Alois Kufner · Anna-Margarete Sändig

Some Applications of Weighted Sobolev Spaces

UB Augsburg

This book is a free continuation of the book about weighted Sobolev spaces which appeared as Volume 31 of the series TEUBNER-TEXTE zur Mathematik. It deals with some applications of these spaces to the solution of boundary value problems. - Part one deals with elliptic boundary value problems in domains whose boundaries have conical corner points and edges; the weighted spaces make it possible to describe in more detail the qualitative properties of the solution including its regularity. One chapter is devoted to the finite element method. - Part two deals mainly with existence theorems for two types of boundary value problems: elliptic problems with "bad behaving" right hand sides, and equations which are degenerate-elliptic or whose coefficients admit some singularities. It is shown how the weighted spaces can be used to overcome these difficulties. Also nonlinear problems are shortly dealt with.

ISBN 978-3-663-11386-7   ISBN 978-3-663-11385-0 (eBook)
DOI 10.1007/978-3-663-11385-0

Dieses Buch ist eine freie Fortsetzung des als Band 31 der
Reihe TEUBNER-TEXTE zur Mathematik erschienenen Buches über
gewichtete Sobolev-Räume. Es werden Anwendungen dieser Räume
zur Lösung von Randwertaufgaben behandelt. - Teil 1 ist elliptischen Randwertproblemen auf Gebieten gewidmet, deren Rand koni-
sche Eckpunkte oder Kanten aufweist. Gewichtete Räume ermögli-
chen eine ausführliche Beschreibung der qualitativen Eigenschaften der Lösungen bis zu Regularitätsaussagen. Ein Kapitel ist der
Methode der finiten Elemente gewidmet. - Teil 2 befaßt sich
hauptsächlich mit Existenzaussagen für zwei Typen von Randwertproblemen: für elliptische Randwertprobleme, deren rechte Seiten
gewisse "schlechte" Eigenschaften haben können, und für Glei-
chungen, die ausarten oder deren Koeffizienten gewisse Singula-
ritäten aufweisen. Es wird gezeigt, wie man die entstehenden
Schwierigkeiten mit Hilfe gewichteter Räume überwinden kann. Es
werden auch kurz nichtlineare Probleme behandelt.

Ce volume représente une suite libre au livre sur les espaces
de Sobolev avec poids, paru comme volume 31 de la série TEUBNER-
TEXTE zur Mathematik. On considère ici les applications de ces
espaces à la résolution des problèmes aux limites. - La première
partie est consacrée aux problèmes aux limites elliptiques
sur des domaines dont les frontières contiennent des points angu-
laires coniques ou des arêtes; les espaces avec poids permettent
de décrire en detail les propriétés qualitatives des solutions,
y compris leur régularité. Un chapitre est consacré à la méthode
des éléments finis. - La deuxième partie s'occupe en principe des
théorèmes d'existence pour deux types de problèmes aux limites:
pour les problèmes aux limites elliptiques dont les seconds
membres peuvent avoir certaines "mauvaises" propriétés et pour
les équations soit elliptiques-dégénérées, soit celles dont les
coefficients présentent certaines singularités. On montre comment
on peut surmonter les difficultés qui y surgissent à l'aide des
espaces avec poids. On traite aussi brièvement des problèmes
non-linéaires.

Настоящая книга представляет собой вольное продолжение книги о весовых пространствах С. Л. Соболева, опубликованной как том 31 се-
рии TEUBNER-TEXTE zur Mathematik. В ней рассматриваются применения весовых пространств к решению краевых задач. - Часть 1 посвящена
эллиптическим краевым задачам для областей, граница которых кони-
ческие угловие точки или ребра. С помощью весовых пространств воз-
можно провести подробное исследование качественных свойств реше-
ний включая утверждения о регулярности решений.Одна глава книги
посвящена методу конечных элементов. - В части 2 исследуются
в основном теоремы о существовании решения для двух типов краевых
задач : для эллиптических задач с некоторыми "нехорошими" правы-
ми сторонами, и для вырождающихся уравнений или уравнений, коэффи-
циенты которых обладают сингулярностью. Указано, как можно преодо-
леть возникающие проблемы с помощью весовых пространств, и коротко
рассмотрены также нелинейные уравнения.

C O N T E N T S

This book is in fact a free continuation of the book of the first author *Weighted Sobolev Spaces*, which appeared in 1980 as Volume 31 of the series *TEUBNER-TEXTE zur Mathematik* and, as the second edition, in Wiley & Sons Publishing House in the year 1985 (in the sequel , this book is refered to as [I]).

In the above mentioned book some fundamental properties of Sobolev spaces with weights were established. In a motivating introduction, several possibilities of application of these spaces were indicated : solution of boundary value problems for partial differential equations with nonstandard domains (i.e., domains with a more complicated geometrical structure) or nonstandard differential operators (coefficients of the equation or of its right hand side or of the boundary values make it impossible to use "current" methods). The book [I] touched only briefly the possibilities of exploiting the weighted spaces, and therefore, the present publication is an attempt to acquaint an interested reader in a little wider framework with the possibilities which the weighted spaces offer when applied to the solution of boundary value problems.

This book was written by two authors and consists of two parts. Both parts are self-contained and can be studied independently. Let us briefly mention their contents.

Part O n e , whose author is A.-M. SÄNDIG, concerns the first of the above mentioned domains of practicability. Here elliptic boundary value problems for domains with conical corners and with edges are studied. In this case the weight functions make it possible to describe in more detail the qualitative properties of the solution, first of all as concerns its regularity. This field, in which a pioneering work was done by V. A. KONDRAT'EV in the sixties, has attracted the interest of quite a number of authors, concerning analytical as well as numerical methods. The application of weighted spaces assumes here a very immediate character also in numerical methods, which is demonstrated by a modification of the popular finite element method. The account presented in this book is an attempt to give a survey of analytical results of V. G. MAZ'JA and B. A. PLAMENEVSKIĬ and of their application in the finite element methods. It was especially the last field to which the author herself has contributed by her own results. The restriction to two types of "singular boun-

daries" - that is, corners and edges - is caused by the technical difficulties with which the investigation meets; in a book of the given extent and
destination it was not possible to present many further existing results.

Part T w o is devoted to rather more theoretical applications, namely
to existence theorems for elliptic differential equations (in this aspect it
is tied up with [I], where these problems were studied for the Dirichlet problem), and further for problems of the type of degenerate equations and equations with singular coefficients. The aim is to show that even here the weighted spaces can provide a useful tool enlarging the scope of boundary value
problems solvable by functional-analytical methods. The author, A. KUFNER,
included in it primarily the results he has lately obtained together with his
colleagues.

The authors do hope that the book will arouse the reader's interest in
weighted spaces and convince him (at least a little) of the usefulness of these
mathematical objects. They welcome any comments which could help them to improve further work in the field, and they use the opportunity to extend their
thanks to all who in any way took part in the preparation of this book. Among
them, at least four names should be mentioned explicitly : Dr. Jiří JARNÍK who
improved the authors' English, Dr. Jiří RÁKOSNÍK who drew the figures, Mrs.
Růžena PACHTOVÁ who carefully typed the manuscript, and Dr. Renate MÜLLER from
the TEUBNER Publishing House who by her patient support has eventually succeeded in making the authors complete the text.

Prague/Rostock 1984 - 1987

A.-M. S.
A. K.

<u>0. P r e l i m i n a r i e s</u>

<u>0.1. THE DOMAIN OF DEFINITION</u>. In what follows we shall work with functions
u = u(x) defined on an (in general arbitrary) measurable set

$$\Omega \subset \mathbb{R}^N .$$

In most cases Ω will be a *domain*, i.e. an open and connected set and we will
suppose that the *boundary* $\partial\Omega$ of Ω will satisfy certain regularity conditions.
Mainly we will work with domains of the class

$$C^{0,1}$$

what means that the boundary can be locally described by a Lipschitz-continuous
function of N – 1 variables (for details, see [I], Chapter 4, or A. KUFNER,
O. JOHN, S. FUČÍK [1], Sections 5.5.6 and 6.2.2). Such a boundary can contain
conical points or *edges*.

Let us give two typical examples of domains considered in Part one of
this book.

<u>0.2. EXAMPLES</u>. (i) Let Ω be a domain in $\mathbb{R}^N$ with one or more conical
points O_i (i = 1,...,s) on $\partial\Omega$. Here, $O \in \partial\Omega$ is a *conical point* if there
exists a neighbourhood U of O such that $U \cap \Omega$ is diffeomorphic to a cone
with vertex at O . For a more detailed explanation see Subsection 5.1 (ii).

(ii) Let Ω be a domain with an *edge* M . This means that M is a
smooth (N–2)-dimensional manifold on
$\partial\Omega$ which divides $\partial\Omega$ into two disjoint parts Γ^+ and Γ^- . E.g., an
infinite roof can serve for $\partial\Omega$, M
being the ridge of the roof, or the
figure from the Fig. 0.

For u = u(x) with $x \in \Omega$ and
for $\alpha = (\alpha_1,...,\alpha_N)$ a multiindex,
we will denote by

$$D^\alpha u$$

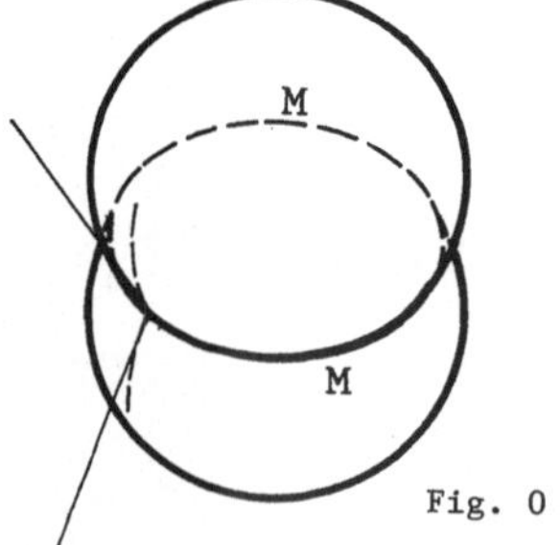

Fig. 0

the derivative $\partial^{\alpha_1+...+\alpha_N} u/\partial x_1^{\alpha_1}...\partial x_N^{\alpha_N}$ of u *in the sense of distributions*.

<u>0.3. CLASSICAL SOBOLEV SPACES</u>. (i) For $1 \leq p \leq \infty$, let

$$(0.1) \qquad L^p(\Omega) = \left\{ u = u(x); \ \|u; L^p(\Omega)\| = \left(\int_\Omega |u(x)|^p \ dx \right)^{1/p} < \infty \right.$$

$$\left. \text{for} \ \ 1 \leq p < \infty , \ \ \|u; L^\infty(\Omega)\| = \sup_{x \in \Omega} \text{ess} \ |u(x)| < \infty \right\} .$$

(ii) For $k \in \mathbb{N}$ and $1 \leq p \leq \infty$, let

$$W^{k,p}(\Omega)$$

denote the *Sobolev space* of all functions $u \in L^p(\Omega)$ whose derivatives $D^\alpha u$ of order $|\alpha| \leq k$ again belong to $L^p(\Omega)$. It is a Banach space if equipped with the norm

$$(0.2) \qquad \|u; W^{k,p}(\Omega)\| = \left(\sum_{|\alpha| \leq k} \int_\Omega |D^\alpha u(x)|^p \ dx \right)^{1/p} .$$

(iii) We denote by

$$W^{k,p}_{loc}(\Omega)$$

the set of all functions $u = u(x)$ defined on Ω which satisfy

$$u \in W^{k,p}(\Omega_0) \ \ \text{for every bounded set} \ \ \Omega_0 \subset \overline{\Omega}_0 \subset \Omega .$$

(iv) We denote by

$$W^{k,p}_0(\Omega)$$

the closure of the set $C^\infty_0(\Omega)$ with respect to the norm (0.2).

(v) The space of *traces* of functions $u \in W^{k,p}(\Omega)$ on $\partial\Omega$ will be denoted by

$$W^{k-1/p,p}(\partial\Omega)$$

and defined as the factor-space

$$W^{k,p}(\Omega)/W^{k,p}_0(\Omega)$$

equipped with the corresponding factor-norm.

0.4. WEIGHTED SOBOLEV SPACES.

(i) Let us denote by

$$(0.3) \qquad W(\Omega)$$

the set of all *weight functions* $w(x)$, i.e. $w(x)$ is measurable and positive almost everywhere (a.e.) in Ω.

(ii) For $1 \leq p < \infty$ and $w \in W(\Omega)$ we denote

$$L^p(\Omega;w) = \left\{ u = u(x); \ \|u; L^p(\Omega;w)\| = \left(\int_\Omega |u(x)|^p \ w(x) \ dx \right)^{1/p} < \infty \right\} .$$

(iii) For

$$(0.4) \qquad S = \{ w_\alpha \in W(\Omega); \ |\alpha| \leq k \}$$

a given collection of weight functions, we denote by

$$(0.5) \qquad W^{k,p}(\Omega;S)$$

the set of all functions $u = u(x)$ defined on Ω whose derivatives $D^\alpha u$ of order $|\alpha| \le k$ belong to $L^p(\Omega;w_\alpha)$. It is a normed linear space if equipped with the norm

$$(0.6) \qquad \|u; W^{k,p}(\Omega;S)\| = \left(\sum_{|\alpha| \le k} \int_\Omega |D^\alpha u(x)|^p \, w_\alpha(x) \, dx \right)^{1/p} .$$

    (iv) If

$$(0.7) \qquad w_\alpha^{-1/(p-1)} \in L^1_{loc}(\Omega) \quad \text{for} \quad |\alpha| \le k ,$$

then the space $W^{k,p}(\Omega;S)$ is a *Banach space* (see A. KUFNER, B. OPIC [4], [6]).

    (v) Let us suppose that

$$(0.8) \qquad w_\alpha \in L^1_{loc}(\Omega) \quad \text{for} \quad |\alpha| \le k .$$

Then all functions from $C_0^\infty(\Omega)$ belong to $W^{k,p}(\Omega;S)$ and it is meaningfull to introduce the space

$$(0.9) \qquad W_0^{k,p}(\Omega;S)$$

as the closure of $C_0^\infty(\Omega)$ with respect to the norm (0.6). It is again a Banach space if additionally (0.7) is satisfied.

<u>0.5.</u> <u>SPACES WITH POWER WEIGHTS</u>. Now, let us consider some special cases of the weighted spaces just introduced.

    For M a subset of $\partial\Omega$, we introduce the following special collection S:

$$(0.10) \qquad S = \{d_M^{\varepsilon(|\alpha|)}; \ |\alpha| \le k, \ \varepsilon(|\alpha|) \in \mathbb{R}\}$$

with $d_M(x) = \text{dist}(x,M)$. In this case, both conditions (0.7), (0.8) are satisfied. Two special cases will be of importance:

    (i) $\varepsilon(|\alpha|)$ is independent of $|\alpha|$, $\varepsilon(|\alpha|) = \varepsilon$ for every $|\alpha| \le k$. Then the corresponding weighted Sobolev spaces (0.5) and (0.9) will be denoted by

$$(0.11) \qquad W^{k,p}(\Omega;d_M,\varepsilon) \quad \text{and} \quad W_0^{k,p}(\Omega;d_M,\varepsilon) ,$$

respectively. Let us remain that according to (0.6), the norm in the spaces (0.11) is given by

$$(0.12) \qquad \|u; W^{k,p}(\Omega;d_M,\varepsilon)\| = \left(\sum_{|\alpha| \le k} \int_\Omega |D^\alpha u(x)|^p \, d_M^\varepsilon(x) \, dx \right)^{1/p} .$$

    (ii) For a fixed $\varepsilon \in \mathbb{R}$, let

$$\varepsilon(|\alpha|) = \varepsilon - (k - |\alpha|)p .$$

Then the corresponding weighted Sobolev space $W^{k,p}(\Omega;S)$ will be denoted by

$$(0.13) \qquad H^{k,p}(\Omega;d_M,\varepsilon) \ .$$

According to (0.6), we have

$$(0.14) \qquad \|u; H^{k,p}(\Omega;d_M,\varepsilon)\| = \left(\sum_{|\alpha|\le k} \int_\Omega |D^\alpha u(x)|^p \, d_M^{\varepsilon-(k-|\alpha|)p}(x) \ dx \right)^{1/p} \ .$$

(iii) Clearly, weight functions of the type (0.10) influence the behaviour of the functions from the corresponding weighted space only in a neighbourhood of the set M . Therefore, it is reasonable to introduce the space

$$W_M^{k,p}(\Omega;d_M,\varepsilon)$$

as the closure of the set

$$(0.15) \qquad C_M^\infty(\Omega) = \left\{ v \in C^\infty(\overline{\Omega}); \ \text{supp } v \ \text{bounded}, \ \text{supp } v \cap \overline{M} = \emptyset \right\}$$

with respect to the norm (0.12). Analogously, we define the space

$$(0.16) \qquad H_M^{k,p}(\Omega;d_M,\varepsilon)$$

as the closure of $C_M^\infty(\Omega)$ with respect to the norm (0.14).

(iv) Sometimes, we will deal with spaces denoted by

$$(0.16^*) \qquad W^{k,p}\big(\Omega; \ s(d_M)\big) \quad \text{and} \quad W_0^{k,p}\big(\Omega; \ s(d_M)\big) \ .$$

These spaces are in fact the weighted Sobolev spaces (0.5) and (0.9) with a special choice of the collection S , namely

$$(0.16^{**}) \qquad S = \left\{ s\big(d_M(x)\big) \ \text{for all} \ |\alpha| \le k \right\} \ ,$$

i.e., $w_\alpha(x) = s\big(d_M(x)\big)$ for all multiindices α , where $s = s(t)$ is a (continuous and positive) function defined on $(0,\infty)$, and d_M is again the distance function $\text{dist}(x,M)$, $M \subset \partial\Omega$. Let us recall that according to (0.6), the norm in the spaces (0.16^*) is given by

$$(0.16^{***}) \qquad \|u; W^{k,p}\big(\Omega;s(d_M)\big)\| = \left(\sum_{|\alpha|\le k} \int_\Omega |D^\alpha u(x)|^p \, s\big(d_M(x)\big) \ dx \right)^{1/p} \ .$$

The spaces (0.11) are special cases of the spaces (0.16^*) : we obtain them if we set $s(t) = t^\varepsilon$.

0.6. **FURTHER SPACES WITH POWER WEIGHTS.** (i) In the literature (see, e.g., V. G. MAZ'JA, B. A. PLAMENEVSKIĬ [1]) often spaces with weight functions of the type (0.10) occur, which are denoted by

$$(0.17) \qquad V^{k,p}(\Omega;\beta) \ , \quad \beta \in \mathbb{R} \ ,$$

and coincide with the spaces introduced in (0.16) by the formula

$$(0.18) \qquad V^{k,p}(\Omega;\beta) = H_M^{k,p}(\Omega;d_M,\beta p) \ .$$

The symbol d_M is omitted in the notation of the V-spaces since it will be clear from the context what the set M is.

(ii)   In Part one of this book, we will deal with a set M which consists of a finite number of boundary points (e.g. conical points) : $M = \{O_i;\ i = 1,\ldots,s\}$. In this case, we consider the following collection of weight functions

$$S = \{w_\alpha(x) = \prod_{i=1}^{s} |x - O_i|^{p(\beta_i - k + |\alpha|)} \,, \quad \beta_i \in \mathbb{R}\}$$

and define the space

(0.19) $V^{k,p}(\Omega;\vec{\beta})$, $\vec{\beta} = (\beta_1,\ldots,\beta_s)$,

as the closure of the set $C_M^\infty(\Omega)$ from (0.15) with respect to the norm (0.6).

(iii) Let us consider a bounded domain Ω with an edge M , which is an (N-2)-dimensional manifold on $\partial\Omega$. Further, let U be a neighbourhood of M with the following property : For every $x \in U$ there is a uniquely determined point $\xi = \xi(x) \in M$ such that

$$d_M(x) = |x - \xi(x)| \ .$$

We introduce a real-valued smooth function $\beta = \beta(\xi)$ defined on M and define – in analogy to the foregoing spaces – the space

$$V^{k,p}(\Omega;\beta(\cdot))$$

as the closure of $C_M^\infty(\Omega)$ with respect to the norm

$$\|u;\ V^{k,p}(\Omega;\beta(\cdot))\| = \left(\sum_{|\alpha|\le k} \int_U |D^\alpha u(x)|^p |x - \xi(x)|^{p(\beta\xi(x)-k+|\alpha|)}\ dx \right.$$
$$\left. + \int_{\Omega-U} |D^\alpha u(x)|^p\ dx \right)^{1/p} \ .$$

Obviously, this is a "local definition" depending on the chosen neighbourhood U , but it is clear that norms obtained for various U's are equivalent.

Some other weighted spaces and modifications of the spaces just introduced can be found later at the beginning of the chapter in which they will be used.

0.7. TRACE SPACES. Analogously as in the case of the classical Sobolev spaces – see Section 0.2 (v) – we can introduce the space of traces of functions from $W^{k,p}(\Omega;S)$ on $\partial\Omega$ as the factor-spaces

$$W^{k,p}(\Omega;S)/W_0^{k,p}(\Omega;S)$$

and denote it by

$$W^{k-1/p,p}(\partial\Omega;S)$$

provided the definition is meaningfull.

Similarly we can proceed in the case of the V-spaces introduced in

Section 0.6. E.g., we have

$$V^{k-1/p,p}(\partial\Omega;\beta) = V^{k,p}(\Omega;\beta)/V_0^{k,p}(\Omega;\beta)$$

where $V_0^{k,p}(\Omega;\beta)$ is the closure of $C_0^\infty(\Omega)$ with respect to the norm of $V^{k,p}(\Omega;\beta)$. This approach is typical for domains with conical points.

Here, we have a *global* description of the trace on the whole $\partial\Omega$ (or, more precisely, a.e. on $\partial\Omega$). The definition of a trace can be modified in the sense that we consider traces *separately* on disjoint parts of the boundary $\partial\Omega$. For example, let us consider the domain Ω with edge M mentioned in Example 0.2 (ii) and let us denote by Γ^+ and Γ^- the remaining two components of $\partial\Omega$. We denote by $V_0^{k,p}(\Omega;\Gamma^\pm,\beta)$ the closure of $C_{\Gamma^\pm}^\infty(\Omega)$ with respect to the norm of $V^{k,p}(\Omega;\beta)$ and introduce the space of traces on Γ^+ or Γ^- , respectively, as the factor-space

$$(0.20) \qquad V^{k-1/p,p}(\Gamma^\pm;\beta) = V^{k,p}(\Omega;\beta)/V_0^{k,p}(\Omega;\Gamma^\pm,\beta) \ .$$

<u>0.8.</u> <u>SPACES WITH "DERIVATIVES OF NEGATIVE ORDER"</u>. (i) Let us consider the space $V^{k,p}(\Omega;\beta)$ introduced by formula (0.18). Here, k was a non-negative integer. Now, for *negative* integers $k = -1,-2,\ldots$ we define the space

$$V^{k,p}(\Omega;\beta) \ , \quad k \in \mathbf{Z} \ , \quad k < 0 \ ,$$

as the closure of the set $C_M^\infty(\Omega)$ from (0.15) with respect to the following norm :

$$(0.21) \qquad \|u;V^{k,p}(\Omega;\beta)\| = \sup_v \left| \int_\Omega u(x)\ \overline{v(x)}\ dx \right| \Big/ \|v;\ V^{-k,q}(\Omega;-\beta)\|$$

with $q = p/(p-1)$, the supremum being taken over all $v \in V^{-k,q}(\Omega;-\beta)$, $v \neq 0$.

(ii) In fact, it follows from (0.21) that $V^{k,p}(\Omega;\beta)$ is, for $k < 0$, the *dual space* to $V^{-k,q}(\Omega;-\beta)$. Similarly we can proceed for the trace spaces, defining the space $V^{k+1/q,p}(\partial\Omega;\beta)$ for $k \in \mathbf{Z}$, $k < 0$, as the dual space to $V^{-k-1/q,q}(\partial\Omega;-\beta)$.

<u>0.9.</u> <u>SPACES OF ROĬTBERG–BEREZANSKIĬ TYPE</u>. Let m be a positive integer, k an arbitrary integer. In connection with the investigation of differential operators of order $2m$, the following spaces are of importance, which are subspaces of the spaces $V^{k,p}(\Omega;\beta)$ introduced in Sections 0.6 (for $k \geq 0$) and 0.8 (for $k < 0$).

(i) Let Ω be a domain with one conical point O and let us suppose that $\partial\Omega\setminus O$ is sufficiently smooth. We introduce the space

$$(0.22) \qquad \tilde{V}^{k,p}(\Omega;\beta)$$

as the closure of $C^{\infty}_{\{0\}}(\Omega)$ with respect to the norm

$$(0.23) \qquad \|u; \ V^{k,p}(\Omega;\beta)\| + \sum_{j=0}^{2m-1} \|\frac{\partial^j u}{\partial n^j} \ ; \ V^{k-j-1/p,p}(\partial\Omega;\beta)\|$$

where n is the exterior normal to $\partial\Omega$ and the trace spaces in the second term in (0.23) are considered in the sense of Sections 0.7 and 0.8.

It is

$$\tilde{V}^{k,p}(\Omega;\beta) = V^{k,p}(\Omega;\beta) \quad \text{if} \quad k \geq 2m$$

since it follows from the definition that

$$(0.24) \qquad \|\frac{\partial^j u}{\partial n^j} \ ; \ V^{k-j-1/p,p}(\partial\Omega;\beta)\| \leq c\|u; \ V^{k,p}(\Omega;\beta)\|$$

for $j = 0,1,\ldots,2m-1$.

(ii) For a domain Ω with an edge M such that $\partial\Omega = \Gamma^+ \cup \Gamma^- \cup M$ with $\Gamma^{\pm}$ sufficiently smooth (see Sections 0.2 (ii) and 0.7), we introduce the space

$$\tilde{V}^{k,p}(\Omega;\beta)$$

as the closure of $C^{\infty}_M(\Omega)$ with respect to the norm

$$(0.25) \qquad \|u; \ V^{k,p}(\Omega;\beta)\| + \sum_{+,-} \sum_{j=0}^{2m-1} \|\frac{\partial^j u}{\partial n^j} \ ; \ V^{k-j-1/p,p}(\Gamma^{\pm};\beta)\|$$

(for the last term, see (0.20)).

(iii) Non-weighted analoga of the above spaces have been introduced by J. M. BEREZANSKIĬ, J. A. ROĬTBERG [1] and by J. A. ROĬTBERG [1]; the weighted spaces have been introduced and used by J. ROSSMANN [1].

<u>0.10. SOME PROPERTIES OF WEIGHTED SPACES.</u> For X , Y two Banach spaces, the symbol

$$X \subsetneq Y$$

means that there is $X \subset Y$ and that there exists a constant $c > 0$ such that

$$\|u;Y\| \leq c\|u;X\| \quad \text{for every} \quad u \in X \ .$$

We shall say that X is (continuously)· *imbedded into* Y .

(i) Now, the following imbeddings follow directly from the definition of the spaces considered :

$$(0.26) \qquad V^{k,p}(\Omega;\beta) \subsetneq V^{k-1,p}(\Omega; \beta-1) \ ,$$

$$(0.27) \qquad V^{k-1/p,p}(\partial\Omega;\beta) \subsetneq V^{k-1-1/p,p}(\partial\Omega; \beta-1) \ ,$$

$$(0.28) \qquad V^{k-1/p,p}(\Gamma^{\pm};\beta) \subsetneq V^{k-1-1/p,p}(\Gamma^{\pm}; \beta-1) \ .$$

(ii) For a *bounded* domain Ω , we have the imbedding

(0.29) $\qquad V^{k,p}(\Omega;\beta) \subsetneqq V^{k,p}(\Omega;\beta_1)$ if $\beta_1 > \beta$.

It follows from the fact that, for a bounded domain Ω , the function $d_M^{\beta_1-\beta}$ is bounded, too.

$\qquad$ (iii) For a *bounded* domain Ω , we have the imbedding

(0.30) $\qquad V^{k,p}(\Omega;\beta) \subsetneqq V^{k,p_1}(\Omega;\beta_1)$

provided that

(0.31) $\qquad p \geq p_1$ and $\beta + \dfrac{N-m}{p} < \beta_1 + \dfrac{N-m}{p_1}$

hold with $m = \dim M$. The imbedding (0.30) follows directly from the Hölder inequality - see [I], Proposition 6.9.

0.11. <u>THE HARDY INEQUALITY AND RELATED IMBEDDINGS</u>. (i) Let us remind the *Hardy inequality*

(0.32) $\qquad \displaystyle\int_0^\infty |f(t)|^p \, t^{\varepsilon-p} \, dt \leq \left(\frac{p}{|\varepsilon-p+1|} \right)^p \int_0^\infty |f'(t)|^p \, t^\varepsilon \, dt$

which holds

$\qquad\qquad$ for $\varepsilon > p - 1$ if $f(\infty) = 0$,
$\qquad\qquad$ for $\varepsilon < p - 1$ if $f(0) = 0$

(see, e.g., G. H. HARDY, J. E. LITTLEWOOD, G. Pólya [1], Theorem 330). This inequality is a usefull tool to derive imbedding theorems for weighted spaces with power type weights. In particular, we have :

$\qquad$ (ii) Let Ω be a bounded domain of class $C^{0,1}$. Then it is

(0.33) $\qquad W^{k,p}(\Omega;d_M,\varepsilon) \subsetneqq H^{k,p}(\Omega;d_M,\varepsilon)$

if

(0.34) $\qquad \varepsilon > kp + m - N$;

in particular we have for these ε the imbedding

(0.35) $\qquad W^{k,p}(\Omega;d_M,\varepsilon) \subsetneqq L^p(\Omega;\, d_M,\, \varepsilon-kp)$.

Further, we have

(0.36) $\qquad W_M^{k,p}(\Omega;d_M,\varepsilon) \subsetneqq H_M^{k,p}(\Omega;d_M,\varepsilon)$

for

(0.37) $\qquad \varepsilon \neq jp + m - N$, $j = 1,\ldots,k$.

Here again $m = \dim M$. For the proof of the above imbeddings see [I].

0.12. <u>REMARK</u>. Imbedding (0.36) together with the trivial imbedding

$\qquad H_M^{k,p}(\Omega;d_M,\varepsilon) \subsetneqq W_M^{k,p}(\Omega;d_M,\varepsilon)$

which holds for *every* ε provided Ω is bounded (compare with the imbedding (0.29)!) yields the identity

$$H_M^{k,p}(\Omega;d_M,\varepsilon) = W_M^{k,p}(\Omega;d_M,\varepsilon)$$

which holds for $\varepsilon \neq jp + m - N$, $j = 1,\ldots,k$. In view of (0.18) we have an interesting connection between the W-spaces and V-spaces, namely

$$V^{k,p}(\Omega;\beta) = W_M^{k,p}(\Omega;d_M,\beta p)$$

for $\beta p \neq jp + m - N$, i.e., for

$$\beta \neq j + \frac{m - N}{p} , \quad j = 1,\ldots,k .$$

<u>0.13. ONE MORE IMBEDDING.</u> Let Ω be a bounded domain of the class $C^{0,1}$, i.e. with a Lipschitzian boundary $\partial\Omega$. Then we have for every $\beta \in \mathbb{R}$

$$(0.38) \qquad V^{1,p_1}(\Omega;\beta) \subsetneqq L^p(\Omega;\beta)$$

with

$$(0.39) \qquad p \geq \frac{N - p_1}{Np_1} \text{ if } p_1 < N , \quad p \geq 1 \text{ if } p_1 \geq N .$$

The result follows from the classical Sobolev imbedding theorem $W^{1,p_1}(\Omega) \subsetneqq L^p(\Omega)$ used for the function $d_M^\beta \cdot u$.

By a repeated use of the imbedding (0.38) we have under the conditions (0.39) for every positive integer k the imbedding

$$(0.40) \qquad V^{k,p_1}(\Omega;\beta) \subsetneqq V^{k-1,p}(\Omega;\beta) .$$

PART ONE

ELLIPTIC BOUNDARY VALUE PROBLEMS IN NON SMOOTH DOMAINS

The theory of elliptic boundary value problems in smooth domains is well developed. One of the main results of the linear theory is the following regularity result: *If the right hand sides of our boundary value problem, the coefficients of the differential operators and the domain are sufficiently smooth and a solution exists, then it is smooth, too.* Such kind of regularity results in Sobolev or Hölder spaces play, among others, an important role in error estimates for the numerical solution of an elliptic boundary value problem.

However, a majority of elliptic boundary value problems which are of interest in practice are formulated in non smooth domains such as polygonal or polyhedral ones, and/or involve mixed boundary conditions. *Unfortunately, no regularity result of the above mentioned type generally holds in these cases.* This situation gives rise to difficulties both in analytical and numerical investigations. Thus we have to study the solvability of such boundary value problems in appropriate (weighted) spaces, to investigate the regularity of the solution and to clarify some numerical effects.

Let us shortly describe the contents of the individual chapters:

In Chapter I we deal with the theory of elliptic boundary value problems in domains with conical points. Mixed boundary conditions are included. First we study the solvability in weighted Sobolev spaces. Then we formulate regularity theorems in the framework of these weighted Sobolev spaces and give expansions of the solutions near the conical points. The coefficients of the singular functions occuring in these expansions are calculated. We closely follow the method of V. A. KONDRAT'EV ($p = 2$) and the results of V. G. MAZ'JA and B. A. PLAMENEVSKIĬ.

Chapter II is concerned with one of the numerical methods of solving boundary value problems, namely with the Finite Element Method (shortly FEM). The results of the first chapter yield error estimates of classical FEM's in domains with conical points as well as for mixed boundary conditions both in the norms of Sobolev spaces and in the maximum norms. The well known polluting effect is qualitatively described. Further, it is shown that a modified FEM, the so-called Dual Singular Function Method, overcomes some numerical difficulties. This method was first proposed by M. DOBROWOLSKI and H. BLUM. We give error estimates for this modified FEM using again the results of the first chapter.

Chapter III deals with elliptic boundary value problems in domains with edges. The content is similar to Chapter I and includes solvability theorems, regularity theorems, expansions of the solution near an edge and the calculation of the singular terms appearing in these expansions. We again use appropriate weighted Sobolev spaces and follow the ideas of V. A. KONDRAT'EV, V. G. MAZ'JA, B. A. PLAMENEVSKIĬ and J. ROSSMANN. We underline that only one of the theories for elliptic boundary value problems in non smooth domains is studied here. Other ideas are developed, e.g., by S. REMPEL and B.-W. SCHULZE [1] and by B.-W. SCHULZE [1], [2], [3], or by P. GRISVARD [2], [3].

Chapter I

ELLIPTIC BOUNDARY VALUE PROBLEMS IN DOMAINS WITH CONICAL POINTS

Section 1 : *Introducing examples*

Let us begin with various examples of boundary value problems arising in practice in which singularities occur. We demonstrate how the geometrical singularities of the boundary or the mixed boundary conditions influence the regularity of the solution and illustrate the method of V. A. KONDRAT'EV [1] with the help of these examples.

§ 1 . T h e   D i r i c h l e t   p r o b l e m   f o r   t h e   L a p l a c e   o p e r a t o r

1.1. <u>FORMULATION OF THE PROBLEM.</u> Let Ω be a polygonal domain in R^2 (see Fig. 1). The problem is the following: *For a given function $f \in L^2(\Omega)$ investigate the smoothness of the solution u of the Dirichlet problem*

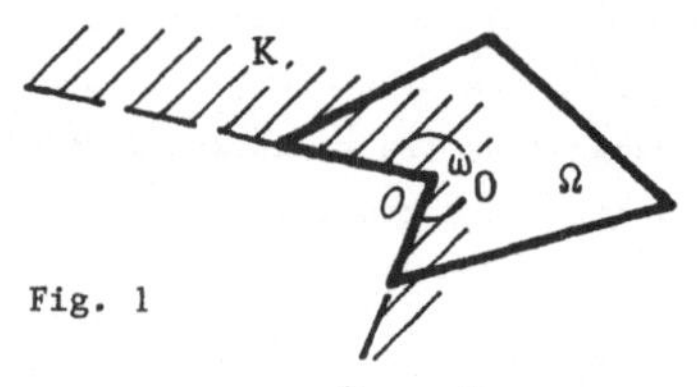

Fig. 1

$$-\Delta u = -\frac{\partial^2 u}{\partial x_1^2} - \frac{\partial^2 u}{\partial x_2^2} = f(x_1, x_2) \quad \textit{for} \quad x = (x_1, x_2) \in \Omega \ ,$$

(1.1)

$$u(x) = 0 \qquad\qquad \textit{for} \quad x \in \partial\Omega \ .$$

The existence of a weak solution $u \in W_0^{1,2}(\Omega)$ (see 15.2 of Chapter II) follows from the Lax-Milgram theorem 15.5. Furthermore, it is well known that $u \in W_{loc}^{2,2}(\Omega)$ and if the domain Ω is smooth, then even $u \in W^{2,2}(\Omega)$ (cf. e.g. J. WLOKA [1]). We now have to study what influence have the sizes of the angles of our polygonal domain on whether $u \in W^{2,2}(\Omega)$ or not.

1.2. THE BOUNDARY VALUE PROBLEM IN AN INFINITE CONE K.

The investigation of the regularity is a local problem. Therefore we choose the origin O at one of the corner points of Ω with the angle ω_0 and multiply the solution $u \in W_0^{1,2}(\Omega)$ of (1.1) by a cut-off function $\eta(|x|) = \eta(r)$, where $0 \leq \eta(r) \leq 1$,

$$(1.2) \qquad \eta(r) = \begin{cases} 1 & \text{for } 0 \leq r \leq \delta , \\ 0 & \text{for } r \geq 2\delta , \end{cases}$$

and $\eta(r) \in C^\infty(R^1)$. The number δ is so small that no other corner point of Ω lies in the circle $\{x : |x| \leq 2\delta\}$.

We denote $u_1 = \eta u$. Let K be the infinite cone with the vertex O and the angle ω_0 (see Fig. 1). Then we have

$$(1.3) \qquad \begin{aligned} - \Delta u_1 &= - \Delta(\eta u) = f_1 && \text{in } K , \\ u_1 &= 0 && \text{on } \partial K , \end{aligned}$$

where $f_1 = - \eta\Delta u - u\Delta\eta - 2\eta_{x_1} u_{x_1} - 2\eta_{x_2} u_{x_2} \in L^2(K)$. We remark that even $f_1 \in L^2(K,\beta)$ for $\beta \geq 0$, where $L^2(K,\beta)$ is the weighted Sobolev space defined by (0.18). The behavior of $u_1 = \eta u$ near O characterizes the regularity of u in a neighborhood of the corner point O of Ω .

1.3. THE SOLUTION OF THE BOUNDARY VALUE PROBLEM IN K.

This point is the most important for our example. The method used was developed by V. A. KONDRAT'EV [1] and demands a lot of theoretical knowledge. Nevertheless, for the sake of brevity and clarity we will only quote the corresponding results here and refer to Section 2 for more information.

First we recall the technique for solving boundary value problems in smooth domains (see e.g. J. WLOKA [1]), namely: view the problem locally, flatten the boundary by a change of variables, freeze the coefficients and use partial Fourier transforms.

(i) For the second step "flatten the boundary" we have to find an alternative in our case. To this end we rewrite (1.3) using polar coordinates (r,ω) . We get

$$- \left(\frac{\partial^2 \bar{u}_1}{\partial r^2} + \frac{1}{r} \frac{\partial \bar{u}_1}{\partial r} + \frac{1}{r^2} \frac{\partial^2 \bar{u}_1}{\partial \omega^2} \right) = \bar{f}_1(r,\omega) \quad \text{in } \tilde{S} ,$$

$$(1.4) \qquad \bar{u}_1(r,0) = 0 ,$$

$$\bar{u}_1(r,\omega_0) = 0 ,$$

where $\tilde{S}$ is an infinite half-strip in the r,ω-plane (see Fig. 2), $\overline{u}_1(r,\omega) = u_1(x_1,x_2)$, $\overline{f}_1(r,\omega) = f_1(x_1,x_2)$. We now put $r = e^\tau$ in (1.4) thus obtaining

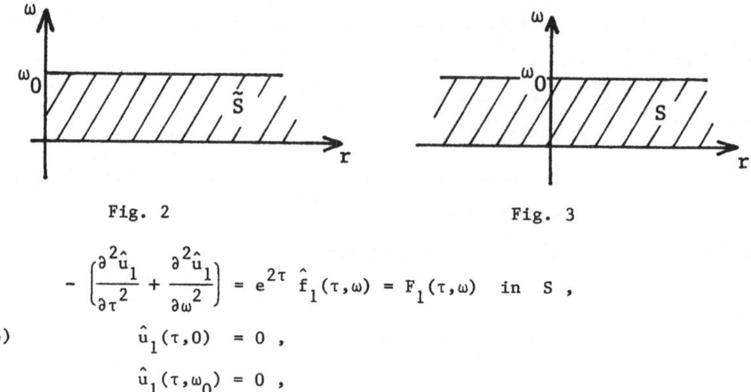

Fig. 2   Fig. 3

$$- \left(\frac{\partial^2 \hat{u}_1}{\partial \tau^2} + \frac{\partial^2 \hat{u}_1}{\partial \omega^2} \right) = e^{2\tau} \hat{f}_1(\tau,\omega) = F_1(\tau,\omega) \quad \text{in} \quad S ,$$

$$(1.5) \qquad \hat{u}_1(\tau,0) = 0 ,$$

$$\hat{u}_1(\tau,\omega_0) = 0 ,$$

where S is an infinite strip in the τ,ω-plane (see Fig. 3), $\hat{u}_1(\tau,\omega) = u_1(x_1,x_2)$, $\hat{f}_1(\tau,\omega) = f_1(x_1,x_2)$. In this way the alternative "flattening process" yields an infinite strip.

(ii) We now use the complex Fourier transform with respect to τ :

$$(1.6) \qquad \mathcal{F}(F)(\lambda) = \tilde{F}(\lambda) = \frac{1}{\sqrt{2\pi}} \int_{-\infty}^{+\infty} e^{-i\lambda\tau} F(\tau)\, d\tau ,$$

where $\lambda \in \mathbb{C}$. Applying (1.6) to the problem (1.5) we get

$$\lambda^2 \tilde{u}_1 - \frac{\partial^2}{\partial\omega^2} \tilde{u}_1 = \tilde{F}_1(\lambda,\omega) \quad \text{for} \quad 0 < \omega < \omega_0 ,$$

$$(1.7) \qquad \tilde{u}_1(\lambda,0) = 0 ,$$

$$\tilde{u}_1(\lambda,\omega_0) = 0 ,$$

where $\tilde{u}_1 = \mathcal{F}(\hat{u}_1)$, $\tilde{F}_1 = \mathcal{F}(\hat{F}_1)$.

(iii) We denote by $\mathcal{U}_0(\lambda)$ the operator of the problem (1.7) :

$$(1.8) \qquad \mathcal{U}_0(\lambda) : W^{2,2}(G) \to L^2(G) \times \mathbb{C} \times \mathbb{C} ,$$

where $G = (0,\omega_0)$. The eigenvalues of $\mathcal{U}_0(\lambda)$ are $\lambda_k = i\,\frac{k\pi}{\omega_0}$, $k = \pm 1, \pm 2, \ldots$, and the corresponding eigenfunctions are $\tilde{e}_k = \sin\frac{k\pi}{\omega_0}\,\omega$. If λ is no eigenvalue of $\mathcal{U}_0(\lambda)$ then for any function $\tilde{F}_1 \in L^2(G)$ there exists a uniquely determined solution $\tilde{u}_1 \in W^{2,2}(G)$ of (1.7). We write

$$(1.9) \qquad \tilde{u}_1 = \mathcal{U}_0^{-1}(\lambda)\,[\tilde{F}_1,0,0]\ ,$$

where $\mathcal{U}_0^{-1}(\lambda)$ denotes the inverse operator (resolvent operator) :

$$\mathcal{U}_0^{-1}(\lambda)\ :\ L^2(G) \times \mathbb{C} \times \mathbb{C} \to W^{2,2}(G)\ .$$

Finally, the inverse Fourier transform of $\tilde{u}_1$ yields the solution $\hat{u}_1(\tau,\omega)$ $= u_1(x_1,x_2)$ of (1.3).

(iv) In order to reach our goal, namely to find regularity theorems and expansions of the solution u_1 or u we need some information about

- the properties of the complex Fourier transform (1.6),

- the properties of the parameter boundary value problem (1.7), especially the properties of $\mathcal{U}_0^{-1}(\lambda)$.

Let us provide a rough survey of the information needed :

(v)   -   The *complex Fourier transform*

$$\mathcal{F}(F)(\lambda) = \hat{F}(\lambda) = \frac{1}{\sqrt{2\pi}} \int\limits_{-\infty}^{+\infty} e^{-i\lambda\tau}\, F(\tau)\ d\tau$$

defines an isomorphic mapping

$$(1.10) \qquad \{F(\tau)\ :\ \int\limits_{-\infty}^{+\infty} |F(\tau)|^2\, e^{2h\tau}\, d\tau < \infty\} \to L^2(\mathbb{R}+ih)$$

for $\lambda = s + ih$, $h = $ const., $\mathbb{R} + ih = \{\lambda \in \mathbb{C}\ :\ \mathrm{Im}\ \lambda = h\}$.

- The following Parseval identity holds :

$$(1.11) \qquad \int\limits_{-\infty}^{+\infty} |F(\tau)|^2\, e^{2h\tau}\, d\tau = \int\limits_{-\infty+ih}^{\infty+ih} |\tilde{F}(\lambda)|^2\, d\lambda\ .$$

- The inverse Fourier transform is given by

$$(1.12) \qquad F(\tau) = \frac{1}{\sqrt{2\pi}} \int\limits_{-\infty+ih}^{\infty+ih} e^{i\lambda\tau}\, \tilde{F}(\lambda)\ d\lambda\ .$$

- We have

$$(1.13) \qquad \mathcal{F}\left(\frac{d^j}{d\tau^j}\, F(\tau)\right)(\lambda) = (i\lambda)^j\, \mathcal{F}\left(F(\tau)\right)(\lambda)\ .$$

- If $h < h_1$, $\displaystyle\int\limits_{-\infty}^{+\infty} |F(\tau)|^2\, e^{2h\tau}\, d\tau < \infty$ and $\displaystyle\int\limits_{-\infty}^{+\infty} |F(\tau)|^2\, e^{2h_1\tau}\, d\tau < \infty$

then

(1.14) $\tilde{F}(\lambda)$ is holomorphic in the strip $h < \text{Im } \lambda < h_1$.

(vi) Let us investigate whether the right hand side $F_1(\tau,\omega)$ of the problem (1.5) is Fourier transformable in this sense. We have

$$\int_{-\infty}^{+\infty} |F_1(\tau,\omega)|^2 \, e^{2h\tau} \, d\tau = \int_{-\infty}^{+\infty} e^{4\tau} \, |\hat{f}_1(\tau,\omega)|^2 \, e^{2h\tau} \, d\tau$$

$$= \int_{-\infty}^{+\infty} e^{2(\tau+\beta\tau)} \, |\hat{f}_1(\tau,\omega)|^2 \, d\tau \quad \text{for} \quad h = \beta - 1 \ , \quad \beta \geq 0 \ .$$

Since $f_1 \in L^2(K,\beta)$ (see (1.3)) and therefore

$$\int_K |x|^{2\beta} \, |f_1(x)|^2 \, dx = \int_S e^{2(\tau+\beta\tau)} \, |\hat{f}_1(\tau,\omega)|^2 \, d\tau \, d\omega < \infty \ ,$$

the Fourier transform of $F_1(\tau,\omega)$ is meaningful in the half plane $\text{Im } \lambda \geq -1$ for almost all $\omega \in G$.

(vii) The following regularity result holds for our example : If no eigenvalues of $\mathcal{U}_0(\lambda)$ lie on the line $\text{Im } \lambda = h = \beta - 1$, $\beta \geq 0$, then the inverse Fourier transform

$$(1.15) \qquad \hat{u}_{1,h}(\tau,\omega) = \frac{1}{\sqrt{2\pi}} \int_{-\infty+ih}^{\infty+ih} \tilde{u}_1(\lambda,\omega) \, e^{i\lambda\tau} \, d\lambda$$

exists and

$$\hat{u}_{1,h}(\tau,\omega) = u_{1,h}(x) \in V^{2,2}(K,\beta) \ ,$$

where $V^{2,2}(K,\beta)$ is the weighted Sobolev space defined by (0.18). $u_{1,h}(x)$ is the uniquely determined solution of (1.3) from $V^{2,2}(K,\beta)$.

Let us illustrate this situation by two special examples : For $h = -1$ we have

$$(1.16) \qquad u_{1,-1}(x) = \hat{u}_{1,-1}(\tau,\omega) = \frac{1}{\sqrt{2\pi}} \int_{-\infty-i}^{\infty-i} \tilde{u}_1(\lambda,\omega) \, e^{i\lambda\tau} \, d\lambda = w_1(x) \ ,$$

which is the uniquely determined solution of (1.3) in $V^{2,2}(K,0)$. For $h = 0$ we get

$$(1.17) \qquad u_{1,0}(x) = \hat{u}_{1,0}(\tau,\omega) = \frac{1}{\sqrt{2\pi}} \int_{-\infty}^{+\infty} \tilde{u}_1(\lambda,\omega) \, e^{i\lambda\tau} \, d\lambda \ ,$$

which is the uniquely determined solution of (1.3) in $V^{2,2}(K,1)$.

We now go back to our starting point, namely to the solution $u_1 = \eta u$

of (1.3), where $u \in W_0^{1,2}(\Omega)$. It is evident that $u_1 \in V^{1,2}(K,0)$. A regularity result proved by V. A. KONDRAT'EV [1] yields that $u_1 \in V^{2,2}(K,1)$. Therefore we have $u_1(x) = u_{1,0}(x)$.

(viii) Let us derive an *expansion of* $u_1(x)$ *in* K , calculating the integral (1.17). The Cauchy theorem implies that

$$\hat{u}_{1,0}(\tau,\omega) = \lim_{N\to\infty} \frac{1}{\sqrt{2\pi}} \int_{-N}^{+N} \tilde{u}_1(\lambda,\omega) \, e^{i\lambda\tau} \, d\lambda$$

$$(1.18) \qquad = \frac{1}{\sqrt{2\pi}} \lim_{N\to\infty} \left(\int_{-N}^{-N-i} \tilde{u}_1(\lambda,\omega) \, e^{i\lambda\tau} \, d\lambda + \int_{-N-i}^{N-i} \tilde{u}_1 \, e^{i\lambda\tau} \, d\lambda + \int_{N-i}^{N} \tilde{u}_1 \, e^{i\lambda\tau} \, d\lambda \right)$$

$$+ \frac{1}{\sqrt{2\pi}} \, 2\pi i \sum_{\substack{-1 \,<\, \mathrm{Im}\,\lambda \,<\, 0 \\ |\mathrm{Re}\,\lambda| \,<\, N}} \mathrm{Res} \ \tilde{u}_1(\lambda,\omega) \, e^{i\lambda\tau} \ .$$

(See Fig. 4.)

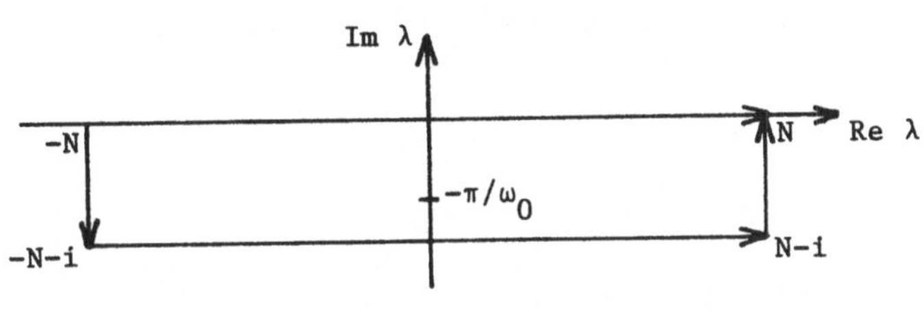

Fig. 4

The first and third integrals of (1.18) tend to 0 ; the second to integral (1.16). We have to calculate the residua of $\tilde{u}_1(\lambda,\omega)e^{i\lambda\tau}$ and therefore to look at the singularities of $\tilde{u}_1(\lambda,\omega) = \mathcal{O}_0^{-1}(\lambda)[\tilde{F}_1,0,0]$. $\mathcal{O}_0^{-1}(\lambda)$ is a meromorphic operator function of λ with poles at $\lambda_k = i \frac{k\pi}{\omega_0}$, $k = \pm 1, \pm 2, \ldots$. For the calculation of the residuum only the pole λ_k with $-1 < \mathrm{Im}\,\lambda_k < 0$ plays a role, which means that, if $\omega_0 > \pi$, then $\lambda_{-1} = -\frac{i\pi}{\omega_0}$ lies in this strip.

In a neighborhood of λ_{-1} the resolvent operator $\mathcal{O}_0^{-1}(\lambda)$ has the form

$$(1.19) \qquad \mathcal{O}_0^{-1}(\lambda) = \frac{P_1}{\lambda - \lambda_{-1}} + \Gamma(\lambda) \ ,$$

where P_1 and $\Gamma(\lambda)$ map $L^2(G) \times \mathbb{C} \times \mathbb{C}$ into $W^{2,2}(G)$. The image of P_1 is the eigenspace of $\mathcal{O}_0(\lambda)$ with respect to λ_{-1} , and $\Gamma(\lambda)$ is holomorphic.

Further, (1.14) and (vi) imply that $\tilde{F}_1(\lambda,\omega)$ is holomorphic with respect to λ in the strip $-1 < \mathrm{Im}\,\lambda < 0$. Therefore we can write

$$(1.20) \qquad [\tilde{F}_1, 0, 0] = \sum_{\ell=0}^{\infty} a_\ell(\omega)(\lambda - \lambda_{-1})^\ell \, ,$$

where the coefficients $a_\ell(\omega) \in L^2(G) \times \mathbb{C} \times \mathbb{C}$.

Finally,

$$(1.21) \qquad e^{i\lambda\tau} = e^{i\lambda_{-1}\tau}\left[1 + i(\lambda - \lambda_{-1})\tau + \ldots + \frac{[i(\lambda - \lambda_{-1})\tau]^n}{n!} + \ldots \right] \, .$$

From (1.19), (1.20) and (1.21) it follows that

$$\tilde{u}_1(\lambda,\omega)e^{i\lambda\tau} = \mathcal{U}_0^{-1}(\lambda)[\tilde{F}_1, 0, 0]\, e^{i\lambda\tau}$$

$$= \sum_{\ell=0}^{\infty}\left[P_1 a_\lambda(\omega)\frac{(\lambda - \lambda_{-1})^\ell}{\lambda - \lambda_{-1}} + \Gamma(\lambda)a_\ell(\omega)(\lambda - \lambda_{-1})^\ell\right]e^{i\lambda_{-1}\tau}[1 + \ldots] \, .$$

Therefore

$$(1.22) \qquad \mathrm{Res}\left(\tilde{u}_1(\lambda,\omega)e^{i\lambda\tau}\right)\Big|_{\lambda=\lambda_{-1}} = e^{i\lambda_{-1}\tau}P_1 a_0(\omega) = e^{\pi\tau/\omega_0}c_1\sin\frac{\pi}{\omega_0}\omega$$

where c_1 is a complex constant. For $\omega_0 > \pi$ the formula (1.18) yields

$$(1.23) \qquad u_1(x) = \hat{u}_{1,0}(\tau,\omega) = c\, e^{\pi\tau/\omega_0}\sin\frac{\pi}{\omega_0}\omega + w_1(x)$$

$$= c\, r^{\pi/\omega_0}\sin\frac{\pi}{\omega_0}\omega + w_1(x) \, ,$$

where $w_1(x) \in V^{2,2}(K,0)$. The constant c can be calculated (see § 12) from

$$c = \frac{1}{\pi}\int_K f_1\, r^{-\pi/\omega_0}\sin\frac{\pi}{\omega_0}\omega\, dx \, .$$

Let us *summarize* our investigations : If $\omega_0 > \pi$ then the solution $u_1 \in V^{1,2}(K,0)$ of the boundary value problem (1.3) has the form

$$u_1(x) = c\, r^{\pi/\omega_0}\sin\frac{\pi}{\omega_0}\omega + w_1(x) \, ,$$

where $w_1(x) \in V^{2,2}(K,0)$ and $c = \frac{1}{\pi}\int_K f_1\, r^{-\pi/\omega_0}\sin\frac{\pi}{\omega_0}\omega\, dx$. If $\omega_0 \leq \pi$, then $u_1(x) = w_1(x)$.

1.4. THE SOLUTION OF THE BOUNDARY VALUE PROBLEM IN THE POYGONAL DOMAIN Ω .

(i) Let Ω be a polygonal domain, O_j its corner points and ω_j the angles at O_j , $j = 1,\ldots,J_0$. Assume that $\omega_j > \pi$ for $j = 1,2,\ldots,J$. We denote $r_j = |x - O_j|$ and $\eta_j(x) = \eta(r_j)$, where η is defined by (1.2). Assume δ is so small that O_j is the only corner point in the ball $\{x : |x - O_j| \leq 2\delta\}$, $j = 1,\ldots,J$.

We consider the weak solution $u \in W_0^{1,2}(\Omega)$ of (1.1). We have

$$(1.24) \qquad u = \sum_{j=1}^{J} n_j^2 u + u - \sum_{j=1}^{J} n_j^2 u = \sum_{j=1}^{J} n_j^2 u + \left(1 - \sum_{j=1}^{J} n_j^2\right) u .$$

It follows from (1.23) that

$$(1.25) \qquad n_j u = c_j \, r_j^{\pi/\omega_j} \sin \frac{\pi}{\omega_j} \omega^j + w_j(x) ,$$

where $w_j(x) \in V^{2,2}(K_j,0)$, K_j is the infinite cone with the vertex O_j and the angle ω_j , $j = 1,\ldots,J$, and ω^j is the variable polar angle to O_j . Inserting (1.25) into (1.24) we get

$$(1.26) \qquad u = \sum_{j=1}^{J} c_j \, n_j \, r_j^{\pi/\omega_j} \sin \frac{\pi}{\omega_j} \omega^j + w(x) ,$$

where

$$w(x) = \sum_{j=1}^{J} n_j \, w_j(x) + \left(1 - \sum_{j=1}^{J} n_j^2\right) u \in W^{2,2}(\Omega) .$$

(ii) Now we are able to give the answer to our problem, formulated at the beginning in 1.1 : *If Ω has J reentrant corner points ($\omega_j > \pi$, $j = 1,\ldots,J$), then the weak solution $u \in W_0^{1,2}(\Omega)$ has the form (1.26). If Ω is a convex polygonal domain, then $u \in W^{2,2}(\Omega)$.*

§ 2 . A m i x e d b o u n d a r y v a l u e p r o b l e m f o r t h e L a p l a c e o p e r a t o r

2.1. <u>FORMULATION OF THE PROBLEM.</u> Let Ω be a polygonal domain in $\mathbb{R}^2$. (See Fig. 5.) Let M be the following set of boundary points of Ω :

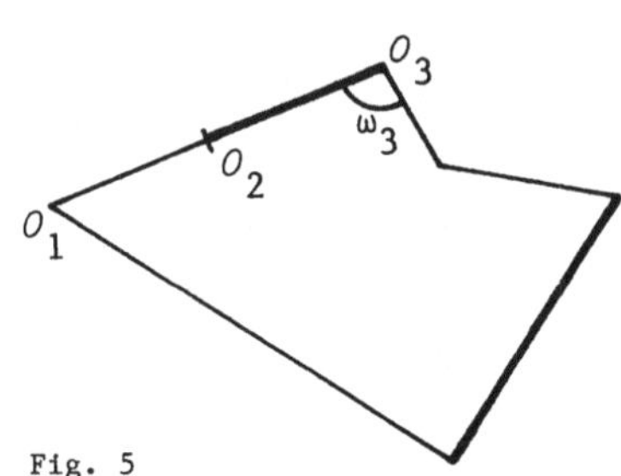

$M = \{O_j\}_{j=1,\ldots,J_0}$ consists of corner points ($\omega_j \neq \pi$) and of points where boundary conditions change their type ($\omega_j = \pi$ is possible).

The problem is the following :

For a given $f \in L^2(\Omega)$ investigate the smoothness of the solution u of the mixed problem:

Fig. 5

$$- \Delta u = - \frac{\partial^2 u}{\partial x_1^2} - \frac{\partial^2 u}{\partial x_2^2} = f(x_1,x_2) \quad \textit{for} \quad x = (x_1,x_2) \in \Omega ,$$

$$(2.1) \qquad u = 0 \qquad\qquad \textit{for} \quad x \in \Gamma_D ,$$

$$\frac{\partial u}{\partial n} = 0 \qquad\qquad \textit{for} \quad x \in \Gamma_N ,$$

where $\Gamma_D \cup \Gamma_N = \partial\Omega$, $\Gamma_N = \bigcup_{j \in J_1} \Gamma_j$, $\Gamma_D = \bigcup_{j \in J_2} \Gamma_j$, $\Gamma_D \cap \Gamma_N = \emptyset$, Γ_j
$= [O_j, O_{j+1})$, $O_{J_0+1} = O_1$, $j = 1, \ldots, J_0$.

The existence of a weak solution $u \in W^{1,2}(\Omega)$ again follows from the Lax-Milgram theorem 15.5 (see 15.2 of Chapter II). In what follows we will study the problem how regular is the weak solution $u \in W^{1,2}(\Omega)$ with respect to the angles ω_j .

We have to consider three cases for O_j :

$O_j \in \overline{\Gamma}_D \smallsetminus \overline{\Gamma}_N$; this means $\Gamma_{j-1} \cup \Gamma_j \subset \Gamma_D$ and we speak of $D - D$ conditions; or

$O_j \in \overline{\Gamma}_N \smallsetminus \overline{\Gamma}_D$; this means $\Gamma_{j-1} \cup \Gamma_j \subset \Gamma_N$ and we speak of $N - N$ conditions; or

$O_j \in \overline{\Gamma}_N \cap \overline{\Gamma}_D$; this means e.g. $\Gamma_{j-1} \subset \Gamma_D$ and $\Gamma_j \subset \Gamma_N$ and we speak of
$\qquad\qquad$ D - N conditions.

The first case has already been studied in § 1.

2.2. N - N BOUNDARY CONDITIONS IN K .

(i) Let O_j be a point at which $N - N$ conditions appear. We choose the origin O at this point and multiply the solution $u \in W^{1,2}(\Omega)$ of (2.1) by the cut-off function η defined by (1.2). Let K be the infinite cone with the vertex O and the angle $\omega_j = \omega_0$. Assume that one side of K coincides with the x_1- axis. For $u_1 = \eta u$ we get, analogously to (1.3), a boundary value problem in K :

$$-\Delta u_1 = f_1 \quad \text{in} \quad K ,$$
$$(2.2)$$
$$\frac{\partial u_1}{\partial n} = g_1 \quad \text{on} \quad \partial K ,$$

where $g_1 \in V^{1,2}(K,\beta)$ for $\beta \geq 0$, and $g_1(x) = 0$ for $|x| < \delta$ and $|x| \geq 2\delta$ and $x \in \partial K$.

(ii) Changing the variables as in 1.3 and applying the Fourier transform (1.6) we obtain

$$\lambda^2 \tilde{u}_1 - \frac{\partial^2}{\partial\omega^2} \tilde{u}_1 = \tilde{F}_1(\lambda,\omega) \quad \text{for} \quad 0 < \omega < \omega_0 ,$$
$$(2.3)$$
$$\frac{\partial \tilde{u}_1}{\partial\omega} (\lambda,0) = \tilde{G}_1(\lambda,0) ,$$
$$\frac{\partial \tilde{u}_1}{\partial\omega} (\lambda,\omega_0) = \tilde{G}_1(\lambda,\omega_0) ,$$

where $\tilde{u}_1 = \mathcal{F}(\hat{u}_1)$, $\tilde{F}_1 = \mathcal{F}(\hat{F}_1) = \mathcal{F}(e^{2\tau} \hat{f}_1)$, $\tilde{G}_1 = \mathcal{F}(e^{\tau} \hat{g}_1)$. We denote by $\mathcal{U}_0(\lambda)$ the operator of the boundary value problem (2.3)

$$\mathcal{U}_0(\lambda) : W^{2,2}(G) \rightarrow L^2(G) \times \mathbb{C} \times \mathbb{C} .$$

The eigenvalues of $\mathcal{U}_0(\lambda)$ are $\lambda_k = \dfrac{ik\pi}{\omega_0}$, $k = 0, \pm1, \pm2, \ldots$, the corresponding eigenfunctions are $\tilde{e}_k = \cos \dfrac{k\pi}{\omega_0} \omega$. Analogously to (vii) of 1.3 we have: If no eigenvalues of $\mathcal{U}_0(\lambda)$ are on the line $\operatorname{Im} \lambda = h = \beta - 1$, $\beta \geq 0$, then the inverse Fourier transform

$$\hat{u}_{1,h}(\tau,\omega) = \frac{1}{\sqrt{2\pi}} \int_{-\infty+ih}^{\infty+ih} \tilde{u}_1(\lambda,\omega) \, e^{i\lambda\tau} \, d\lambda$$

exists. Here $\tilde{u}_1(\lambda,\omega) = \mathcal{U}_0^{-1}(\lambda)\left[\tilde{F}_1,\tilde{G}_1(0),\tilde{G}_1(\omega_0)\right]$. Further, $\tilde{u}_{1,h}(\tau,\omega) = u_{1,h}(x)$ is the solution of (2.3) from $V^{2,2}(K,\beta)$.

(iii) Let us now consider the solution $u_1 = \eta u$ of (2.2), where $u \in W^{1,2}(\Omega)$ is the weak solution of the problem (2.1). It was proved by V. A. KONDRAT'EV [1] that $u_1 \in V^{2,2}(K, 1 + \varepsilon)$, where ε is a small positive number. Since the solution from this weighted Sobolev space is uniquely determined we have

$$u_1(x) = u_{1,\varepsilon}(x) = \frac{1}{\sqrt{2\pi}} \int_{-\infty+i\varepsilon}^{\infty+i\varepsilon} \tilde{u}_1(\lambda,\omega) \, e^{i\lambda\tau} \, d\lambda .$$

Calculating this integral in the same way as in 1.3, (viii), (in this case in the rectangle with the corner points $- N + i\varepsilon$, $- N - i$, $N - i$, $N + i\varepsilon$) we get the following expansion of u_1 for $\omega_0 > \pi$:

$$(2.4) \qquad u_1(x) = 1 \cdot c_1 + c_2 \, r^{\pi/\omega_0} \cos \frac{\pi}{\omega_0} \omega + w_1(x) ,$$

where $w_1(x) \in V^{2,2}(K,0)$.

<u>2.3. N – D BOUNDARY CONDITIONS IN K .</u> (i) Let O_j be a boundary point, where $N - D$ conditions appear. Choosing the origin O at O_j and multiplying the solution $u \in W^{1,2}(\Omega)$ of (2.1) with η we get the following boundary value problem in the infinite cone K (K has the vertex O , the angle $\omega_j = \omega_0$ and the side with Neumann conditions coincides with the x_1 – axis) :

$$- \Delta u_1 = f_1 \quad \text{in } K ,$$

$$(2.5) \qquad \frac{\partial u_1}{\partial n} = g_1 \quad \text{for } \omega = 0 ,$$

$$u_1 = 0 \quad \text{for } \omega = \omega_0 ,$$

where $u_1 = \eta u$, f_1 and g_1 are described by (2.2) and (1.3).

(ii) Changing the variables and applying the complex Fourier transform as in the preceding examples we get

$$\lambda^2 \tilde{u}_1 - \frac{\partial^2}{\partial\omega^2} \tilde{u}_1 = \tilde{F}_1(\lambda,\omega) \quad \text{for} \quad 0 < \omega < \omega_0 \,,$$

$$(2.6) \qquad \frac{\partial \tilde{u}_1}{\partial\omega}(\lambda,0) = \tilde{G}_1(\lambda,0) \,,$$

$$\tilde{u}_1(\lambda,\omega_0) = 0 \,.$$

The eigenvalues of the corresponding operator $\mathcal{U}_0(\lambda)$ are $\lambda_k = i(k + \frac{1}{2})\frac{\pi}{\omega_0}$, $k = 0, \pm 1, \pm 2, \ldots$ and the eigenfunctions are $\tilde{e}_k(\omega) = \cos\left(\frac{k + 1/2}{\omega_0}\right)\pi\omega$. Again we have : If no eigenvalues of $\mathcal{U}_0(\lambda)$ lie on the line $\operatorname{Im}\lambda = h = \beta - 1$, $\beta \geq 0$, then

$$\hat{u}_{1,h}(\tau,\omega) = \frac{1}{\sqrt{2\pi}} \int_{-\infty+ih}^{\infty+ih} \tilde{u}_1(\lambda,\omega)\, e^{i\lambda\tau}\, d\lambda = u_{1,h}(x) \in V^{2,2}(K,\beta)$$

and $u_{1,h}(x)$ is the uniquely determined solution from $V^{2,2}(K,\beta)$ of the problem (2.5).

(iii) Since $u_1 = \eta u \in V^{2,2}(K, 1 + \varepsilon)$ we have

$$(2.7) \qquad u_1(x) = u_{1,\varepsilon}(x) = \frac{1}{\sqrt{2\pi}} \int_{-\infty+i\varepsilon}^{\infty+i\varepsilon} \tilde{u}_1(\lambda,\omega)\, e^{i\lambda\tau}\, d\lambda \,,$$

where $\varepsilon > 0$ is sufficiently small ($\varepsilon < \frac{1}{8}$). We calculate the integral (2.7) analogously as for the D-D condition and the N-N condition, choosing the rectangle with the corner points $-N + i\varepsilon$, $-N - i$, $N - i$, $N + i\varepsilon$. Since the eigenvalues $\lambda_{-1} = -\frac{i}{2}\frac{\pi}{\omega_0}$ for $\frac{\pi}{2} < \omega_0 < \frac{3}{2}\pi$ and both eigenvalues λ_{-1} and $\lambda_{-2} = -\frac{3}{2}i\frac{\pi}{\omega_0}$ for $\omega_0 > \frac{3}{2}\pi$ lie in this rectangle, we get

$$(2.8) \qquad u_1(x) = c\, r^{\pi/2\omega_0} \cos\frac{\pi\omega}{2\omega_0} + w_1(x) \quad \text{for} \quad \frac{\pi}{2} < \omega_0 < \frac{3}{2}\pi \,,$$

$$(2.9) \qquad u_1(x) = c_1\, r^{\pi/2\omega_0} \cos\frac{\pi\omega}{2\omega_0} + c_2\, r^{3\pi/2\omega_0} \cos\frac{3\pi}{2\omega_0}\omega + w_1(x)$$

$$\text{for} \quad \omega_0 > \frac{3}{2}\pi \,,$$

where $w_1(x) \in V^{2,2}(K,0)$.

(iv) <u>REMARK.</u> If we choose the coordinate system (x_1,x_2) in such a way that the Dirichlet condition is concentrated on that side of K which coincides with the x_1-axis, i.e.

$$- \Delta u_1 = f_1 \quad \text{in} \quad K \,,$$

$$u_1 = 0 \quad \text{for} \quad \omega = 0 \,, \quad \frac{\partial u_1}{\partial n} = g_1 \quad \text{for} \quad \omega = \omega_0 \,,$$

we get the same eigenvalues λ_k of the corresponding operator $\mathcal{O}_0(\lambda)$ but the eigenfunctions are $\sin (k + \frac{1}{2}) \frac{\pi}{\omega_0} \omega$.

2.4. THE SOLUTION IN THE POLYGONAL DOMAIN Ω .

(i) Let us consider the following points of M :

- Points from $\overline{\Gamma}_D \setminus \overline{\Gamma}_N$ where $\omega_j > \pi$. We denote by J_D the corresponding index set.

- Points from $\overline{\Gamma}_N \setminus \overline{\Gamma}_D$ where $\omega_j > \pi$. We denote by J_N the corresponding index set.

- Points from $\overline{\Gamma}_N \cap \overline{\Gamma}_D$ where $\frac{\pi}{2} < \omega_j < \frac{3}{2} \pi$. We denote by J_{ND_1} the corresponding index set.

- Points from $\overline{\Gamma}_N \cap \overline{\Gamma}_D$ where $\omega_j > \frac{3}{2} \pi$. Let J_{ND_2} be the corresponding index set.

(ii) Let $u \in W^{1,2}(\Omega)$ be the weak solution of (2.1). We investigate its regularity using the same method as in 1.4. We write the solution u in the form

$$(2.10) \qquad u = \sum_{j \in J_D} \eta_j^2 u + \sum_{j \in J_N} \eta_j^2 u + \sum_{j \in J_{ND_1}} \eta_j^2 u + \sum_{j \in J_{ND_2}} \eta_j^2 u$$

$$+ \left(1 - \sum_{j \in J_0 \cup J_N \cup J_{ND_1} \cup J_{ND_2}} \eta_j^2 u\right) .$$

Then we use the expansions (1.25), (2.4), (2.8) and (2.9) obtaining

$$(2.11) \qquad u = \sum_{j \in J_D} c_j \eta_j r_j^{\pi/\omega_j} \sin \frac{\pi}{\omega_j} \omega^j + \sum_{j \in J_N} c_j \eta_j r_j^{\pi/\omega_j} \cos \frac{\pi}{\omega_j} \omega^j$$

$$+ \sum_{j \in J_{ND_1}} c_j \eta_j r_j^{\pi/2\omega_j} \cos \frac{\pi}{2\omega_j} \omega^j$$

$$+ \sum_{j \in J_{ND_2}} \left(c_j \eta_j r_j^{\pi/2\omega_j} \cos \frac{\pi}{2\omega_j} \omega^j + \tilde{c}_j \eta_j r_j^{3\pi/2\omega_j} \cos \frac{3\pi}{2\omega_j} \omega^j\right) + w(x) ,$$

where $w(x) \in W^{2,2}(\Omega)$. ω^j denotes the locally defined variable angle, $0 < \omega^j < \omega_j$. In § 13 we will give the formulae for the coefficients c_j and $\tilde{c}_j$.

Formula (2.11) completely describes the regularity of u .

§3. The Dirichlet problem for the bi-harmonic operator

3.1. <u>FORMULATION OF THE PROBLEM.</u> Let Ω be a polygonal domain in R^2 . *For a given* $f \in L^2(\Omega)$ *investigate the smoothness of the solution* u *of the Dirichlet problem*

$$\Delta^2 u = \frac{\partial^4 u}{\partial x_1^4} + 2 \frac{\partial^4 u}{\partial x_1^2 \partial x_2^2} + \frac{\partial^4 u}{\partial x_2^4} = f \quad \text{for} \quad x = (x_1, x_2) \in \Omega ,$$

(3.1)

$$u = \frac{\partial u}{\partial n} = 0 \quad \text{for} \quad x \in \partial\Omega .$$

The Lax-Milgram theorem 15.5 implies that there exists a weak solution $u \in W_0^{2,2}(\Omega)$. In what follows we study how the regularity of u depends on the sizes of the angles of our polygonal domain Ω .

3.2. <u>THE BOUNDARY VALUE PROBLEM IN AN INFINITE CONE K .</u> (i) We choose the origin O at one corner point of Ω with the angle ω_0 and consider the infinite cone K with the vertex O and the angle ω_0 . Multiplying the solution $u \in W_0^{2,2}(\Omega)$ of (3.1) by the cut-off function η defined by (1.2), we get a boundary value problem for $u_1 = \eta u$ in K , namely

$$\Delta^2 u_1(x) = f_1(x) \quad \text{for} \quad x = (x_1, x_2) \in K ,$$

(3.2)

$$u_1(x) = \frac{\partial u_1(x)}{\partial n} = 0 \quad \text{for} \quad x \in \partial K ,$$

where $f_1(x) = \sum\limits_{|\alpha|=4} a_\alpha D^\alpha u_1 = \sum\limits_{|\alpha|=4} a_\alpha D^\alpha \eta u = \eta \Delta^2 u + \sum\limits_{|\alpha|=4} a_\alpha \sum\limits_{\gamma < \alpha} D^{\alpha-\gamma} \eta D^\gamma u$, $\alpha = (\alpha_1, \alpha_2)$, $\gamma = (\gamma_1, \gamma_2)$, and $\alpha > \gamma$ means $\alpha_1 > \gamma_1$ and $\alpha_2 \geq \gamma_2$ or $\alpha_1 \geq \gamma_1$ and $\alpha_2 > \gamma_2$. It is evident that $f_1 \in L^2(K, \beta)$ for $\beta \geq 0$.

(ii) We change the variables in (3.2) in the same way as in 1.3 and use the complex Fourier transform 1.6. We obtain the following boundary value problem depending on the parameter λ for $\tilde{u}_1 = \tilde{u}_1(\lambda, \omega)$:

$$A_0(\lambda)\tilde{u}_1 = \frac{\partial^4 \tilde{u}_1}{\partial \omega^4} + 2(-\lambda^2 - 2i\lambda + 2) \frac{\partial^2 \tilde{u}_1}{\partial \omega^2} + (-\lambda^2)(i\lambda - 2)^2 \tilde{u}_1 = \tilde{F}_1(\lambda, \omega)$$
$$\text{for} \quad \omega \in G = (0, \omega_0) ,$$

(3.3)

$$B_{01}(\lambda)\tilde{u}_1 = \tilde{u}_1(\lambda, \omega) = 0 \quad \text{for} \quad \omega \in \partial G ,$$

$$B_{02}(\lambda)\tilde{u}_1 = \frac{\partial \tilde{u}_1}{\partial \omega}(\lambda, \omega) = 0 \quad \text{for} \quad \omega \in \partial G ,$$

where $\tilde{F}_1(\lambda, \omega) = \tilde{\mathcal{F}}(e^{4\tau} \hat{f}_1(\tau, \omega))$, $\tilde{u}_1 = \mathcal{F}(\hat{u}_1(\tau, \omega))$. Analogously to § 1 we can show that $F_1(\tau, \omega) = e^{4\tau} \hat{f}_1(\tau, \omega)$ is Fourier transformable in the half-space $\text{Im } \lambda \geq -3$ ($h = \beta - 3$, $\beta \geq 0$).

The operator $\mathcal{Ol}_0(\lambda) = \left(A_0(\lambda), B_{01}(\lambda), B_{02}(\lambda)\right)$ maps $W^{4,2}(G)$ into $L^2(G) \times W^{2,3+1/2}(\partial G) \times W^{2,2+1/2}(\partial G)$.

(iii) Let us say that $\lambda = \lambda_0$ is a *generalized eigenvalue* of $\mathcal{Ol}_0(\lambda)$ if there exists a nontrivial solution $\tilde{e}(\lambda_0, \omega) \in W^{2,4}(G)$ of $\mathcal{Ol}_0(\lambda)\,\tilde{e}(\lambda, \omega) = (0,0,0)$ (shortly we also speak of an eigenvalue of $\mathcal{Ol}_0(\lambda)$); $\tilde{e}(\lambda_0, \omega)$ is the eigenfunction of $\mathcal{Ol}_0(\lambda)$ with respect to λ_0 .

The calculation of the eigenvalues and the corresponding eigenfunctions of $\mathcal{Ol}_0(\lambda)$ is not so simple as in § 1 and § 2 . It was done by H. MELZER and R. RANNACHER [1], H. BLUM and R. RANNACHER [1] and V. G. MAZ'JA and B. A. PLAMENEVSKIĬ [7]. Let us shortly describe the method :

The general solutions of the equation
$$A_0(\lambda)\tilde{e} = 0$$

are of the following form :

(3.4) $\quad \tilde{e}(\lambda, \omega) = c_1 \cos (i\lambda\omega) + c_2 \sin (i\lambda\omega) + c_3 \cos (i\lambda-2)\omega$
$$\qquad\qquad + c_4 \sin (i\lambda-2)\omega \quad \text{for} \quad \lambda \neq 0 , \quad \lambda \neq -2i , \quad \lambda \neq -i ,$$

(3.5) $\quad \tilde{e}(\lambda, \omega) = c_1 + c_2\omega + c_3 \sin 2\omega + c_4 \cos 2\omega \quad \text{for} \quad \lambda = 0 \quad \text{and} \quad \lambda = -2i,$

(3.6) $\quad \tilde{e}(\lambda, \omega) = c_1 \sin \omega + c_2\omega \sin \omega + c_3 \cos \omega + c_4\omega \cos \omega \quad \text{for} \quad \lambda = -i$.

We determine the coefficients c_1 , c_2 , c_3 and c_4 in such a way that the boundary conditions $B_{01}\tilde{e} = 0$ and $B_{02}\tilde{e} = 0$ are satisfied. This leads to a linear system of four equations which admits a nontrivial solution if and only if the determinant $D(\lambda)$ of the corresponding matrix of coefficients vanishes. We get

(3.7) $\quad D(\lambda) = \sin^2(i\lambda-1)\omega_0 - (i\lambda-1)^2 \sin^2\omega_0 = 0 \quad \text{for} \quad \lambda \neq 0 , \quad \lambda \neq -i ,$
$$\qquad\qquad\qquad\qquad\qquad\qquad\qquad\qquad\qquad\qquad\qquad\qquad \lambda \neq -2i ,$$

(3.8) $\quad D(\lambda) = \sin \omega_0(-\sin \omega_0 + \omega_0 \cos \omega_0) = 0 \quad \text{for} \quad \lambda = 0 , \quad \lambda = -2i ,$

(3.9) $\quad D(\lambda) = \sin^2\omega_0 - \omega_0^2 = 0 \qquad\qquad\qquad \text{for} \quad \lambda = -i$.

Since $\omega_0 > 0$, $\lambda = -i$ is not an eigenvalue. If $\sin \omega_0 \neq \omega_0 \cos \omega_0$ then the eigenvalues of $\mathcal{Ol}_0(\lambda)$ are the zeros of $D(\lambda)$ given by (3.7). We give a description of Im λ in Fig. 6. (Compare H. MELZER and R. RANNACHER [1].) If $\sin \omega_0 = \omega_0 \cos \omega_0$, then $\lambda = 0$ and $\lambda = -2i$ are eigenvalues.

(iv) The following regularity result holds (cf. 1.3 (vii)) : If no eigenvalues of $\mathcal{Ol}_0(\lambda)$ lie on the line Im $\lambda = h = \beta - 3$, $\beta \geq 0$, then the inverse Fourier transform

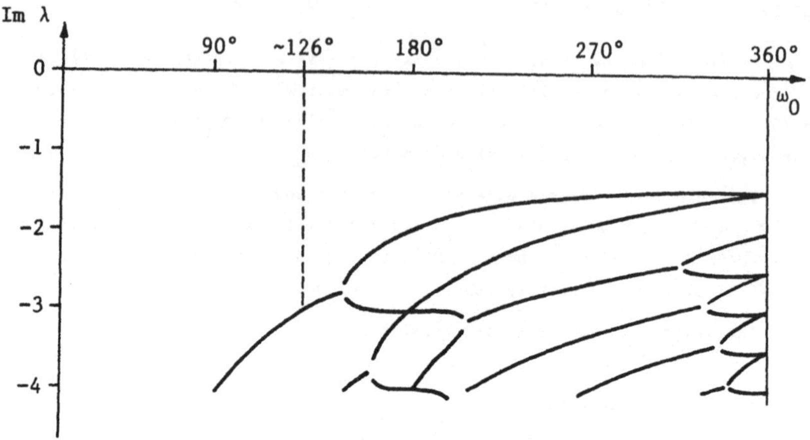

Fig. 6

$$\hat{u}_{1,h}(\tau,\omega) = \frac{1}{\sqrt{2\pi}} \int\limits_{-\infty+ih}^{\infty+ih} \tilde{u}_1(\lambda,\omega)e^{i\lambda\tau}\, d\lambda$$

exists and $\hat{u}_{1,h}(\tau,\omega) = u_{1,h}(x)$ is the uniquely determined solution of (3.2) from $V^{2,4}(K,\beta)$.

Let us consider the solution $u_1 = \eta u$ of (3.2), where $u \in W_0^{2,2}(\Omega)$. V. A. KONDRAT'EV [1] has proved that $u_1 \in V^{4,2}(K,2)$. Therefore

$$(3.10) \qquad u_1(x) = \hat{u}_{1,-1}(\tau,\omega) = \frac{1}{\sqrt{2\pi}} \int\limits_{-\infty-i}^{\infty-i} \tilde{u}_1(\lambda,\omega)e^{i\lambda\tau}\, d\lambda \ .$$

(v) In order to get an expansion of $u_1(x)$ analogous to (1.23) we have to calculate the integral (3.10) via the Cauchy theorem. We get

$$(3.11) \qquad u_1(x) = \frac{1}{\sqrt{2\pi}} \lim_{N\to\infty} \Big(\int\limits_{-N-i}^{-N-3i} \tilde{u}_1\, e^{i\lambda\tau}\, d\lambda + \int\limits_{-N-3i}^{N-3i} \tilde{u}_1\, e^{i\lambda\tau}\, d\lambda + \int\limits_{N-3i}^{N-i} \tilde{u}_1\, e^{i\lambda\tau}\, d\lambda \Big)$$

$$+ \frac{1}{\sqrt{2\pi}} 2\pi i \sum_{-3 < \text{Im } \lambda < -1} \text{Res } \tilde{u}_1(\lambda,\omega)e^{i\lambda\tau} \ .$$

The first and third integrals tend to zero for $N \to \infty$. The second integral yields a function from $V^{4,2}(K,0)$. The residua are calculated in the same manner as in 1.3, (viii), provided the corresponding zeros of $D(\lambda)$ (the eigenvalues of $\mathcal{U}_0(\lambda)$) are simple. If the multiplicity of the zeros of $D(\lambda)$ is two (a higher multiplicity is impossible) then the calculation of the corresponding residuum is more complicated.

Let us show that zeros of $D(\lambda)$ of multiplicity two exist. We consider $D(\lambda)$ defined by (3.7). $D'(\lambda) = 0$ if and only if

$$(3.12) \qquad \omega_0 \sin (i\lambda-1)\omega_0 \cos (i\lambda-1)\omega_0 = (i\lambda-1) \sin^2\omega_0 \; .$$

If $\dfrac{\sin \omega_0}{\omega_0} = \cos (i\lambda-1)\omega_0$ and $\tan (i\lambda-1)\omega_0 = (i\lambda-1)\omega_0$ then (3.12) is valid (e.g. for $\omega_0 \cong 0.406 \; \pi$, cf. H. MELZER and R. RANNACHER [1]).

We now describe the calculation of the residua for zeros of $D(\lambda)$ of multiplicity two. We have to distinguish two cases :

1° The rank of the corresponding coefficient matrix is two for $\lambda = \lambda_0$. Then two linearly independent eigenfunctions $\tilde{e}_{01}$ and $\tilde{e}_{02}$ exist. In a neighborhood of λ_0 the operator $\mathcal{U}_0^{-1}(\lambda)$ has the form

$$\mathcal{U}_0^{-1}(\lambda) = \frac{P_1}{\lambda - \lambda_0} + \Gamma(\lambda)$$

and the image of P_1 is contained in the eigenspace of λ_0 . Analogous to 1.3, (viii) it follows that

$$(3.13) \qquad \mathrm{Res} \; \tilde{u}_1(\lambda,\omega)e^{i\lambda\tau}\big|_{\lambda=\lambda_0} = r^{i\lambda_0}(c_1\tilde{e}_{01} + c_2\tilde{e}_{02}) \; .$$

2° The rank of the corresponding coefficient matrix is three for $\lambda = \lambda_0$. Then there is one eigenfunction $\tilde{e}_{01}$ and one so-called associate function $\tilde{k}_{01}$ with respect to λ_0 . (For the definition of an associate function see § 7.) In this case we have

$$\mathcal{U}_0^{-1}(\lambda) = \frac{P_2}{(\lambda - \lambda_0)^2} + \frac{P_1}{(\lambda - \lambda_0)} + \Gamma(\lambda) \; ,$$

where the images of P_2 and P_1 are contained in the one dimensional spaces of the eigenfunctions and of the associate functions, respectively. We get

$$(3.14) \qquad \mathrm{Res} \; \tilde{u}_1(\lambda,\omega)e^{i\lambda\tau}\big|_{\lambda=\lambda_0} = r^{i\lambda_0}[c_1\tilde{e}_{01} + c_2\tilde{k}_{0,1}] + r^{i\lambda_0} i \ln r \; c_3 \tilde{e}_{01} \; .$$

Let us *summarize* these results denoting in the strip $-3 < \mathrm{Im} \; \lambda < -1$ by $\lambda_1,\dots,\lambda_{N_1}$ the simple zeros of $D(\lambda)$, by $\lambda_{N_1+1},\dots,\lambda_{N_2}$ the zeros of $D(\lambda)$ of multiplicity two provided the rank of the corresponding coefficient matrix is two and by $\lambda_{N_2+1},\dots,\lambda_{N_3}$ the zeros of $D(\lambda)$ of the multiplicity two provided the rank of the corresponding coefficient matrix is three. The equations (3.11), (3.13) and (3.14) imply that *the following expansion holds for* $u_1 = \eta u$, $u \in W_0^{2,2}(\Omega)$:

$$(3.15) \qquad u_1(x) = \sum_{j=1}^{N_1} c_j \, r^{i\lambda_j} \tilde{e}_j + \sum_{j=N_1+1}^{N_2} r^{i\lambda_j}(c_{j1}\tilde{e}_{j1} + c_{j2}\tilde{e}_{j2}) +$$

$$+ \sum_{j=N_2+1}^{N_3} r^{i\lambda_j} [c_{j1}\tilde{e}_{j1} + c_{j2}\tilde{k}_{j1} + (i \ln r)c_{j3}\tilde{e}_{j1}] + w_1(x) \ ,$$

where $w_1(x) \in V^{4,2}(K,0)$; c_j , c_{j1} , c_{j2} , c_{j3} are constants, $\tilde{e}_j$, $\tilde{e}_{j1}$, $\tilde{e}_{j2}$ are eigenfunctions with respect to λ_j , $\tilde{k}_{j1}$ are the associate functions.

<u>3.3.</u> <u>THE SOLUTION IN THE POLYGONAL DOMAIN Ω .</u> (i) For the sake of simplicity we assume that the polygonal domain Ω has only one corner point O with an angle $\omega_0 > 126°$. In fact, it follows from formula (3.15) (see Fig.6) that for $\omega_0 \leq 126°$ no eigenvalues of $\mathcal{U}_0(\lambda)$ lie in the strip $-3 < \operatorname{Im} \lambda < -1$ and therefore $u_1(x) = w_1(x) \in V^{4,2}(K,0)$. In this case the solution $u \in W_0^{2,2}(\Omega)$ of the boundary value problem (3.1) is "regular" in a neighborhood of such a corner point.

We write again
$$u = \eta^2 u + (1 - \eta^2)u \ ,$$
where η is defined by (1.2). Using the expansion (3.15) we get

$$(3.16) \qquad u(x) = \sum_{j=1}^{N_1} c_j \eta r^{i\lambda_j} \tilde{e}_j + \sum_{j=N_1+1}^{N_2} \eta r^{i\lambda_j} (c_{j1}\tilde{e}_{j1} + c_{j2}\tilde{e}_{j2})$$

$$+ \sum_{j=N_2+1}^{N_3} \eta r^{i\lambda_j} [c_{j1}\tilde{e}_{j1} + c_{j2}\tilde{k}_{j1} + (i \ln r)c_{j3}\tilde{e}_{j1}] + w(x)$$

where $w(x) \in W^{4,2}(\Omega)$.

(ii) <u>REMARK.</u> Other boundary value problems for the biharmonic operator, among them those with mixed boundary conditions, were investigated by H. MELZER and R. RANNACHER [1] and H. BLUM and R. RANNACHER [1].

§ 4 . A N a v i e r - S t o k e s e q u a t i o n

<u>4.1.</u> <u>FORMULATION OF THE PROBLEM.</u> Let Ω be a two dimensional, simply connected domain which has only one corner point O with the angle ω_0 (see Fig. 7).

Fig. 7

Assume there is a neighborhood $U(0)$ of 0 such that $U(0) \cap \overline{\Omega} = \{x \in \mathbb{R}^2 : |x| < \delta , \ 0 \le \omega \le \omega_0\}$ where ω again denotes the polar angle.

We consider the two dimensional field equation for a steady, viscous incompressible flow

$$\gamma \Delta^2 u + \frac{\partial u}{\partial x_1} \frac{\partial \Delta u}{\partial x_2} - \frac{\partial u}{\partial x_2} \frac{\partial \Delta u}{\partial x_1} = \operatorname{div} \vec{F} = f \quad \text{in} \quad \Omega ,$$

(4.1)

$$u = \frac{\partial u}{\partial n} = 0 \qquad\qquad \text{on} \quad \partial \Omega ,$$

where u is the so-called stream function of the flow and $\vec{F}$ is the mass force density ; $f \in L^2(\Omega)$; γ is a constant. We consider the weak solution $u \in W_0^{2,2}(\Omega)$. It was proved by L.A.OGANESJAN [1] that $u \in C^1(\overline{\Omega})$. Furthermore, the inclusion $u \in V^{4,2}(\Omega,2)$ holds, see V. A. KONDRAT'EV [1], [2]. Hence

$$\gamma \Delta^2 u = f + \frac{\partial u}{\partial x_2} \frac{\partial \Delta u}{\partial x_1} - \frac{\partial u}{\partial x_1} \frac{\partial \Delta u}{\partial x_2} = g ,$$

where $g \in L^2(\Omega,1)$. Now we consider the following problem: *How does the smoothness of the solution* u *depend on the size of the angle* ω_0 ?

4.2. <u>SOLUTION OF THE PROBLEM.</u> We apply the results of § 3 to the new linear problem

(4.2) $\qquad \gamma \Delta^2 u = g \qquad \text{in} \quad \Omega ,$

$$u = \frac{\partial u}{\partial n} = 0 \quad \text{on} \quad \partial \Omega .$$

Using the notation from formula (3.15) we obtain

$$(4.3) \qquad u(x) = \sum_{j=1}^{N_1} c_j \eta r^{i\lambda_j} \tilde{e}_j + \sum_{j=N_1+1}^{N_2} \eta r^{i\lambda_j} (c_{j1}\tilde{e}_{j1} + c_{j2}\tilde{e}_{j2})$$

$$+ \sum_{j=N_2+1}^{N_3} \eta r^{i\lambda_j} \left[c_{j1}\tilde{e}_{j1} + c_{j2}\tilde{k}_{j1} + (i \ln r)c_{j3}\tilde{e}_{j1} \right] + w(x)$$

where $-2 < \operatorname{Im} \lambda_j < -1$ and $w(x) \in V^{4,2}(\Omega,1)$.

Section 2 : *A s p e c i a l   b o u n d a r y   v a l u e   p r o b l e m   i n   a n   i n f i n i t e   c o n e   K*

§ 5 .   F o r m u l a t i o n   o f   s o m e   b o u n d a r y   v a l u e   p r o b l e m s

5.1. <u>THE DOMAINS.</u> (i) An *infinite cone* $K \subset \mathbb{R}^N$ with the vertex 0 is defined by its surface-equation

$$x_N^{2p} = \sum_{i_1+\ldots+i_{N-1}=2p} a_{i_1\ldots i_{N-1}} x_1^{i_1}\ldots x_{N-1}^{i_{N-1}} + p(x) \ , \ x = (x_1,\ldots,x_N),$$

where $\displaystyle\sum_{i_1+\ldots+i_{N-1}=2p} a_{i_1\ldots i_{N-1}} x_1^{i_1}\ldots x_{N-1}^{i_{N-1}} \geq 0$, $p(x)$ is a smooth function

such that $|p(x)| = o\big((x_1^2+\ldots+x_{N-1}^2)^p\big)$, $p > 0$.

(ii) Let $\Omega \subset \mathbb{R}^N$ be an open subset of $\mathbb{R}^N$ with the compact closure $\overline{\Omega}$ whose boundary $\partial\Omega$ is an $(N-1)$-dimensional manifold. A point $O \in \partial\Omega$ is called a *conical point*, if there is a neighborhood $U_\varepsilon(O)$ of O such that $U_\varepsilon(O) \cap \Omega$ is diffeomorphic to a cone K intersected with the unit ball. The intersection of K with the unit sphere is a domain G with a smooth boundary ∂G . (See Fig. 8 for the special case when $U_\varepsilon(O)$ is the unit ball.)

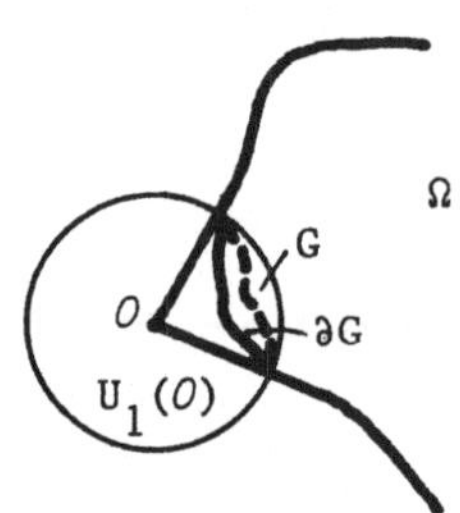

If $N = 2$, then the conical points are corner points (with the angle $\omega \neq \pi$).

Assume that there is a finite number of conical points $\{O_i\}_{i=1,\ldots,I}$ at $\partial\Omega$ and that $\partial\Omega \setminus \{O_i\}_{i=1,\ldots,I}$ is smooth.

Fig. 8

5.2. THE DIFFERENTIAL OPERATORS.

(i) We consider the linear differential operators

$$(5.1) \qquad A(x,D_x) = \sum_{|\alpha|\leq 2m} a_\alpha(x) D_x^\alpha \qquad \text{defined for } x \in \overline{\Omega} \ ,$$

$$(5.2) \qquad B_j(x,D_x) = \sum_{|\alpha|\leq m_j} b_{j,\alpha}(x) D_x^\alpha \qquad \text{defined for } x \in \partial\Omega \setminus \{O_i\} \ ,$$

$$j = 1,2,\ldots,m \ ,$$

using the notation $\displaystyle D_x^\alpha = (-i)^{|\alpha|} \frac{\partial^\alpha}{\partial x^\alpha} = (-i)^{|\alpha|} \frac{\partial^{\alpha_1+\ldots+\alpha_N}}{\partial x_1^{\alpha_1}\ldots\partial x_N^{\alpha_N}}$.

Assume that A is elliptic in Ω and $\{B_1,\ldots,B_m\}$ is a normal system on $\partial\Omega \setminus \{O_i\}$ which covers A . (For more details see J. WLOKA [1].) Suppose that the coefficients a_α and $b_{j,\alpha}$ are sufficiently smooth on $\overline{\Omega}$ and $\partial\Omega \setminus \{O_i\}$, respectively.

(ii) If $N = 2$ and *various types of boundary conditions* occur, then we add the boundary points where the boundary conditions change the type and at

which the corresponding angle equals π to the set of corner points. We have
then a bigger set $M = \{O_i\}_{i=1,\ldots,T} \subset \partial\Omega$ of "singular points" and we consider the operators (5.2) on $\partial\Omega \setminus M$. (See Fig. 9, where O_2 is a corner point in the sense just mentioned.)

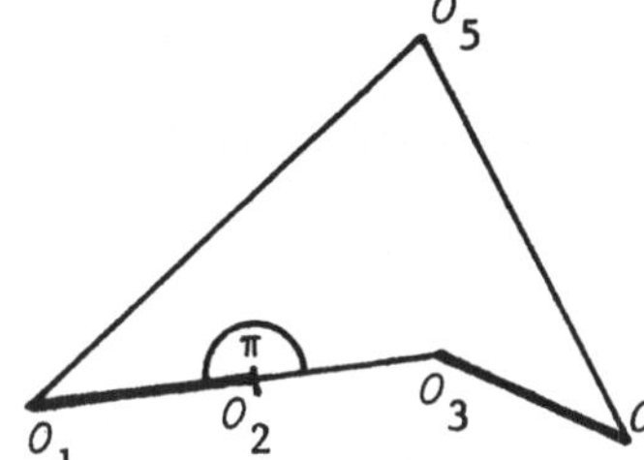

$$M = \{O_i\}_{i=1,\ldots,5}$$

Fig. 9

<u>5.3. BOUNDARY VALUE PROBLEMS IN Ω AND K </u>. (i) Now the boundary value problem in Ω is : *Investigate the solvability of*

$$(5.3) \qquad \begin{aligned} A(x,D_x)u &= f(x) &&in\ \ \Omega\ , \\ B_j(x,D_x)u &= g_j(x) &&on\ \ \partial\Omega \setminus M\ ,\quad j = 1,2,\ldots,m\ , \end{aligned}$$

in the weighted spaces $V^{K,p}(\Omega,\vec{\beta})$ *defined by* (0.19) *and study the behavior of the solutions* u *near the set* M .

If we introduce the operator $\mathcal{U}(x,D_x) = \{A(x,D_x),B_1(x,D_x),\ldots,B_m(x,D_x)\}$

$$(5.4) \qquad \mathcal{U}(x,D_x) : V^{2m+\ell,p}(\Omega,\vec{\beta}) \to V^{\ell,p}(\Omega,\vec{\beta}) \times \sum_{j=1}^{m} V^{\ell+2m-m_j-1/p,p}(\partial\Omega,\vec{\beta})$$

(for the definition of the trace spaces see 0.7), then we can formulate the boundary value problem as follows:
Investigate the properties of the operator (5.4); *for example, is this operator a Fredholm operator* ?

For this purpose we reduce our problem – using certain well-known localization principles (see § 9) – to a boundary value problem in a special domain, namely in the infinite cone K .

   (ii) Let $K \subset \mathbb{R}^N$ be the cone defined by 5.1 (i) and let $A(x,D_x)$ and $B_j(x,D_x)$ be differential operators of the type (5.1) and (5.2), defined for $x \in K$ and $x \in \partial K \setminus \{0\}$, respectively. We consider the following boundary value problem :

$$(5.5) \qquad \begin{aligned} A(x,D_x)u &= \sum_{|\alpha|\leq 2m} a_\alpha(x)D_x^\alpha u = f(x) &&in\ \ K\ , \\ B_j(x,D_x)u &= \sum_{|\alpha|\leq m_j} b_{j,\alpha}(x)D_x^\alpha u = g_j(x) &&on\ \ \partial K \setminus \{0\},\quad j = 1,\ldots,m\ . \end{aligned}$$

We denote by $\mathcal{U}_K(x,D_x) = \{A(x,D_x),B_1(x,D_x),\ldots,B_m(x,D_x)\}$ the corresponding operator

$$(5.6) \qquad \mathscr{A}_K(x,D_x) : V^{\ell+2m,p}(K,\gamma) \to V^{\ell,p}(K,\gamma) \times \sum_{j=1}^{m} V^{\ell+2m-m_j-1/p,p}(\partial K,\gamma) .$$

The properties of $\mathscr{A}_K(x,D_x)$ in a neighborhood of 0 are determined by the properties of a special operator $\mathscr{A}_0(D_x)$ given by the principal parts of $A(x,D_x)$ and $B_j(x,D_x)$, $j = 1,2,\ldots,m$, with frozen coefficients. Therefore we first consider the special boundary value problem assuming it is well defined

$$(5.7) \qquad \begin{aligned} A_0(0,D_x)u &= A_0(D_x)u = \sum_{|\alpha|=2m} a_\alpha(0)D_x^\alpha u = f \qquad \text{in } K , \\[2mm] B_{j,0}(0,D_x)u &= B_{j,0}(D_x)u = \sum_{|\alpha|=m_j} b_{j,\alpha}(0)D_x^\alpha u = g_j \quad \text{on } \partial K , \\[2mm] & \qquad j = 1,2,\ldots,m . \end{aligned}$$

The corresponding operator $\mathscr{A}_0(D_x) = \{A_0(D_x),B_{1,0}(D_x),\ldots,B_{m,0}(D_x)\}$ maps

$$(5.8) \qquad V^{2m+\ell,p}(K,\beta) \to V^{\ell,p}(K,\beta) \times \sum_{j=1}^{m} V^{2m+\ell-m_j-1/p,p}(\partial K,\beta) .$$

We will deal with the special boundary value problem (5.7) in the next sections.

§ 6. Solvability of the special problem in $V^{\ell+2m,p}(K,\beta)$

<u>6.1.</u> <u>A BOUNDARY VALUE PROBLEM DEPENDING ON A PARAMETER.</u> (i) Formulation. Let (r,ω) be the spherical coordinates in R^N . We write the operators of problem (5.7) as follows :

$$\begin{aligned} A_0(D_x) &= r^{-2m}L(\omega,D_\omega,rD_r) , \\[2mm] B_{j,0}(D_x) &= r^{-m_j} M_j(\omega,D_\omega,rD_r) , \quad j = 1,\ldots,m , \text{ where } D_r = \frac{1}{i}\frac{\partial}{\partial r} . \end{aligned}$$

The calculation of the operators L and M_j is generally not very easy (see § 3), but it is very useful **since we can** substitute $r = e^\tau$ and apply the complex Fourier transform (1.6). In fact, transforming (5.7) in this manner we get

$$(6.1) \qquad \begin{aligned} L(\omega,D_\omega,\lambda)\tilde{u}(\lambda,\omega) &= \tilde{F}(\lambda,\omega) \qquad \text{for } \omega \in G , \\[2mm] M_j(\omega,D_\omega,\lambda)\tilde{u}(\lambda,\omega) &= \tilde{G}_j(\lambda,\omega) \quad \text{for } \omega \in \partial G , \end{aligned}$$

where $\tilde{u} = \mathscr{F}(\hat{u})$, $\hat{u}(\tau,\omega) = u(x)$, $\tilde{F} = \mathscr{F}(\hat{F})$, $\hat{F}(\tau,\omega) = e^{2m\tau}\hat{f}(\tau,\omega)$, $\hat{f}(\tau,\omega) = f(x)$, $\tilde{G}_j = \mathscr{F}(\hat{G}_j)$, $\hat{G}_j(\tau,\omega) = e^{m_j\tau}\hat{g}_j(\tau,\omega)$, $\hat{g}_j(\tau,\omega) = g_j(x)$. We denote by $\mathscr{A}_0(\lambda) = \{L(\lambda),M_1(\lambda),\ldots,M_j(\lambda)\}$ the operator of the boundary value problem (6.1)

$$(6.2) \qquad \mathcal{U}_0(\lambda) : W^{2m+\ell,p}(G) \to W^{\ell,p}(G) \times \prod_{j=1}^{m} W^{2m+\ell-m_j-1/p,p}(\partial G) .$$

(ii) Solvability. We have to answer the following question : *For which λ has the problem (6.1) a uniquely determined solution ?* The answer was given by M. S. AGRANOVIČ and M. I. VIŠIK [1] for $p = 2$ and was formulated by V. G. MAZ'JA and B. A. PLAMENEVSKIĬ [4]. Before proceeding to these results we introduce a definition :

(6.3) <u>DEFINITION.</u> A complex number λ_0 is an eigenvalue of $\mathcal{U}_0(\lambda)$ if there is a function $u(\lambda) = u(\lambda,\omega) \in \mathcal{D}\left(\mathcal{U}_0(\lambda)\right) = W^{2m+\ell,p}(G)$ which is holomorphic at λ_0 , $u(\lambda_0) \neq 0$, and $\mathcal{U}_0(\lambda_0)u(\lambda_0) = 0$, $u_0 = u(\lambda_0)$ is an eigenfunction of $\mathcal{U}_0(\lambda)$ with respect to λ_0 .

(6.4) <u>LEMMA.</u> *Assume that λ is situated in a double angle containing the real axis and that $|\lambda| > a_0$ where a_0 is sufficiently large (see Fig. 10). Then there is a uniquely determined solution $\tilde{u}$ of (6.1) and*

$$(6.5) \qquad \sum_{\nu=0}^{\ell+2m} |\lambda|^\nu \, \|\tilde{u}; \, W^{\ell+2m-\nu,p}(G)\| \leq C\Big[\sum_{\nu=0}^{\ell} |\lambda|^\nu \, \|\tilde{F}; \, W^{\ell-\nu,p}(G)\|$$

$$+ \sum_{j=1}^{m} \{\|\tilde{G}_j; \, W^{2m+\ell-m_j-1/p,p}(\partial G)\| + |\lambda|^{\ell+2m-m_j-1/p} \, \|\tilde{G}_j; \, L^p(\partial G)\|\}\Big] ,$$

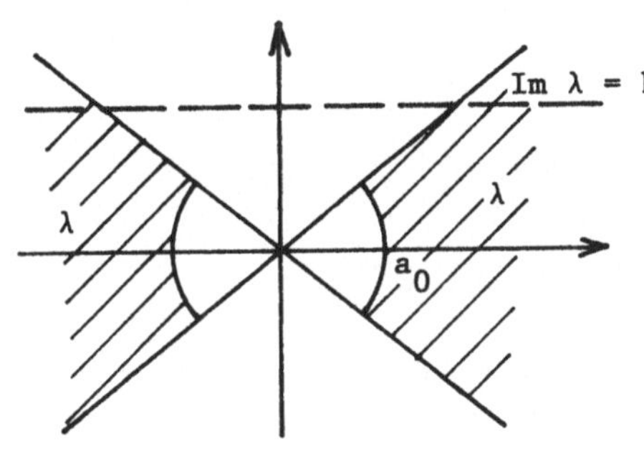

Fig. 10

where $C > 0$ is a constant independent of $\tilde{u}$ and λ .

This Lemma has the following consequence : Let ℓ_h be the line, where $\mathrm{Im}\,\lambda = h$. If no eigenvalue of $\mathcal{U}_0(\lambda)$ lies on ℓ_h, then there is a uniquely determined solution $\tilde{u}$ of (6.1) for all $\lambda \in \ell_h$ and

$$(6.6) \qquad \sum_{\nu=0}^{\ell+2m} |\lambda|^\nu \, \|\tilde{u}; \, W^{\ell+2m-\nu,p}(G)\|$$

$$\leq C(h)\Big[\sum_{j=1}^{m} \{\|\tilde{G}_j; \, W^{2m+\ell-m_j-1/p,p}(\partial G)\| + |\lambda|^{\ell+2m-m_j-1/p} \, \|\tilde{G}_j; \, L^p(\partial G)\|\}$$

$$+ \sum_{\nu=0}^{\ell} |\lambda|^\nu \, \|\tilde{F}; \, W^{\ell-\nu,p}(G)\|\Big] \quad \forall \lambda \in \ell_h.$$

Using Lemma (6.4) we are able to prove the following theorem about the solvability of the special problem (5.7) in the infinite cone K .

6.2. THEOREM. *The problem (5.7) has a uniquely determined solution* u $\in V^{\ell+2m,p}(K,\beta)$ *for every* $f \in V^{\ell,p}(K,\beta)$ *and* $g_j \in V^{\ell+2m-m_j-1/p,p}(\partial K,\beta)$, $j = 1,2,\ldots,m$, *if and only if the line* $\mathrm{Im}\,\lambda = h = \beta + \frac{N}{p} - \ell - 2m$ *contains no eigenvalue of the operator* $\mathcal{O}_0(\lambda)$.

If the latter condition is valid, then the solution u *satisfies the estimate*

$$(6.7) \qquad \| u;\ V^{\ell+2m,p}(K,\beta) \|$$

$$\leq C \Big[\| f;\ V^{\ell,p}(K,\beta) \| + \sum_{j=1}^{m} \| g_j;\ V^{\ell+2m-m_j-1/p,p}(\partial K,\beta) \| \Big] ,$$

where $C > 0$ *is independent of* u .

P r o o f : We prove the theorem only for $p = 2$, following the ideas of V. A. KONDRAT'EV [1]. We refer to V. G. MAZ'JA and B. A. PLAMENEVSKIĬ [4] for the case $1 < p < \infty$. First we show that the condition is sufficient.

(i) Assume that the line $\mathrm{Im}\,\lambda = h = \beta + \frac{N}{2} - \ell - 2m$ contains no eigenvalue of the operator $\mathcal{O}_0(\lambda)$. We verify that the transformed right hand sides $\hat{F}(\tau,\omega)$ and $\hat{G}_j(\tau,\omega)$, $j = 1,2,\ldots,m$, given by (6.1), are Fourier transformable for $\lambda = s + ih$, $s \in \mathbb{R}$, $h = \beta + \frac{N}{2} - \ell - 2m$, in the sense of (1.10). Since

$$\int_K |x|^{2(\beta-\ell)} |f(x)|^2 \, dx = \int_{-\infty}^{+\infty} \int_G e^{2\tau(\beta-\ell+N/2)} |\hat{f}(\tau,\omega)|^2 \, d\omega \, d\tau < \infty$$

it follows that

$$\int_{-\infty}^{+\infty} |\hat{F}(\tau,\omega)|^2 e^{2h\tau} \, d\tau = \int_{-\infty}^{+\infty} e^{4m\tau} |\hat{f}(\tau,\omega)|^2 e^{2h\tau} \, d\tau < \infty$$

for a.e. $\omega \in G$. An analogous result is valid for $\hat{G}_j(\tau,\omega)$, $j = 1,2,\ldots,m$. Lemma (6.4) and formula (1.12) imply that there exists a uniquely determined solution

$$(6.8) \qquad \hat{u}(\tau,\omega) = \frac{1}{\sqrt{2\pi}} \int_{-\infty+ih}^{\infty+ih} e^{i\lambda\tau} \mathcal{O}_0^{-1}(\lambda) \big[\tilde{F}, \tilde{G}_j \big] \, d\lambda = \frac{1}{\sqrt{2\pi}} \int_{-\infty+ih}^{\infty+ih} e^{i\lambda\tau} \tilde{u}(\lambda,\omega) \, d\lambda$$

$$= u(x)$$

of the problem (5.7).

(ii) We show that $\hat{u}(\tau,\omega) = u(x) \in V^{\ell+2m,\beta}(K)$ and that it satisfies (6.7). Using (1.11), (1.13) and (6.6) we get

40

$$\|u;\ V^{\ell+2m,2}(K,\beta)\|^2$$

$$\leq c_1 \sum_{|\alpha_1|+|\alpha_2|\leq \ell+2m} \int_{-\infty}^{+\infty}\iint_G \left|\left|\frac{\partial^{\alpha_1+|\alpha_2|}\hat{u}}{\partial\tau^{\alpha_1}\partial\omega^{\alpha_2}}\right|\right|^2 e^{2(\beta+N/2-2m-\ell)}\ d\omega\ d\tau$$

$$= c_1 \sum_{|\alpha_1|+|\alpha_2|\leq \ell+2m} \int_G \int_{-\infty+ih}^{\infty+ih} \left|\mathcal{F}\left(\frac{\partial^{\alpha_1+|\alpha_2|}\hat{u}}{\partial\tau^{\alpha_1}\partial\omega^{\alpha_2}}\right)\right|^2 d\lambda\ d\omega$$

$$\leq c_2 \sum_{\nu=0}^{\ell+2m} \int_{-\infty+ih}^{\infty+ih} |\lambda|^{2\nu}\ \|\tilde{u};\ W^{\ell+2m-\nu,2}(G)\|^2\ d\lambda$$

$$\leq c_3(h)\left[\sum_{\nu=0}^{\ell} \int_{-\infty+ih}^{\infty+ih} |\lambda|^{2\nu}\ \|\tilde{F};\ W^{\ell-\nu,2}(G)\|^2\ d\lambda\right.$$

$$+ \sum_{j=1}^{m} \int_{-\infty+ih}^{\infty+ih} \{\|\tilde{G}_j;\ W^{2m+\ell-m_j-1/2,2}(\partial G)\|^2$$

$$+ |\lambda|^{2(\ell+2m-m_j-1/2)}\ \|\tilde{G}_j;\ L^2(\partial G)\|^2\}\ d\lambda\Bigg]$$

$$= c_3(h)\left[\sum_{\nu=0}^{\ell} \int_{-\infty}^{+\infty} \left\|\frac{\partial^\nu}{\partial\tau^\nu}\hat{F}(\tau,\omega);\ W^{\ell-\nu,2}(G)\right\|^2 e^{2h\tau}\ d\tau\right.$$

$$+ \sum_{j=1}^{m} \int_{-\infty}^{+\infty} \{\|\hat{G}_j(\tau,\omega);\ W^{2m+\ell-m_j-1/2}(G)\|\ e^{2h\tau}$$

$$+ \left\|\frac{\partial^{\ell+2m-m_j}}{\partial\tau^{\ell+2m-m_j}}\hat{G}_j(\tau,\omega);\ L^2(\partial G)\right\|^2 e^{2h\tau}\}\ d\tau\Bigg]$$

$$\leq c_4(h)\left[\sum_{\nu=0}^{\ell} \int_{-\infty}^{+\infty} \left\|\frac{\partial^\nu}{\partial\tau^\nu}\hat{f}(\tau,\omega);\ W^{\ell-\nu,2}(G)\right\|^2 e^{2(\beta-\ell+N/2)}\ d\tau\right.$$

$$+ \sum_{j=1}^{m} \int_{-\infty}^{+\infty} \{\|\hat{g}_j(\tau,\omega);\ W^{2m+\ell-m_j-1/2}(\partial G)\|^2\ e^{2(\beta-2m-\ell+m_j+N/2)}$$

$$+ \left\|\frac{\partial^{\ell+2m-m_j}}{\partial\tau^{\ell+2m-m_j}}\hat{g}_j;\ L^2(\partial G)\right\|^2 e^{2(\beta-2m-\ell+m_j+N/2)}\}\ d\tau\Bigg]$$

$$\leq c_5(h)\left[\|f;\ V^{\ell,2}(K,\beta)\|^2 + \sum_{j=1}^{m} \|g_j;\ V^{2m+\ell-m_j-1/2,2}(\partial K,\beta)\|^2\right].$$

We now prove the necessity : Assume that there is an eigenvalue λ_0 of $\mathcal{U}_0(\lambda)$ on the line $\operatorname{Im}\lambda = \beta + \frac{N}{2} - \ell - 2m$. Let $u_0(\omega)$ be a corresponding eigenfunction. We consider the sequence of functions

$$\left(u_n = \frac{r^{i\lambda_0 + 1/n} \, u_0(\omega)}{\| r^{i\lambda_0 + 1/n} \, u_0(\omega); \, V^{\ell+2m,2}(K,\beta) \|} \right)_{n=1,2,\ldots} .$$

We have $\mathcal{U}_0(D_x)u_n = r^{i\lambda_0 + 1/n} \, \mathcal{U}_0(\lambda_0 - i \tfrac{1}{n}) \, u_0(\omega) \cdot \dfrac{1}{\| r^{i\lambda_0 + 1/n} \, u_0(\omega); V^{\ell+2m,2}(K,\beta) \|}$

and therefore $\lim\limits_{n\to\infty} \mathcal{U}_0(D_x)u_n = 0$. This contradicts the assumption that $\mathcal{U}_0(D_x)$ is an isomorphism.

6.3. REMARKS (increase of smoothness). (i) Assume, moreover, that $f \in V^{\ell+\ell_0,p}(K,\,\beta+\ell_0)$ and $g_j \in V^{2m+\ell+\ell_0-1/p,p}(\partial K,\,\beta+\ell_0)$, $j = 1,2,\ldots,m$, where ℓ_0 is a positive integer (f and g_j are the right hand sides of (5.7)). Further, assume that no eigenvalue of $\mathcal{U}_0(\lambda)$ lies on the line $\mathrm{Im}\,\lambda = h = \beta + N/p - \ell - 2m$. Since $h = \beta + N/p - \ell - 2m = \beta + \ell_0 + N/p - \ell - \ell_0 - 2m$ and $V^{\ell+\ell_0,p}(K,\,\beta+\ell_0) \subset V^{\ell,p}(K,\beta)$ the solution $u \in V^{2m+\ell,p}(K,\beta)$ of (5.7) is an element of $V^{2m+\ell+\ell_0,p}(K,\,\beta+\ell_0)$, too.

(ii) We now assume that $f \in V^{\ell,p}(K,\beta) \cap V^{\ell,p_1}(K,\beta_1)$ and $g_j \in V^{2m+\ell-m_j-1/p,p}(\partial K,\beta) \cap V^{2m+\ell-m_j-1/p_1,p_1}(\partial K,\beta_1)$, $j = 1,\ldots,m$, where $\beta_1 = \beta + \dfrac{N}{p} - \dfrac{N}{p_1}$. Further, we assume that the line $\mathrm{Im}\,\lambda = h = \beta + \dfrac{N}{p} - \ell - 2m$ contains no eigenvalue of $\mathcal{U}_0(\lambda)$. Then the solution $u \in V^{2m+\ell,\beta}(K,\beta)$ of the problem (5.7) is an element of $V^{2m+\ell,p_1}(K,\beta_1)$ too, because $h = \beta + \dfrac{N}{p} - \ell - 2m = \beta_1 + \dfrac{N}{p_1} - \ell - 2m$.

§ 7 . R e g u l a r i t y   a n d   t h e   e x p a n s i o n   o f   t h e   s o l u t i o n   o f   t h e   s p e c i a l   p r o b l e m

The above Remarks 6.3 about the increase of smoothness are based on the fact that we consider the same line $\mathrm{Im}\,\lambda = h$ for various β_1 , p_1 and ℓ_0 . In this section we study what happens if we have different lines $\mathrm{Im}\,\lambda = h$ and $\mathrm{Im}\,\lambda = h_1$, $h_1 < h$, associated with the spaces $V^{\ell+2m,p}(K,\beta)$ and $V^{\ell_1+2m,p_1}(K,\beta_1)$, respectively. We have already investigated this problem in the introducing examples. We know that the eigenvalues of $\mathcal{U}_0(\lambda)$ in the strip $h_1 < \mathrm{Im}\,\lambda < h$ play an important role for regularity results. Further, we have seen that the corresponding eigen- and "associate" functions of $\mathcal{U}_0(\lambda)$ occur in the expansion of the solution u . Therefore let us start with some definitions.

7.1. SOME DEFINITIONS AND RESULTS CONCERNING OPERATOR-VALUED FUNCTIONS.

(i) <u>DEFINITION</u> (see I. I. GOCHBERG, E. I. SIGAL [1]). Let λ_0 be an eigenvalue of $\mathfrak{A}_0(\lambda)$ and $u(\lambda)$ the function appearing in Definition (6.3). $u(\lambda)$ is called a *root function* of $\mathfrak{A}_0(\lambda)$ at the point λ_0 . The multiplicity of the zero of $\mathfrak{A}_0(\lambda)u(\lambda)$ at λ_0 is called the *multiplicity of* $u(\lambda)$, *rang* u_0 denotes the maximum of the multiplicities of all root functions u with $u(\lambda_0) = u_0$.

(ii) <u>DEFINITION.</u> Let $u(\lambda) = \sum\limits_{j=0}^{\infty} (\lambda - \lambda_0)^j u_j$ be a root function of a multiplicity s at λ_0 . The functions $u_1,\ldots,u_{s-1}$, $u_i = u_i(\omega)$, $i = 1, 2,\ldots,s-1$, are called the *associate functions* of $\mathfrak{A}_0(\lambda)$ at λ_0 . $(u_0,\ldots,u_{s-1})$ is called a *Jordan chain* of the length s of $\mathfrak{A}_0(\lambda)$ with respect to λ_0 .

(iii) <u>REMARK.</u> For operator-valued functions depending on a complex variable λ , properties analogous to those of ordinary functions are valid (see e.g. G. BACHMAN, L. NARIČI [1]). Therefore, since $\mathfrak{A}_0(\lambda)u(\lambda)$ vanishes at λ_0 of the order s , we have

$$\mathfrak{A}_0(\lambda)u(\lambda)\big|_{\lambda=\lambda_0} = \mathfrak{A}_0(\lambda_0)u_0 = 0 \, ,$$

$$\frac{\partial}{\partial\lambda}\left(\mathfrak{A}_0(\lambda)u(\lambda)\right)\big|_{\lambda=\lambda_0} = \mathfrak{A}_0'(\lambda_0)u_0 + \mathfrak{A}_0(\lambda_0)u_1 = 0 \, ,$$

$$\vdots$$

$$\frac{\partial^{s-1}}{\partial\lambda^{s-1}}\left(\mathfrak{A}_0(\lambda)u(\lambda)\right)\big|_{\lambda=\lambda_0} = \mathfrak{A}_0^{s-1}(\lambda_0)u_0 + \ldots + (s-1)!\,\mathfrak{A}_0(\lambda_0)u_{s-1} = 0,$$

or shortly

$$(7.1) \qquad \sum_{\mu=0}^{k} \frac{1}{\mu!} \frac{\partial^\mu \mathfrak{A}_0(\lambda)}{\partial\lambda^\mu} u_{k-\mu} = 0 \quad \text{for } k = 0,\ldots,s-1 \, .$$

(iv) <u>DEFINITION.</u> Let $I = \dim \ker \mathfrak{A}_0(\lambda_0)$ and $\max\limits_{u_0 \in \ker \mathfrak{A}_0(\lambda_0)} \text{rang } u_0 = \kappa_1 < \infty$. A system $\{u^{0,1},\ldots,u^{0,I}\} \subset \ker \mathfrak{A}_0(\lambda_0)$ is a canonical system of eigenfunctions of $\mathfrak{A}_0(\lambda)$ with respect to λ_0 , if it is linearly independent and $\kappa_1 \geq \kappa_2 \geq \ldots \geq \kappa_I$, where $\kappa_i = \text{rang } u^{0,i}$, $i = 1,2,\ldots,I$. The corresponding system of Jordan chains is called the *canonical system of Jordan chains* of $\mathfrak{A}_0(\lambda)$ with respect to the eigenvalue λ_0 .

(v) <u>LEMMA</u> (S. G. KREJN, V. P. TROFIMOV [1]). *Let* $\mathfrak{A}_0^{-1}(\lambda)$ *be the inverse operator of* $\mathfrak{A}_0(\lambda)$ *(given by (6.2)).* $\mathfrak{A}_0^{-1}(\lambda)$ *is a meromorphic operator-valued function with poles which are the eigenvalues of* $\mathfrak{A}_0(\lambda)$. *The order* s *of a pole* λ_0 *is the largest of the lengths of the Jordan chains corresponding to* λ_0 . *There is an expansion of* $\mathcal{C}_0^{-1}(\lambda)$ *in a neighborhood of the pole* λ_0 *of the following form :*

$$(7.2) \qquad \mathcal{C}_0^{-1}(\lambda) = \frac{P_s}{(\lambda - \lambda_0)^s} + \ldots + \frac{P_1}{\lambda - \lambda_0} + \Gamma(\lambda) \ .$$

P_j , $j = 1,\ldots,s$, *are finite-dimensional operators not depending on* λ , $\Gamma(\lambda)$ *is holomorphic; the operator* P_s *acts into the space of eigenfunctions of* $\mathcal{C}_0(\lambda)$ *corresponding to* λ_0 , *while the operators* $P_{s-1},\ldots,P_1$ *act into the subspaces of the corresponding associate functions.*

<u>7.2. THE HOMOGENEOUS PROBLEM.</u> Let us consider the problem

$$(7.3) \qquad \begin{aligned} A_0(D_x)u &= 0 \quad \text{in} \ \ K \ , \\ B_{j,0}(D_x)u &= 0 \quad \text{on} \ \ \partial K \ , \quad j = 1,\ldots,m \ , \end{aligned}$$

i.e. the homogeneous analogue of the problem (5.7). The following lemma was proved by V. G. MAZ'JA, B. A. PLAMENEVSKIĬ [4]:

<u>LEMMA.</u> *The functions*

$$(7.4) \qquad u(x) = \bar{u}(r,\omega) = r^{i\lambda_0} \sum_{k=0}^{\ell} \frac{1}{k!} (i \log r)^k u_{\ell-k}(\omega) \ , \quad 0 \le \ell \le s-1 \ ,$$

are solutions of (7.3) *if and only if* λ_0 *is an eigenvalue of* $\mathcal{C}_0(\lambda)$ *and* $(u_0,\ldots,u_\ell)$ *is a Jordan chain of the length* $\ell+1$ *corresponding to* λ_0 .

P r o o f : We again write the operators in spherical coordinates (r,ω) :

$$\begin{aligned} A_0(D_x) &= r^{-2m} L(\omega,D_\omega,rD_r) \ , \\ B_{j,0}(D_x) &= r^{-m_j} M_j(\omega,D_\omega,rD_r) \ . \end{aligned}$$

Since

$$(rD_r)^j \bar{u}(r,\omega) = (rD_r)^j \Big[r^{i\lambda_0} \sum_{k=0}^{\ell} \frac{1}{k!} (i \log r)^k u_{\ell-k}(\omega) \Big]$$

$$= r^{i\lambda_0} \sum_{k=0}^{\ell} \frac{1}{k!} (i \log r)^k \sum_{\mu=0}^{\ell-k} \binom{j}{\mu} \lambda_0^{j-\mu} u_{\ell-k-\mu}(\omega)$$

we obtain

$$(7.5) \qquad L(\omega,D_\omega,rD_r)\bar{u}(r,\omega) = r^{i\lambda_0} \sum_{k=0}^{\ell} \frac{1}{k!} (i \log r)^k \sum_{\mu=0}^{\ell-k} \frac{1}{\mu!} \frac{\partial^\mu L(\lambda_0)}{\partial\lambda^\mu} u_{\ell-k-\mu}(\omega) \ ,$$

$$(7.6) \qquad M_j(\omega,D_\omega,rD_r)\bar{u}(r,\omega) = r^{i\lambda_0} \sum_{k=0}^{\ell} \frac{1}{k!} (i \log r)^k \sum_{\mu=0}^{\ell-k} \frac{1}{\mu!} \frac{\partial^\mu M_j(\lambda_0)}{\partial\lambda^\mu} u_{\ell-k-\mu}(\omega) \ ,$$

$$j = 1,\ldots,m \ .$$

The homogeneous equations $L(\omega,D_\omega,rD_r)\bar{u}(r,\omega) = 0$, $M_j(\omega,D_\omega,rD_r)\bar{u}(r,\omega) = 0$

are satisfied, if and only if

$$(7.7) \qquad \sum_{\mu=0}^{q} \frac{1}{\mu!} \frac{\partial^{\mu} L(\lambda_0)}{\partial \lambda^{\mu}} u_{q-\mu}(\omega) = 0 \quad \text{for} \quad \omega \in G ,$$

$$(7.8) \qquad \sum_{\mu=0}^{q} \frac{1}{\mu!} \frac{\partial^{\mu} M_j(\lambda_0)}{\partial \lambda^{\mu}} u_{q-\mu}(\omega) = 0 \quad \text{for} \quad \omega \in \partial G ,$$

$$j = 1,2,\ldots,m , \qquad q = 0,1,\ldots,\ell , \qquad 0 \leq \ell \leq s-1 .$$

Together with (7.1) we get the assertion.

The following theorem describes the expansion and the regularity of the solution of the special problem (5.7).

7.3. **THEOREM.** *Let* $u \in V^{2m+\ell,p}(K,\beta)$ *be a solution of the problem* (5.7), $f \in V^{\ell,p}(K,\beta) \cap V^{\ell,p}(K,\beta_1)$ *and* $g_j \in V^{\ell+2m-m_j-1/p,p}(\partial K,\beta)$ $\cap V^{\ell+2m-m_j-1/p,p}(\partial K,\beta_1)$, $j = 1,\ldots,m$, $\beta_1 < \beta$, $1 < p < \infty$. *Assume that no eigenvalue of* $\mathfrak{A}_0(\lambda)$ *lies on the lines* $\mathrm{Im}\,\lambda = h_1 = \beta_1 + \frac{N}{p} - \ell - 2m$ *and* $\mathrm{Im}\,\lambda = h = \beta + \frac{N}{p} - \ell - 2m$. *Let* $\lambda_1,\ldots,\lambda_{N_0}$ *be the eigenvalues of* $\mathfrak{A}_0(\lambda)$ *in the strip* $h_1 < \mathrm{Im}\,\lambda < h$.

Then

$$(7.9) \qquad u(x) = \bar{u}(r,\omega) = \sum_{\mu=1}^{N_0} \sum_{\sigma=1}^{I_\mu} \sum_{k=0}^{\kappa_{\mu_\sigma}-1} c_{\mu,\sigma,k} u_{\mu,\sigma,k}(r,\omega) + \bar{w}(r,\omega)$$

$$= \sum_{\gamma \in I} c_\gamma u_\gamma + w ,$$

where $u_\gamma = u_{\mu,\sigma,k}(r,\omega) = r^{i\lambda_\mu} \sum_{q=0}^{k} \frac{1}{q!} (i \log r)^q u_\mu^{k-q,\sigma}(\omega)$, $\gamma = (\mu,\sigma,k)$,

$I = \{(\mu,\sigma,k)\}_{\substack{\mu=1,\ldots,N_0 \\ \sigma=1,\ldots,I_\mu}}$, $k = 0,\ldots,\kappa_{\mu_\sigma}-1$, $I_\mu = \dim \ker \mathfrak{A}_0(\lambda_\mu)$, *and*

$$(7.10) \qquad \left\{ \begin{pmatrix} u_\mu^{0,1} \\ \cdot \\ \cdot \\ \cdot \\ u_\mu^{\kappa_{\mu_1}-1,1} \end{pmatrix} , \begin{pmatrix} u_\mu^{0,2} \\ \cdot \\ \cdot \\ \cdot \\ u_\mu^{\kappa_{\mu_2}-1,2} \end{pmatrix} , \ldots , \begin{pmatrix} u_\mu^{0,I_\mu} \\ \cdot \\ \cdot \\ \cdot \\ u_\mu^{\kappa_{\mu_{I_\mu}}-1,I_\mu} \end{pmatrix} \right\}$$

is a canonical system of Jordan chains of $\mathfrak{A}_0(\lambda)$ *with respect to* λ_μ ; c_γ *are constants,* $w(x) = \bar{w}(r,\omega) \in V^{\ell+2m,p}(K,\beta_1)$ *is a solution of* (5.7) *and*

$$(7.11) \qquad \|w;V^{\ell+2m,p}(K,\beta_1)\| \leq c \Big[\|f;V^{\ell,p}(K,\beta_1)\| + \sum_{j=1}^{m} \|g_j;V^{\ell+2m-m_j-1/p,p}(\partial K,\beta_1)\|\Big]$$

P r o o f : We only prove this theorem for $p = 2$ and refer to V. G. MAZ'JA, B. A. PLAMENEVSKIĬ [4] for $p \neq 2$.

(i) It follows from Theorem 6.2 that the solution $u \in V^{\ell+2m,2}(K,\beta)$ of (5.7) is uniquely determined and given by formula (6.8) :

$$u(x) = \hat{u}(\tau,\omega) = \frac{1}{\sqrt{2\pi}} \int_{-\infty+ih}^{\infty+ih} e^{i\lambda\tau} \mathcal{U}_0^{-1}(\lambda) [\tilde{F},\tilde{G}_j] \, d\lambda \ .$$

By the Cauchy theorem (see Fig. 11) we obtain

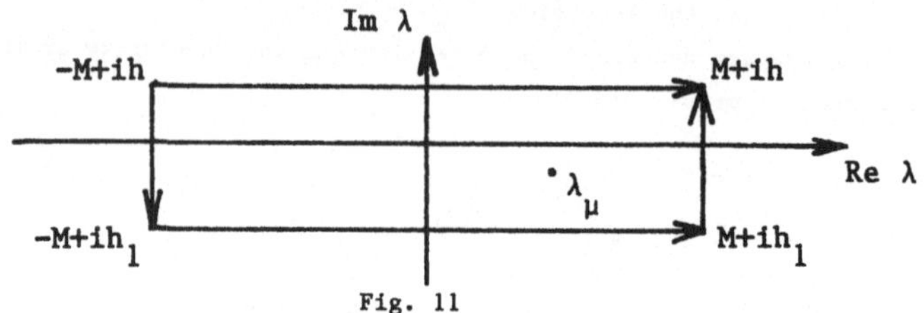

Fig. 11

$$(7.12) \qquad \hat{u}(\tau,\omega) = \frac{1}{2\pi} \lim_{M\to\infty} \left[\int_{-M+ih}^{-M+ih_1} e^{i\lambda\tau} \mathcal{U}_0^{-1}(\lambda) [\tilde{F},\tilde{G}_j] \, d\lambda \right.$$

$$+ \int_{-M+ih_1}^{M+ih_1} e^{i\lambda\tau} \mathcal{U}_0^{-1}(\lambda) [\tilde{F},\tilde{G}_j] \, d\lambda + \int_{M+ih_1}^{M+ih} e^{i\lambda\tau} \mathcal{U}_0^{-1}(\lambda) [\tilde{F},\tilde{G}_j] \, d\lambda \left. \right]$$

$$+ \frac{1}{2\pi} \, 2\pi i \sum_{\mu=1}^{N_0} \text{Res} \left(e^{i\lambda\tau} \mathcal{U}_0^{-1}(\lambda) [\tilde{F},\tilde{G}_j] \right) \big|_{\lambda=\lambda_\mu} \ .$$

The first and third integrals of (7.12) tend to 0 for $M \to \infty$ (see V. A. KONDRAT'EV [1]). The second integral

$$(7.13) \qquad \frac{1}{\sqrt{2\pi}} \int_{-\infty+ih_1}^{\infty+ih_1} e^{i\lambda\tau} \mathcal{U}_0^{-1}(\lambda) [\tilde{F},\tilde{G}_j] \, d\lambda = \hat{w}(\tau,\omega) = w(x)$$

is the uniquely determined solution $w \in V^{\ell+2m,2}(K,\beta_1)$ of (5.7), and the estimate (7.11) is valid. This assertion follows from Theorem 6.2.

(ii) It remains to calculate the residua in (7.12). To this end we write

$$[\tilde{F},\tilde{G}_j] = \sum_{\ell=0}^{\infty} a_\ell(\omega)(\lambda - \lambda_\mu)^\ell$$

in a neighborhood of λ_μ , which is possible·by (1.14). The coefficients $a_\ell(\omega)$ are elements of $W^{\ell,2}(G) \times \prod_{j=1}^{m} W^{\ell+2m-m_j-1/2,2}(\partial G)$. Further, $e^{i\lambda\tau}$ $= e^{i\lambda_\mu\tau} \left[1 + i(\lambda - \lambda_\mu)\tau + \dots + [i(\lambda - \lambda_\mu)\tau]^n/n! + \dots \right]$ and therefore in a neighborhood of λ_μ we have

$$e^{i\lambda\tau}\,\mathcal{O}_0^{-1}(\lambda)\,[\tilde{F},\tilde{G}_j] = e^{i\lambda_\mu\tau}\Big[1 + \ldots + \frac{[i(\lambda - \lambda_\mu)\tau]^n}{n!} + \ldots\Big]$$

$$\cdot\Big[p_{\kappa_{\mu_1}}(\lambda - \lambda_\mu)^{-\kappa_{\mu_1}} + \ldots + \frac{p_1}{\lambda - \lambda_\mu} + \Gamma(\lambda)\Big]\Big[\sum_{\ell=0}^{\infty} a_\ell(\omega)(\lambda - \lambda_\mu)^\ell\Big]\ .$$

We conclude

$$\operatorname{Res} e^{i\lambda\tau}\,\mathcal{O}_0^{-1}(\lambda)\,[\tilde{F},\tilde{G}_j]\Big|_{\lambda=\lambda_\mu} = e^{i\lambda_\mu\tau}\Big[p_1 a_1(\omega) + \ldots + p_{\kappa_{\mu_1}} a_{\kappa_{\mu_1}}(\omega)\Big]$$

$$+ e^{i\lambda_\mu\tau}\frac{(i\tau)^1}{1!}\Big[p_2 a_1(\omega) + \ldots + p_{\kappa_{\mu_1}} a_{\kappa_{\mu_1}-1}(\omega)\Big] + \ldots$$

$$+ e^{i\lambda_\mu\tau}\frac{(i\tau)^{\kappa_{\mu_1}-1}}{(\kappa_{\mu_1}-1)!}\Big[p_{\kappa_{\mu_1}} a_1(\omega)\Big]\ .$$

Substituting $e^\tau = r$ and applying Lemmas 7.1 (v) and 7.2 we obtain (7.9).

From Theorem 7.3 we get very easily a more general theorem.

<u>7.4. THEOREM.</u> *Let $u \in V^{2m+\ell,p}(K,\beta)$ be a solution of the problem (5.7), $f \in V^{\ell,p}(K,\beta) \cap V^{\ell_1,p_1}(K,\beta_1)$ and $g_j \in V^{\ell+2m-m_j-1/p,p}(\partial K,\beta)$ $\cap V^{\ell_1+2m-m_j-1/p_1,p_1}(\partial K,\beta_1)$, $j = 1,\ldots,m$, $\ell_1 \geq \ell$, $h_1 = \beta_1 + \dfrac{N}{p_1} - \ell_1 - 2m$ $< h = \beta + \dfrac{N}{p} - \ell - 2m$, $1 < p < \infty$, $1 < p_1 < \infty$. Assume that no eigenvalues of $\mathcal{O}_0(\lambda)$ be on the lines $\operatorname{Im}\lambda = h$ and $\operatorname{Im}\lambda = h_1$. Let $\lambda_1,\ldots,\lambda_{N_0}$ be the eigenvalues of $\mathcal{O}_0(\lambda)$ in the strip $h_1 < \operatorname{Im}\lambda < h$.*

Then

$$(7.14)\qquad u(x) = \bar{u}(r,\omega) = \sum_{\gamma \in I} c_\gamma u_\gamma + w\ ,$$

where $w \in V^{\ell_1+2m,p_1}(K,\beta_1)$ and

$$(7.15)\qquad \|w;\ V^{\ell_1+2m,p_1}(K,\beta_1)\| \leq c\Big[\|f;\ V^{\ell_1,p_1}(K,\beta_1)\|$$

$$+ \sum_{j=1}^{m}\|g_j;\ V^{\ell_1+2m-m_j-1/p_1,p_1}(\partial K,\beta_1)\|\Big]\ .$$

(For the notation see Theorem 7.3.)

P r o o f : Let $\ell_1 = \ell_0 + \ell$, $\ell_0 \geq 0$. We have $h_1 = \beta_1 + \dfrac{N}{p_1} - \ell_0 - \ell - 2m$ $= \beta_1 + \dfrac{N}{p_1} - \ell_0 - \dfrac{N}{p} + \dfrac{N}{p} - \ell - 2m = \beta_1' + \dfrac{N}{p} - \ell - 2m$, where $\beta_1' = \beta_1 + \dfrac{N}{p_1} - \dfrac{N}{p} -\ell_0$. Using Theorem 7.3 we get the expansion (7.9), where $w \in V^{\ell+2m,p}(K,\beta_1')$ $= V^{\ell+2m,p}(K,\ \beta_1 + \dfrac{N}{p_1} - \dfrac{N}{p} - \ell_0)$. The function w is a solution of (5.7).

Remarks 6.3(i) and (ii) about the increase of smoothness yield that w
$\in V^{\ell+\ell_0+2m,p_1}(K,\beta_1)$. The estimate (7.15) follows from Theorem 6.2.

Now we are able to formulate the following regularity result.

7.5.   THEOREM. *Assume that the assumptions of Theorem 7.4 are satisfied. If no
eigenvalues of* $\mathcal{U}_0(\lambda)$ *are situated in the strip* $h_1 < \mathrm{Im}\,\lambda < h$, *then the
solution* $u \in V^{\ell+2m,p}(K,\beta)$ *is an element of* $V^{\ell_1+2m,p_1}(K,\beta_1)$, *too.*

P r o o f : We obtain from the formula (7.10) that $u = w$.

§ 8 . A g e n e r a l b o u n d a r y v a l u e p r o b l e m i n K

We investigate the more general boundary value problem (5.5) in the infinite cone K , namely,

$$A(x,D_x)u = f \quad \text{in } K ,$$

$$B_j(x,D_x)u = g_j \quad \text{on } \partial K , \quad j = 1,2,\ldots m$$

with the corresponding operator $\mathcal{U}_K(x,D_x)$ defined by (5.6).

8.1.   THE SOLVABILITY IN $V^{\ell+2m,p}(K,\beta)$.   (i) We assume that the coefficients
$a_\alpha(x)$ and $b_{j,\alpha}(x)$ of the operators $A(x,D_x)$ and $B_j(x,D_x)$, $j = 1,\ldots,m$,
respectively, satisfy the following conditions : There exists a $\delta > 0$ such
that

$$(8.1) \quad \begin{aligned}
&\left| r^{|\gamma|} D_x^\gamma \big(a_\alpha(x) - a_\alpha(0)\big)\right| < \delta \quad \text{for} \quad |\gamma| \leq \ell , \quad |\alpha| = 2m , \quad x \in K \setminus \{0\} , \\
&\left| r^{|\gamma|} D_x^\gamma \big(b_{j,\alpha}(x) - b_{j,\alpha}(0)\big)\right| < \delta \quad \text{for} \quad |\gamma| \leq \ell + 2m - m_j , \quad |\alpha| = m_j , \\
&\qquad\qquad\qquad\qquad\qquad\qquad\qquad\qquad x \in \partial K \setminus \{0\} , \quad j = 1,\ldots,m .
\end{aligned}$$

$$(8.2) \quad \begin{aligned}
&\left| r^{|\gamma|+2m-|\alpha|} D_x^\gamma a_\alpha(x)\right| < \delta \quad \text{for} \quad |\gamma| \leq \ell , \quad |\alpha| < 2m , \quad x \in K \setminus \{0\} , \\
&\left| r^{|\gamma|+m_j-|\alpha|} D_x^\gamma b_{j,\alpha}(x)\right| < \delta \quad \text{for} \quad |\gamma| \leq \ell + 2m - m_j , \quad |\alpha| < m_j , \\
&\qquad\qquad\qquad\qquad\qquad\qquad\qquad\qquad x \in \partial K \setminus \{0\} , \quad j = 1,2,\ldots,m .
\end{aligned}$$

If the coefficients a_α and $b_{j,\alpha}$ are sufficiently smooth, then the conditions (8.1) and (8.2) are valid in a neighborhood of the origin.

(ii) **LEMMA.** *Assume the conditions* (8.1) *and* (8.2) *are satisfied for a
certain* δ . *The operator* $\mathcal{U}_K(x,D_x)$ *is an isomorphism if and only if no
eigenvalues of* $\mathcal{U}_0(\lambda)$ *are situated on the line* $\mathrm{Im}\,\lambda = h = \beta + \dfrac{N}{p} - 2m - \ell$.

P r o o f : We have to show that the inverse operator of $\mathcal{U}_K(x,D_x)$ exists.
To this end we write

$$\mathcal{U}_K^{-1}(x,D_x) = \big[\mathcal{U}_K(x,D_x) - \mathcal{U}_0(0,D_x) + \mathcal{U}_0(0,D_x)\big]^{-1} =$$

$$= \left[\mathcal{A}_0(0,D_x)\left(I + \mathcal{A}_0^{-1}(0,D_x)\right)\left(\mathcal{A}_K(x,D_x) - \mathcal{A}_0(0,D_x)\right)\right]^{-1}$$

$$= \left[I + \mathcal{A}_0^{-1}(0,D_x)\left(\mathcal{A}_K(x,D_x) - \mathcal{A}_0(0,D_x)\right)\right]^{-1} \mathcal{A}_0^{-1}(0,D_x)$$

and verify that

$$(8.3) \qquad \left[I + \mathcal{A}_0^{-1}(0,D_x)\left(\mathcal{A}_K(x\ D_x) - \mathcal{A}_0(0,D_x)\right)\right]^{-1}$$

is defined. Let us denote by $\|| \cdot \||$ the operator norm for linear operators acting from $V^{\ell+2m,p}(K,\beta)$ into $V^{\ell,p}(K,\beta) \times \prod\limits_{j=1}^{m} V^{\ell+2m-m_j-1/p,p}(\partial K,\beta)$. The operator (8.3) exists, if

$$(8.4) \qquad \left\||\, \mathcal{A}_0^{-1}(0,D_x)\left(\mathcal{A}_K(x,D_x) - \mathcal{A}_0(0,D_x)\right)\,\right\|| < 1 \;.$$

Let us show that the estimate (8.4) is valid, by using the decomposition

$$\mathcal{A}_K(x,D_x) = \mathcal{A}_0(x,D_x) + \mathcal{A}_1(x,D_x)$$
$$= \mathcal{A}(x,D_x) - \mathcal{A}_0(0,D_x) + \mathcal{A}_0(0,D_x) + \mathcal{A}_1(x,D_x) \;,$$

where $\mathcal{A}_0(x,D_x)$ is the principal part of $\mathcal{A}_K(x,D_x)$. In the remaining operator $\mathcal{A}_1(x,D_x)$ only the derivatives up to the order $2m-1$ or m_j-1 occur, respectively. Property (8.1) implies that

$$\left\||\, \mathcal{A}_0(x,D_x) - \mathcal{A}_0(0,D_x)\,\right\|| < c_0 \delta \;;$$

and property (8.2) implies that

$$\left\||\, \mathcal{A}_1(x,D_x)\,\right\|| < c_1 \delta \;.$$

Therefore $\left\||\, \mathcal{A}_K(x,D_x) - \mathcal{A}_0(0,D_x)\,\right\|| < c\delta$. Choosing $c\delta = \dfrac{1}{\||\mathcal{A}_0^{-1}(0,D_x)\||}$ we obtain the estimate (8.4).

   (iii) <u>COROLLARY.</u> *There is an operator* $\mathcal{A}^\varepsilon(x,D_x)$ *which is an isomorphism of* $V^{\ell+2m,p}(K,\beta)$ *into* $V^{\ell,p}(K,\beta)$ $\times \prod\limits_{j=1}^{m} V^{\ell+2m-m_j-1/p,p}(\partial K,\beta)$ *with the following properties:*

$$(8.5) \qquad \mathcal{A}^\varepsilon(x,D_x) = \begin{cases} \mathcal{A}_K(x,D_x) & \text{if } |x| < \dfrac{\varepsilon}{2} \;, \\[2mm] \mathcal{A}_0(0,D_x) & \text{if } |x| > \varepsilon \;, \end{cases}$$

and

$$\left\||\, \mathcal{A}^\varepsilon(x,D_x) - \mathcal{A}_0(0,D_x)\,\right\|| < \dfrac{1}{2\,\||\mathcal{A}_0^{-1}(0,D_x)\||} \;,$$

provided ε *is sufficiently small.*

<u>8.2. LEMMA (regularity result).</u> *Let* $u \in V^{2m+\ell,p}(K,\beta)$ *be a solution of* (5.5) *with sufficiently small support in a neighborhood of the origin* 0 *. Assume* $f \in V^{\ell,p}(K,\beta) \cap V^{\ell_1,p_1}(K,\beta_1)$ *and* $g_j \in V^{\ell+2m-m_j-1/p,p}(\partial K,\beta) \cap$

$\cap\, V^{\ell+2m-m_j-1/p_1,p_1}(\partial K,\beta_1)$; $\ell_1 \geq \ell$, $h_1 = \beta_1 + \dfrac{N}{p_1} - 2m - \ell_1 \leq \beta + \dfrac{N}{p} - 2m - \ell$
$= h$. *If no eigenvalues of* $\alpha_0(\lambda)$ *are situated in the strip* $h_1 \leq \mathrm{Im}\,\lambda \leq h$,
then $u \in V^{2m+\ell_1,p_1}(K,\beta_1)$, *too.*

P r o o f : (i) We first consider the case $\ell = \ell_1$, $p = p_1$. Let k_1 be an integer such that $\beta_1 \geq \beta - k_1$. We have

$$\alpha_0(0,D_x)u = \alpha_K(x,D_x)u + \big(\alpha_0(0,D_x) - \alpha_K(x,D_x)\big)u$$
$$= \big[f,\{g_j\}_{j=1,\ldots,m}\big] + \big[F,\{G_j\}_{j=1,\ldots,m}\big]$$

where $F \in V^{\ell,p}(K,\beta-1)$, $G_j \in V^{\ell+2m-m_j-1/p,p}(\partial K,\beta-1)$. If $k_1 = 1$ then $f + F \in V^{\ell,p}(K,\beta_1) \cap V^{\ell,p}(K,\beta)$ and $g_j + G_j \in V^{2m+\ell-m_j-1/p,p}(\partial K,\beta_1)$
$\cap\, V^{2m+\ell-m_j-1/p,p}(\partial K,\beta)$. Theorem 7.3 implies that $u \in V^{2m+\ell,p}(K,\beta_1)$. If $k_1 > 1$, then $u \in V^{2m+\ell,p}(K,\beta-1)$ and we repeat this procedure for $u \in V^{2m+\ell,p}(K,\beta-1)$, and so on.   Finally we obtain that $u \in V^{2m+\ell,p}(K,\beta_1)$.

(ii) We now study the case $\ell_1 = \ell + \ell_0 > \ell$, $p = p_1$. Assume that the support of u is contained in the ball with the radius $\dfrac{\varepsilon}{2}$ and the center O , where ε is determined by (8.5). Then we have $\alpha_K(x,D_x)u = \alpha^\varepsilon(x,D_x)u$. It follows from the first part (i) of the proof that $u \in V^{2m+\ell,p}(K,\beta_1')$, where $\beta_1' = \beta_1 - \ell_0$. On the other hand, there is a uniquely determined solution $u_1 \in V^{2m+\ell_1,p}(K,\beta_1)$ of the problem

$$\alpha^\varepsilon(x,D_x)u_1 = \big[f,\{g_j\}_{j=1,\ldots,m}\big] \; .$$

Since $V^{2m+\ell_1,p}(K,\beta_1) \subset V^{2m+\ell,p}(K,\beta_1')$, we obtain $u = u_1$.

(iii) Finally, we consider the case $p \neq p_1$. It follows from parts (i) and (ii) of the proof that $u \in V^{2m+\ell_1,p}(K,\beta_2)$ where $\beta_2 = \beta_1 + \dfrac{N}{p_1} - \dfrac{N}{p}$.
Assume $\underline{p > p_1}$. We have

$$\alpha_0(0,D_x)u = \alpha_K(x,D_x)u + \big[\alpha_0(0,D_x) - \alpha_K(x,D_x)\big]u$$
$$= \big[f,\{g_j\}_{j=1,\ldots,m}\big] + \big[F,\{G_j\}_{j=1,\ldots,m}\big]$$

where $F \in V^{\ell_1,p}(\beta_2-1)$, $G_j \in V^{2m+\ell_1-m_j-1/p,p}(\partial K,\beta_2-1)$. Since F , G_j have bounded supports and $\beta_2 - 1 + \dfrac{N}{p} < \dfrac{N}{p_1} + \beta_1$ we obtain from the imbedding 0.10, formula (0.30), that

$$F \in V^{\ell_1,p_1}(K,\beta_1) \quad \text{and} \quad G_j \in V^{2m+\ell_1-m_j-1/p_1,p_1}(\partial K,\beta_1) \; .$$

Theorem 7.4 implies that $u \in V^{2m+\ell_1,p_1}(K,\beta_1)$.

Assume $p_1 > p$. We again consider the uniquely determined solution u_1 = $[\mathcal{O}^{\varepsilon}]^{-1}[f,\{g_j\}] \in V^{\ell_1+2m,p_1}(K,\beta_1)$, which also has a bounded support. The solution $u \in V^{2m+\ell_1,p}(K,\beta_2)$ is an element of $V^{2m+\ell_1,p}(K, \beta_1 + \frac{N}{p_1} - \frac{N}{p} + \varepsilon_0)$, too, and it is uniquely determined, provided $\varepsilon_0 > 0$ is so small that the strip $h \leq \mathrm{Im}\ \lambda \leq h + \varepsilon_0$ contains no eigenvalues of $\mathcal{O}_0(\lambda)$. Since $\beta_1 + \frac{N}{p_1} < \beta_1 + \frac{N}{p_1} - \frac{N}{p} + \varepsilon_0 + \frac{N}{p}$ we use the imbedding (0.30) thus concluding $u = u_1$.

Section 3 : *T h e   b o u n d a r y   v a l u e   p r o b l e m   i n   a   b o u n d e d   d o m a i n*

We are concerned with the basic problem, namely, with the elliptic boundary value problem (5.3) in a bounded domain with conical points. First we deal
with its solvability and regularity and then we study the expansion of the
solution near a conical point.

§ 9 . S o l v a b i l i t y   i n $V^{\ell+2m,p}(\Omega,\vec{\beta})$ a n d   r e g u l a r i t y

We use the results for boundary value problems in an infinite cone in
order to prove the corresponding results in bounded domains with conical
points. Roughly speaking, the main idea is as follows :
We consider a sufficiently fine covering $\{U_i\}_{i=1,\ldots,Q}$ of the domain Ω and
investigate boundary value problems in U_i . Fitting together the solutions u_i
of these "local" boundary value problems we construct a right (and a left)
regularizer for the original problem (5.3). It follows from the existence of
the regularizers that the operator $\mathcal{O}(x,D_x)$ (formula (5.4)) of the problem
(5.3) is a Fredholm operator. Further, we immediately get regularity results,
multiplying the solution by a cut-off function which has a support in a neighborhood of a conical point and using the regularity results in an infinite
cone.

Let us now give a more detailed account. We begin with some definitions.

9.1. <u>DEFINITIONS AND NOTATION.</u> (i) <u>DEFINITION.</u> Let $\mathcal{O}$ be a linear operator mapping the Banach space E into the Banach space F , i.e. $\mathcal{O} \in \mathcal{L}(E,F)$.
$\mathcal{O}$ is a *Fredholm operator* (or an index operator) if $\mathcal{O}$ is a closed operator
(that means if $x_n \to x$ and $\mathcal{O}x_n \to y$, then $x \in \mathcal{D}(\mathcal{O})$ and $\mathcal{O}x = y$), the
range of $\mathcal{O}$ is closed, both dim ker $\mathcal{O}$ and dim coker $\mathcal{O}$ are finite.

(ii) <u>DEFINITION.</u> Let $\mathcal{O} \in \mathcal{L}(E,F)$, let E and F be Banach spaces.
R_ℓ is a *left regularizer* of $\mathcal{O}$ if $R_\ell \mathcal{O} = I + S$, where S is a compact
operator mapping E into E and   I is the identical operator of E into
E . R_r is a *right regularizer* of $\mathcal{O}$ if $\mathcal{O}R_r = I + T$, where T is a

compact operator of F into F , and I is the corresponding identical operator.

(iii) <u>LEMMA</u> (S. G. KREJN [1]). *Let* $\mathcal{A} \in \mathcal{L}(E,F)$, $\mathcal{A}$ *closed.* $\mathcal{A}$ *is a Fredholm operator if and only if there exist left and right regularizers of* $\mathcal{A}$.

(iv) Assume that the bounded domain Ω has N_1 conical points O_i , $\{O_i\}_{i=1,\ldots,N_1} = M$. (In the case 5.2 (ii) M is bigger.) For every point $O_i \in M$ we consider the infinite cones K_i defined analogously as in 5.1 (ii). Further, we choose $\varepsilon_i > 0$ so small that the operators

$$(9.1) \qquad \mathcal{A}^{\varepsilon_i}(x,D_x) = \begin{cases} \mathcal{A}_{K_i}(x,D_x) & \text{if } |x - O_i| < \dfrac{\varepsilon_i}{2} , \\ \mathcal{A}_0(O_i,D_x) & \text{if } |x - O_i| > \varepsilon_i \end{cases}$$

are isomorphisms of $V^{\ell+2m,P}(K_i,\beta_i)$ into $V^{\ell,P}(K_i,\beta_i)$
$\times \prod\limits_{j=1}^{m} V^{2m+\ell-m_j-1/p,p}(\partial K_i,\beta_i)$ and that only one point from M lies in the ball
$B_{\varepsilon_i/2}(O_i) = \{x : |x - O_i| < \dfrac{\varepsilon_i}{2}\}$. We denote by $\mathcal{A}_0^i(\lambda)$ the operator

$$(9.2) \qquad \mathcal{A}_0^i(\lambda) : W^{2m+\ell,P}(G_i) \to W^{\ell,P}(G_i) \times \prod\limits_{j=1}^{m} W^{2m+\ell-m_j-1/p,p}(\partial G_i)$$

which is defined analogously to (6.2). G_i occurs instead of G .

(v) We describe an *appropriate covering* of Ω : Let $\{U_i\}_{i=1,\ldots,Q}$ be a finite covering of Ω with the properties

$$(9.3) \qquad U_i \subset B_{\varepsilon_i/2}(O_i) \quad \text{for } i = 1, ,\ldots,N_1 ,$$

$$(9.4) \qquad U_i \cap \Omega = U_i , \quad U_i \text{ has a smooth boundary for } i = N_1+1,\ldots,Q .$$

The following theorem concerns the solvability of the problem (5.3) in $V^{\ell+2m,P}(\Omega,\vec{\beta})$, $\vec{\beta} = (\beta_1,\ldots,\beta_{N_1})$.

<u>9.2. THEOREM.</u> *The operator* (5.4), $\mathcal{A}(x,D_x)$, *which maps* $V^{\ell+2m,P}(\Omega,\vec{\beta})$ *into* $V^{\ell,P}(\Omega,\vec{\beta}) \times \prod\limits_{j=1}^{m} V^{\ell+2m-m_j-1/p,p}(\partial\Omega,\vec{\beta})$, *is a Fredholm operator if and only if no eigenvalues of* $\mathcal{A}_0^i(\lambda)$, $i = 1,\ldots N_1$, *lie on the lines* $\mathrm{Im}\,\lambda = h = \beta_i + \dfrac{N}{p} - 2m - \ell$.

P r o o f : We first show that the condition is *sufficient.*

(i) Let us consider a finite covering of Ω with the properties (9.3) and (9.4), and a corresponding partition of unity $\{\phi_i\}_{i=1,\ldots,Q}$. Let ψ_i be smooth functions with supports in U_i , which satisfy the conditions
$$(9.5) \qquad \psi_i \phi_i = \phi_i , \quad i = 1,\ldots,Q .$$

Let $f \in V^{\ell,p}(\Omega,\vec{\beta})$ and $g_j \in V^{\ell+2m-m_j-1/p,p}(\partial\Omega,\vec{\beta})$, $j = 1,\ldots,m$. We consider
the following boundary value problems in $\Omega \cap U_i$, $i = 1,\ldots,N_1$:

$$(9.6) \qquad \mathcal{U}(x,D_x)u = \mathcal{U}^{\varepsilon_i}(x,D_x)u = [\psi_i f, \{\psi_i g_j\}_{j=1,\ldots,m}] .$$

For simplicity we assume that $\Omega \cap U_i = K_i \cap U_i$ (otherwise we use a diffeomorphism) and consider the right hand sides of (9.6) as elements of $V^{\ell,p}(K_i,\beta_i)$
$\times \prod\limits_{j=1}^{m} V^{\ell+2m-m_j-1/p,p}(\partial K_i,\beta_i)$. Since $\mathcal{U}^{\varepsilon_i}(x,D_x)$, defined by (9.1), is an isomorphism we obtain solutions $u_i \in V^{\ell+2m,p}(K_i,\beta)$ of (9.6). We write

$$(9.7) \qquad u_i = R_i[\psi_i f, \{\psi_i g_j\}_{j=1,\ldots,m}] , \quad i = 1,\ldots,N_1 .$$

(ii) We now consider the boundary value problems in the smooth domains
U_i , $i = N_1+1,\ldots,Q$,

$$\mathcal{U}(x,D_x)u = [\psi_i f, \{\psi_j g_j\}_{j=1,\ldots,m}] .$$

It is well known (see e.g. S. AGMON, A. DOUGLIS, L. NIRENBERG [1]) that in
smooth domains U the operator of an elliptic boundary value problem is a
Fredholm operator, mapping $W^{\ell+2m,p}(U)$ into $W^{\ell,p}(U) \times W^{\ell+2m-m_j-1/p,p}(\partial U)$.
Therefore there exist right and left regularizers (Lemma 9.1, (iii)) of
$\mathcal{U}(x,D_x)$ in U_i , $i = N_1+1,\ldots,Q$. Let R_i be right regularizers. We again
write

$$(9.8) \qquad u_i = R_i[\psi_i f, \{\psi_j g_j\}_{j=1,\ldots,m}] , \qquad i = N_1+1,\ldots,Q ,$$
$$(9.9) \qquad \mathcal{U}R_i = I + T_i ,$$

where $u_i \in W^{\ell+2m,p}(U_i)$ and T_i is defined by 9.1 (ii).

(iii) We construct a right regularizer R of $\mathcal{U}(x,D_x)$, setting

$$u = R[f,\{g_j\}_{j=1,\ldots,m}] = \sum_{i=1}^{Q} \phi_i R_i[\psi_i f, \{\psi_i g_j\}_{j=1,\ldots,m}] .$$

Indeed, we have

$$\mathcal{U}R[f, \{g_j\}_{j=1,\ldots,m}] = \sum_{i=1}^{Q} \mathcal{U}(\phi_i u_i) = \sum_{i=1}^{N_1} \mathcal{U}^{\varepsilon_i}(\phi_i u_i) + \sum_{i=N_1+1}^{Q} \mathcal{U}(\phi_i u_i)$$

$$= \sum_{i=1}^{N_1} \{\phi_i[\psi_i f, \{\psi_i g_j\}_{j=1,\ldots,m}] + [\mathcal{U}^{\varepsilon_i},\phi_i]R_i[\psi_i f, \{\psi_i g_j\}_{j=1,\ldots,m}]\}$$

$$(9.10) \qquad + \sum_{i=N_1+1}^{Q} \{\phi_i[\psi_i f, \{\psi_i g_j\}_{j=1,\ldots,m}] + \phi_i T_i[\psi_i f, \{\psi_i g_j\}_{j=1,\ldots,m}]$$

$$+ [\mathcal{U},\phi_i]R_i[\psi_i f, \{\psi_i g_j\}_{j=1,\ldots,m}]\}$$

$$= [f, \{g_j\}_{j=1,\ldots,m}] + T[f, \{g_j\}_{j=1,\ldots,m}] ,$$

where the operator $\mathcal{O}$ (or $\mathcal{O}^{\varepsilon_i}$) acts by the rule $\mathcal{O}(\phi_i u_i) = \phi_i \mathcal{O} u_i + [\mathcal{O}, \phi_i] u_i$. The first term of (9.10) follows from property (9.5), the second one from the remaining three terms. Since $T[f, \{g_j\}_{j=1,\ldots,m}]$
$\in V^{\ell+1,p}(\Omega,\vec{\beta}) \times \prod_{j=1}^{m} V^{\ell+2m-m_j-1/p+1,p}(\partial\Omega,\vec{\beta})$ and this space is compactly imbedded into $V^{\ell,p}(\Omega,\vec{\beta}) \times \prod_{j=1}^{m} V^{\ell+2m-m_j-1/p,p}(\partial\Omega,\vec{\beta})$ (see V. A. KONDRAT'EV [1] for $p = 2$), T is a compact operator.

In a similar way we can construct a left regularizer. Since $\mathcal{O}$ is a bounded operator defined on the whole space $V^{\ell+2m,p}(\Omega,\vec{\beta})$, $\mathcal{O}$ is closed. Lemma 9.1 (iii) yields the assertion.

(iv) Now we deal with the necessity of the assumption following the ideas of V. A. KONDRAT'EV [1]. On the contrary, assume there is an eigenvalue $\lambda_0^{(j)}$ of $\mathcal{O}_0^{(j)}(\lambda)$, $j \in \{1,\ldots,N_1\}$, on the line $\operatorname{Im} \lambda = \frac{N}{p} + \beta_j - \ell - 2m$. Let $u_0^{(j)}(\omega)$ be a corresponding eigenfunction. We consider the sequence of functions

$$\left\{ u_n^{(j)} = \frac{r_j^{i\lambda_0^{(j)}+1/n}\, u_0^{(j)}(\omega^j)\eta_j(r_j)}{\left\| r_j^{i\lambda_0^{(j)}+1/n}\, u_0^{(j)}(\omega^j); \ V^{\ell+2m,p}(\Omega,\vec{\beta}) \right\|} \right\}_{n=1,2\ldots}$$

where ω^j is the polar angle to O_j and

$$(9.11) \qquad \eta_j(r_j) = \begin{cases} 1 & \text{for } r_j = |x - O_j| < \varepsilon_j/4 \\ 0 & \text{for } r_j = |x - O_j| > \varepsilon_j/2 \end{cases}, \quad \eta_j(r_j) \in C_0^{\infty}(R^N).$$

The functions $u_n^{(j)}$ are from $V^{\ell+2m,p}(\Omega,\vec{\beta})$ and $\|u_n^{(j)}; V^{\ell+2m,p}(\Omega,\vec{\beta})\|$ $\geq \text{const} > 0$ for all n, but $\lim_{n\to\infty} u_n^{(j)}(r,\omega) = 0$ a.e. On the other hand, we have

$$\lim_{n\to\infty} \left\| \mathcal{O}(x,D_x)v_n^{(j)}; \ V^{\ell,p}(\Omega,\vec{\beta}) \times \prod_{j=1}^{m} V^{2m+\ell-m_j-1/p,p}(\partial\Omega,\vec{\beta}) \right\| = 0$$

where $v_n^{(j)}(x) = u_n^{(j)}(r,\omega^j)$. This means that the limit of the sequence $\left(\mathcal{O}(x,D_x)v_n^j\right)_{n=1,2\ldots}$ has no original from $V^{\ell+2m,p}(\Omega,\vec{\beta})$ and therefore the image of $\mathcal{O}(x,D_x)$ is not closed. Consequently $\mathcal{O}(x,D_x)$ is not a Fredholm operator.

9.3. **THEOREM** (on regularity). *Let* $u \in V^{\ell+2m,p}(\Omega,\vec{\beta})$ *be a solution of problem* (5.3), *where* $f \in V^{\ell_1,p_1}(\Omega,\vec{\beta}_1) \cap V^{\ell,p}(\Omega,\vec{\beta})$ *and* $g_j \in V^{\ell_1+2m-m_j-1/p_1,p_1}(\partial\Omega,\vec{\beta}_1)$ $\cap V^{\ell+2m-m_j-1/p,p}(\partial\Omega,\vec{\beta})$, $j = 1,\ldots,m$, $\ell_1 \geq \ell$, $h_{1,i} = \frac{N}{p_1} + \beta_{1,i} - \ell_1 - 2m \leq \frac{N}{p} + \beta_i - \ell - 2m = h_i$, $i = 1,\ldots,N_1$, $(\beta_{1,1},\beta_{1,2},\ldots,\beta_{1,N_1}) = \vec{\beta}_1$.

If no eigenvalues of the operators $\mathcal{U}_0^{(i)}(\lambda)$, $i = 1,\ldots,N_1$, *are situated in the strips* $h_{1,i} \leq \operatorname{Im} \lambda \leq h_i$, *then* $u \in V^{\ell_1+2m,P_1}(\Omega,\vec{\beta}_1)$, *too.*

P r o o f . Let η_i be the cut-off functions defined by (9.11) and ε_i the numbers given by (9.1). We write $u = \sum_{i=1}^{N_1} \eta_i u + u - \sum_{i=1}^{N_1} \eta_i u$. The function $u_1 = u - \sum_{i=1}^{N_1} \eta_i u$ vanishes in $B_{\varepsilon_i/4}(O_i)$, $i = 1,\ldots,N_1$. Therefore it follows from the regularity theory for elliptic boundary value problems in smooth domains (see e.g. S. AGMON, A. DOUGLIS, L. NIRENBERG [1]) that $u_1 \in W^{\ell_1+2m,P_1}(\Omega)$ and consequently, $u_1 \in V^{\ell_1+2m,P_1}(\Omega,\vec{\beta}_1)$. Further,

$$(9.12) \qquad \mathcal{U}(\eta_i u) = \mathcal{U}^{\varepsilon_i}(\eta_i u) = \eta_i[f, \{g_j\}_{j=1,\ldots,m}] + [\mathcal{U}^{\varepsilon_i}, \eta_i]u$$

(see (9.1) and 9.2 (iii)). Since only the derivatives of η_i occur in the last term, it vanishes in $B_{\varepsilon_i/4}(O_i)$ and also for all x with $|x - O_i| > \varepsilon_i/2$.

Therefore the right hand side of (9.12) satisfies $\eta_i[f, \{g_j\}_{j=1,\ldots,m}]$

$$+ [\mathcal{U}^{\varepsilon_i},\eta_i]u \in \left[V^{\ell_1,P_1}(K_i,\beta_{1,i}) \times \prod_{j=1}^{m} V^{\ell_1+2m-m_j-1/P_1,P_1}(\partial K_i,\beta_{1,i})\right]$$

$$\cap \left[V^{\ell,P}(K_i,\beta_i) \times \prod_{j=1}^{m} V^{\ell+2m-m_j-1/p,p}(\partial K_i,\beta_i)\right] \quad (\ K_i \ \text{is the infinite cone gene-}$$

rated by the conical point O_i). Lemma 8.2 yields that $\eta_i u \in V^{\ell_1+2m,P_1}(K_i,\beta_{1,i})$ and consequently, $\eta_i u \in V^{\ell_1+2m,P_1}(\Omega,\vec{\beta}_1)$ for $i = 1,2,\ldots,N_1$. It follows that $u = \sum_{i=1}^{N_1} \eta_i u + u_1 \in V^{\ell_1+2m,P_1}(\Omega,\vec{\beta}_1)$.

§ 10 . T h e   e x p a n s i o n   o f   t h e   s o l u t i o n   n e a r
      a   c o n i c a l   p o i n t

Theorem 9.3 describes the regularity of the solution, provided the strips $h_{1,i} \leq \operatorname{Im} \lambda \leq h_i$ do not contain eigenvalues of $\mathcal{U}_0^i(\lambda)$. We now study what happens if eigenvalues of $\mathcal{U}_0^i(\lambda)$ occur in these strips.

$$(10.1) \qquad \begin{array}{l} \text{Assume for simplicity that there is only one conical point } O \text{ at} \\ \partial\Omega \text{ and that there is a neighborhood } B_\varepsilon(O) \text{ where } \Omega \text{ coincides} \\ \text{with an infinite cone } K . \end{array}$$

<u>10.1. THEOREM.</u> *Let* $\Omega \subset \mathbb{R}^N$ *be a bounded domain with the property* (10.1) . *Let* $u \in V^{\ell+2m,P}(\Omega,\beta)$ *be a solution of* (5.3) *where* $f \in V^{\ell_1,P_1}(\Omega,\beta_1)$ *and* $g_j \in V^{\ell_1+2m-m_j-1/P_1,P_1}(\partial\Omega,\beta_1)$, $j = 1,\ldots,m$, $\ell_1 \geq \ell$, $h_1 = \beta_1 + \dfrac{N}{P_1} - 2m - \ell_1$

$< h = \beta + \frac{N}{p} - 2m - \ell$. *Assume the lines* $\operatorname{Im} \lambda = h_1$ *and* $\operatorname{Im} \lambda = h$ *contain no eigenvalues of* $\mathfrak{A}_0(\lambda)$ *and that the number* $\varepsilon > 0$ *in* (10.1) *is so small that* (8.2) *is valid for a certain* δ *in* $B_\varepsilon(0)$. *Let* η *be a cut-off function of the type* (9.11).

Then the expansion

$$(10.2) \qquad \eta u = \sum_{h_1 < \operatorname{Im} \lambda_j < h} \; \sum_{\mu=0}^{\{h-h_1\}} r^{i\lambda_j + \mu} \, P_{j\mu}(\log r) + w$$

holds, where λ_j *are eigenvalues of* $\mathfrak{A}_0(\lambda)$ *in the strip* $h_1 < \operatorname{Im} \lambda_j < h$; $\{h - h_1\}$ *is the biggest integer which is less than* $h - h_1$; $\eta w \in$

$\in V^{\ell_1 + 2m, p_1}(\Omega, \beta_1)$ *and* $P_{j\mu}$ *are polynomial functions with coefficients depending on* ω .

Remarks to the p r o o f : This theorem was proved by V. A. KONDRAT'EV [1] for $p = p_1 = 2$. The ideas of that proof can be applied also for general $p = p_1$, $\ell = \ell_1$. The extension to the more general case $p \neq p_1$, $\ell \neq \ell_1$ is simple and proceeds by using the results of Theorem 9.3 and the fact that the regularity of the remainder w is determined by those eigenvalues λ_w of $\mathfrak{A}_0(\lambda)$ for which $\operatorname{Im} \lambda_w < h_1$. In what follows we shall prove special cases of Theorem 10.1 when the expansion (10.2) has a form analogous to the expansion (7.9) of the solution of the special problem (5.7) in an infinite cone.

<u>10.2.   THEOREM.</u> *Suppose the assumptions of Theorem 10.1 are valid and moreover,*

$$(10.3) \qquad 0 \leq h - h_1 \leq 1$$

or

$$h - h_1 > 1 \; , \quad \mathfrak{A}_0(x, D) = \mathfrak{A}_0(0, D) \quad and$$

$$(10.4) \qquad a_\alpha(x) \equiv 0 \quad for \quad 2m - 1 \geq |\alpha| > 2m - (h - h_1) \; ,$$

$$b_{j,\alpha}(x) \equiv 0 \quad for \quad m_j - 1 \geq |\alpha_j| > m_j - (h - h_1) \; , \quad j = 1, \ldots, m \; .$$

Then

$$(10.5) \qquad \eta u = \sum_{\gamma \in I} c_\gamma u_\gamma + w$$

where $\eta w \in V^{\ell_1 + 2m, p_1}(\Omega, \beta_1)$. *The index set* I *and the functions* u_γ *are defined in Theorem 7.3,* c_γ *are constants.*

P r o o f : (i) Let $\beta_1^* = \beta + \frac{N}{p} - \ell - \frac{N}{p_1} + \ell_1 = h - h_1 + \beta_1$. Since $\beta_1^* \geq \beta_1$ we have $V^{\ell_1, p_1}(\Omega, \beta_1) \subset V^{\ell_1, p_1}(\Omega, \beta_1^*)$ and $V^{\ell_1 + 2m - m_j - 1/p_1, p_1}(\partial\Omega, \beta_1)$ $\subset V^{\ell_1 + 2m - m_j - 1/p_1, p_1}(\partial\Omega, \beta_1^*)$. The line $\operatorname{Im} \lambda = h = \beta + \frac{N}{p} - \ell - 2m = \beta_1^* - 2m$ $- \ell_1 + \frac{N}{p_1}$ contains no eigenvalues of $\mathfrak{A}_0(\lambda)$. Therefore it follows from

Theorem 9.3 that $u \in V^{\ell_1 + 2m, p_1}(\Omega, \beta_1^*)$.

(ii) Let us start with the case $h - h_1 \leq 1$. The function $\eta u \in V^{\ell_1 + 2m, p_1}(\Omega, \beta_1^*)$ fulfils

$$A_0(0, D_x)\eta u = \left[A(x, D_x) + A_0(0, D_x) - A_0(x, D_x) - A_1(x, D_x)\right]\eta u$$

(10.6)
$$= \eta f - \eta A_1(x, D_x)u + \eta\left[A_0(0, D_x) - A_0(x, D_x)\right]u$$

$$+ \sum_{|\alpha| = 2m} a_\alpha(0) \sum_{\gamma < \alpha} \binom{\alpha}{\gamma} D_x^{\alpha - \gamma} \eta D_x^\gamma u = F ,$$

(10.7)
$$B_{j,0}(0, D_x)\eta u = \eta g_j - \eta B_{j,1}(x, D_x)u + \eta\left[B_{j,0}(0, D_x) - B_{j,0}(x, D_x)\right]u$$

$$+ \sum_{|\alpha| = m_j} b_{j,\alpha}(0) \sum_{\gamma < \alpha} \binom{\alpha}{\gamma} D_x^{\alpha - \gamma} \eta D_x^\gamma u = G_j , \quad j = 1,\ldots,m ,$$

where $\gamma = (\gamma_1,\ldots,\gamma_N) < (\alpha_1,\ldots,\alpha_N) = \alpha$ means that $\gamma_i \leq \alpha_i$ for $i = 1,\ldots,N$ and $\gamma_{i_0} < \alpha_{i_0}$ for at least one index i_0 , $1 \leq i_0 \leq N$. The right hand side F of (10.6) is an element of $V^{\ell_1, p_1}(\Omega, \beta_1)$, because $\eta A_1(x, D_x)u \in V^{\ell_1, p_1}(\Omega, \beta_1^* - 1)$, $\eta\left[A_0(0, D_x) - A_0(x, D_x)\right]u \in V^{\ell_1, p_1}(\Omega, \beta_1^* - 1)$,

$\sum\limits_{|\alpha| = 2m} a_\alpha(0) \sum\limits_{\gamma < \alpha} \binom{\alpha}{\gamma} D_x^{\alpha - \gamma} \eta D_x^\gamma u$ vanishes in a neighborhood of 0 and $\beta_1^* - 1 = h - h_1 + \beta_1 - 1 \leq \beta_1$. We obtain analogously that the right hand sides G_j of (10.7) are elements of $V^{\ell_1 + 2m - m_j - 1/p_1, p_1}(\partial\Omega, \beta_1)$, $j = 1, 2,\ldots,m$. Since $\eta u \in V^{2m + \ell_1, p_1}(K, \beta_1^*)$, $F \in V^{\ell_1, p_1}(K, \beta_1) \cap V^{\ell_1, p_1}(K, \beta_1^*)$, $G_j \in V^{\ell_1 + 2m - m_j - 1/p_1, p_1}(\partial K, \beta_1) \cap V^{\ell_1 + 2m - m_j - 1/p_1, p_1}(\partial K, \beta_1^*)$, $\beta_1 \leq \beta_1^*$ we can apply Theorem 7.3 obtaining

$$\eta u = \sum_{\gamma \in I} c_\gamma u_\gamma + w$$

where $w \in V^{2m + \ell_1, p_1}(K, \beta_1)$ and $\eta w \in V^{2m + \ell_1, p_1}(\Omega, \beta_1)$.

(iii) We now assume that conditions (10.4) are satisfied. These conditions guarantee that the right hand sides F and G_j of (10.6) and (10.7) belong to $V^{\ell_1, p_1}(\Omega, \beta_1)$ and $V^{\ell_1 + 2m - m_j - 1/p_1, p_1}(\partial\Omega, \beta_1)$, $j = 1,\ldots,m$, respectively. Therefore we conclude as in the first case that

$$\eta u = \sum_{\gamma \in I} c_\gamma u_\gamma + w ,$$

where $\eta w \in V^{2m + \ell_1, p_1}(\Omega, \beta_1)$.

(iv) <u>COROLLARY</u>. *Let the assumptions of Theorem 10.2 be satisfied. Then*

(10.8)
$$u = \sum_{\gamma \in I} \eta c_\gamma u_\gamma + w_1 ,$$

where $w_1 \in V^{2m+\ell_1, P_1}(\Omega, \beta_1)$.

P r o o f : We have

$$u = \eta^2 u + (1 - \eta^2)u = \eta(\sum_{\gamma \in I} c_\gamma u_\gamma + w) + (1 - \eta^2)u = \sum_{\gamma \in I} c_\gamma \eta u_\gamma + w_1 \ ,$$

where $w_1 = \eta w + (1 - \eta^2)u$. The function $(1 - \eta^2)u$ vanishes in a neighborhood of O .

The following theorem yields an estimate of w_1 .

<u>10.3.</u>   <u>THEOREM.</u>   *Let the assumptions of Theorem 10.2 be satisfied. If* $\mathcal{O}(x, D_x)$ *is an isomorphism of* $V^{\ell+2m, P}(\Omega, \beta)$ *into* $V^{\ell, P}(\Omega, \beta) \times \prod_{j=1}^{m} V^{\ell+2m-m_j+1/p, p}(\partial\Omega, \beta)$, *then*

(10.9) $\|w_1; V^{\ell_1+2m, P_1}(\Omega, \beta_1)\|$

$$\leq c\left[\|f; V^{\ell_1, P_1}(\Omega, \beta_1)\| + \sum_{j=1}^{m} \|g_j; V^{\ell_1+2m-m_j-1/p_1, P_1}(\partial\Omega, \beta_1)\|\right] \ ,$$

where w_1 *is given by* (10.8).

P r o o f :   (i) We first estimate ηw , using (7.11) :

$$\|\eta w; V^{\ell_1+2m, P_1}(\Omega, \beta_1)\| \leq c\|w; V^{\ell_1+2m, P_1}(K, \beta_1)\|$$

(10.10) $\leq c\left[\|F; V^{\ell_1, P_1}(K, \beta_1)\| + \sum_{j=1}^{m} \|G_j; V^{2m+\ell_1-m_j-1/p_1, P_1}(\partial K, \beta_1)\|\right]$

$$\leq c\left[\|F; V^{\ell_1, P_1}(\Omega, \beta_1)\| + \sum_{j=1}^{m} \|G_j; V^{2m+\ell_1-m_j-1/p_1, P_1}(\partial\Omega, \beta_1)\|\right]$$

where F and G_j are determined by (10.6) and (10.7), respectively.

  - If (10.3) is valid, then $\beta_1^* = \beta + \dfrac{N}{p} - \ell - \dfrac{N}{P_1} + \ell_1 = h - h_1 + \beta_1 \leq 1 + \beta_1$.
     Estimating every term of F and G_j we obtain in this case

$$\|\eta w; V^{\ell_1+2m, P_1}(\Omega, \beta_1)\| \leq c\left[\|f; V^{\ell_1, P_1}(\Omega, \beta_1)\|\right.$$

$$+ \sum_{j=1}^{m} \|g_j; V^{2m+\ell_1-m_j-1/p_1, P_1}(\partial\Omega, \beta_1)\| + \|u; V^{2m-1+\ell_1, P_1}(\Omega, \beta_1)\|$$

(10.11) $+ \|u; V^{2m+\ell_1, P_1}(\Omega, \ \beta_1 + 1)\| + \|u; V^{2m+\ell_1, P_1}(\Omega, \beta_1^*)\|\Big]$

$$\leq c\left[\|f; V^{\ell_1, P_1}(\Omega, \beta_1)\| + \sum_{j=1}^{m} \|g_j; V^{2m+\ell_1-m_j-1/p_1, P_1}(\partial\Omega, \beta_1)\|\right.$$

$$+ \|u; V^{2m+\ell_1, P_1}(\Omega, \beta_1^*)\|\Big] \ .$$

$-$ If (10.4) is valid, we obtain, instead of (10.11),

$$\|\eta w; V^{\ell_1+2m,p_1}(\Omega,\beta_1)\| \le c\big[\|f;V^{\ell_1+2m,p_1}(\Omega,\beta_1)\| + \|u;V^{\ell_1+k,p_1}(\Omega,\beta_1)\|$$

$$(10.12) \qquad + \|u;V^{\ell_1+2m,p_1}(\Omega,\beta_1^*)\| + \sum_{j=1}^{m}\|g_j;V^{\ell_1+2m-m_j-1/p_1,p_1}(\partial\Omega,\beta_1)\|\big]\;,$$

where $k = 2m - \{h - h_1\}$ provided $2m - \{h - h_1\} \ge 0$. If $2m - \{h - h_1\}$ < 0 then $A_1(x,D_x) = 0$ and $B_{j,1}(x,D_x) = 0$. Since $\|u;V^{\ell_1+k,p_1}(\Omega,\beta_1)\|$ $\le c\|u;V^{\ell_1+2m,p_1}(\Omega,\ \beta_1 + 2m - k)\| \le c\|u;V^{\ell_1+2m,p_1}(\Omega,\beta_1^*)\|$

we again obtain

$$\|\eta w\| \le c\big[\|f;V^{\ell_1,p_1}(\Omega,\beta_1)\| + \sum_{j=1}^{m}\|g_j;V^{2m+\ell_1-m_j-1/p_1,p_1}(\partial\Omega,\beta_1)\|$$

$$(10.13) \qquad + \|u;V^{2m+\ell_1,p_1}(\Omega,\beta_1^*)\|\big]\;.$$

$-$ Since $\mathcal{A}(x,D_x)$ is an isomorphism of $V^{\ell+2m,p}(\Omega,\beta)$ into $V^{\ell,p}(\Omega,\beta)$

$$\times\ \prod_{j=1}^{m} V^{\ell+2m-m_j+1/p,p}(\partial\Omega,\beta)\;,$$ it follows from Theorems 9.2 and 9.3 that

$\mathcal{A}(x,D_x)$ is also an isomorphism of $V^{\ell_1+2m,p_1}(\Omega,\beta_1^*)$ into $V^{\ell_1,p_1}(\Omega,\beta_1^*)$

$$\times\ \prod_{j=1}^{m} V^{\ell_1+2m-m_j+1/p_1,p_1}(\partial\Omega,\beta_1^*)\;.$$ Therefore

$$\|u;V^{\ell_1+2m,p_1}(\Omega,\beta_1^*)\|$$

$$(10.14) \qquad \le c\big[\|f;V^{\ell_1,p_1}(\Omega,\beta_1^*)\| + \sum_{j=1}^{m}\|g_j;V^{\ell_1+2m-m_j-1/p_1,p_1}(\partial\Omega,\beta_1^*)\|\big]$$

$$\le c\big[\|f;V^{\ell_1,p_1}(\Omega,\beta_1)\| + \sum_{j=1}^{m}\|g_j;V^{\ell_1+2m-m_j-1/p_1,p_1}(\partial\Omega,\beta_1)\|\big]$$

and we conclude

$$(10.15) \qquad \|\eta w\| \le c\big[\|f;V^{\ell_1,p_1}(\Omega,\beta_1)\| + \sum_{j=1}^{m}\|g_j;V^{\ell_1+2m-m_j-1/p_1,p_1}(\partial\Omega,\beta_1)\|\big]\;.$$

(ii) We now estimate $(1 - \eta^2)u$. Since η is independent of u and $(1 - \eta^2)u$ vanishes in a neighborhood of O we have

$$\|(1 - \eta^2)u;\ V^{\ell_1+2m,p_1}(\Omega,\beta_1)\| \le c\|(1 - \eta^2)u;\ V^{\ell_1+2m,p_1}(\Omega,\beta_1^*)\|$$

$$\le c\|u;V^{\ell_1+2m,p_1}(\Omega,\beta_1^*)\|\;.$$

Using (10.14) we obtain

$$(10.16) \qquad \|(1 - \eta^2)u;\ V^{\ell_1+2m,p_1}(\Omega,\beta_1)\| \le$$

$$\leq c \left[\|f; V^{\ell_1, p_1}(\Omega, \beta_1)\| + \sum_{j=1}^{m} \|g_j; V^{\ell_1 + 2m - m_j - 1/p_1, p_1}(\partial\Omega, \beta_1)\| \right] .$$

(10.15) and (10.16) yield the assertion (10.9).

10.4. <u>EXAMPLE</u>. Let $\Omega \subset R^N$ be a bounded domain with the property (10.1). Let A be an elliptic operator of the second order with variable coefficients. We consider the Dirichlet problem

$$Au = f \quad \text{in} \quad \Omega ,$$

$$u = 0 \quad \text{on} \quad \partial\Omega .$$

Assume there is a uniquely determined solution $u \in V^{2,2}(\Omega, 1)$ for every $f \in$ $\in L^2(\Omega)$. We have $h = 1 + \frac{N}{2} - 2$ and choose $h_1 = 0 + \frac{N}{2} - 2$. Then $h - h_1 = 1$ and (10.8) implies

$$u = \sum_{\gamma \in I} \eta c_\gamma u_\gamma + w_1 ,$$

where $w_1 \in V^{2,2}(\Omega, 0)$. Further, $\|w_1; V^{2,2}(\Omega, 0)\| \leq c\|f; L^2(\Omega)\|$.

§ 11 . T h e   c a s e $\ell < 0$

We only formulate some basic results, from which one can derive further solvability and regularity theorems.

Results for $\ell < 0$ are of interest for the study of the behavior of the so-called weak solutions (see 15.2) of problem (5.3) near conical points. The difference as compared with the case $\ell \geq 0$ is the following : The operator $\mathcal{U}(x, D_x)$ maps $\tilde{V}^{\ell+2m, 2}(\Omega, \beta)$ into the space $V^{\ell, 2}(\Omega, \beta)$ $\times \prod_{j=1}^{m} V^{\ell+2m-m_j-1/2, 2}(\partial\Omega, \beta)$, where $\tilde{V}^{\ell+2m, 2}(\Omega, \beta)$ is a subspace of $V^{\ell+2m, 2}(\Omega, \beta)$ defined by (0.22) and $\Omega \subset R^N$ is a bounded domain with only one conical point $O \subset \partial\Omega$.

11.1. <u>THE SPECIAL PROBLEM IN K</u> . Let us formulate two lemmas about solvability and regularity of the special problem (5.7) in an infinite cone K with the vertex O for $\ell < 0$. These lemmas have been proved by J. ROSSMANN [2].

(i) <u>LEMMA</u>. *Let ℓ be an integer. Assume no eigenvalue of $\mathcal{U}_0(\lambda)$ lies on the line* Im $\lambda = \beta + \frac{N}{2} - 2m - \ell$. *Then*

$$\mathcal{U}_0(0, D_x) : \tilde{V}^{\ell+2m, 2}(K, \beta) \to V^{\ell, 2}(K, \beta) \times \prod_{j=1}^{m} V^{2m+\ell-m_j-1/2, 2}(\partial K, \beta)$$

is an isomorphism.

(ii) <u>LEMMA</u>. *Let ℓ be an integer. Assume no eigenvalue of $\mathcal{U}_0(\lambda)$ lies on the lines* Im $\lambda = h = \beta + \frac{N}{2} - 2m - \ell$ *and* Im $\lambda = h_1 = \beta_1 + \frac{N}{2} - 2m - \ell_1$, $\ell_1 \geq \ell$, $h_1 \leq h$. *Let* $u \in \tilde{V}^{\ell+2m, 2}(K, \beta)$ *be a solution of (5.7) with*

$$f \in V^{\ell_1, 2}(K, \beta_1) \quad and \quad g_j \in V^{\ell_1 + 2m - m_j - 1/2, 2}(\partial K, \beta_1) \ .$$

Then

$$u = \sum_{\gamma \in I} c_\gamma u_\gamma + w$$

where $w \in \tilde{V}^{\ell_1 + 2m, 2}(K, \beta_1)$ *and* I *is defined by* (7.9).

<u>11.2. THE GENERAL PROBLEM IN</u> Ω . The following lemma follows from 11.1, Lemma (i).

(i) <u>LEMMA</u>. *Let* $\Omega \subset R^N$ *be a bounded domain with the property* (10.1). *Let* ℓ *be an integer. If the line* $\operatorname{Im} \lambda = \beta + \frac{N}{2} - 2m - \ell$ *contains no eigenvalues of* $\mathcal{U}_0(\lambda)$ *, then the operator of problem* (5.3), *that is,*

$$\mathcal{U}(x, D_x) : \ \tilde{V}^{\ell + 2m, 2}(\Omega, \beta) \to V^{\ell, 2}(\Omega, \beta) \times \prod_{j=1}^{m} V^{2m + \ell - m_j - 1/2, 2}(\partial \Omega)$$

is a Fredholm operator.

Further, a regularity result holds:

(ii) <u>LEMMA</u>. *Let* $\Omega \subset R^N$ *be a bounded domain with the property* (10.1) *and let* ℓ *be an integer. Assume no eigenvalue of* $\mathcal{U}_0(\lambda)$ *lies on the line* $\operatorname{Im} \lambda = h = \beta + \frac{N}{2} - 2m - \ell = \beta_1 + \frac{N}{2} - 2m - \ell_1$ *. Let* $u \in \tilde{V}^{\ell + 2m, 2}(\Omega, \beta)$ *be a solution of the problem* (5.3) *and* $f \in V^{\ell_1, 2}(\Omega, \beta_1)$ *,* $g_j \in V^{\ell_1 + 2m - m_j - 1/2, 2}(\partial \Omega, \beta_1)$ *,* $j = 1, \ldots, m$ *. Then* $u \in \tilde{V}^{\ell_1 + 2m, 2}(\Omega, \beta_1)$ *, too.*

<u>11.3. EXAMPLE.</u> Let $\Omega \subset R^N$ be a bounded domain with the property (10.1). Let A be an elliptic operator of the second order with variable coefficients. We consider the Dirichlet problem

$$Au = f \quad in \ \Omega \ ,$$
$$u = 0 \quad on \ \partial \Omega \ ,$$

where $f \in L^2(\Omega)$. Assume that a uniquely determined "weak" solution $u \in W_0^{1,2}(\Omega)$ exists. Since $W_0^{1,2}(\Omega) \subset \tilde{V}^{1,2}(\Omega, 0)$ we obtain from 11.2 Lemma (ii) that u $\in \tilde{V}^{2,2}(\Omega, 1) = V^{2,2}(\Omega, 1)$ (see 0.9).

Section 4 : *C a l c u l a t i o n   o f   t h e   c o e f f i c i e n t s   i n   t h e   e x p a n s i o n*

We want to derive formulae for the calculation of the coefficients c_γ occuring in (7.9) and (10.5). These coefficients are of interest e.g. in engineering.

§ 12. The coefficient formula for the
       special problem in an infinite cone

We consider the special boundary value problem (5.7). Since its adjoint problem will play an important role, let us start with it.

<u>12.1. THE ADJOINT PROBLEM</u>. (i) Let $\{B_{j,0}\}_{j=1,\ldots,m}$ be a normal system (for the definition, see e.g. J. WLOKA [1] or J. L. LIONS, E. MAGENES [1]) of boundary operators such that Green's formula

$$\int_K A_0 \, u \, \bar{v} \, dx + \sum_{j=1}^{m} \int_{\partial K} B_{j,0} \, u \, \overline{T_{j,0} \, v} \, d\sigma$$

(12.1)

$$= \int_K u \, \overline{A_0^* \, v} \, dx + \sum_{j=1}^{m} \int_{\partial K} S_{j,0} \, u \, \overline{B_{j,0}^* \, v} \, d\sigma$$

holds for all $u \in V^{2m,p}(K, \beta-\ell)$ and $v \in V^{2m,q}(K, -\beta+\ell+2m)$, $\frac{1}{p} + \frac{1}{q} = 1$.
A_0^* , $T_{j,0}$, $S_{j,0}$ and $B_{j,0}^*$ are operators with constant coefficients and of orders $2m$, $2m-m_j-1$, $2m-1-\mu_j$ and μ_j , respectively. Now the adjoint problem to (5.7) is

$$A_0^* v = f \qquad \text{in} \ \ K \ ,$$

(12.2) $\quad B_{j,0}^* v = g_j \quad$ on ∂K , $\ j = 1,2,\ldots,m$.

The corresponding operator $\mathcal{U}_0^*$ has the following property:

$$\mathcal{U}_0^* = \{A_0^*, B_{j,0}^*\}_{j=1,\ldots,m} \ : \ V^{2m,q}(K, -\beta+\ell+2m) \to V^{0,q}(K, -\beta+\ell+2m)$$

(12.3)

$$\times \prod_{j=1}^{m} V^{2m-\mu_j-1/q,q}(\partial K, -\beta+\ell+2m) \ .$$

   (ii) We write $\mathcal{U}_0^*$ in spherical coordinates

$$A_0^* = A_0^*(0,D_x) = r^{-2m} L^*(\omega, D_\omega, rD_r)$$

$$B_{j,0}^* = B_{j,0}^*(0,D_x) = r^{-\mu_j} M_j^*(\omega, D_\omega, rD_r)$$

and consider the boundary value problem analogous to (6.1) depending on the parameter λ^* :

(12.4)

$$L^*(\omega, D_\omega, \lambda^*) \, \tilde{v}(\lambda^*, \omega) = \tilde{F}(\lambda^*, \omega) \quad \text{for} \ \ \omega \in G \ ,$$

$$M_j^*(\omega, D_\omega, \lambda^*) \, \tilde{v}(\lambda^*, \omega) = \tilde{G}_j(\lambda^*, \omega) \quad \text{for} \ \ \omega \in \partial G \ , \ \ j = 1,2,\ldots,m \ .$$

We denote by $\mathcal{U}_0^*(\lambda^*) = \{L^*, M_1^*, \ldots, M_m^*\}$ the corresponding operator,

(12.5) $\quad \mathcal{U}_0^*(\lambda^*) \ : \ W^{2m,q}(G) \to W^{0,q}(G) \times \prod_{j=1}^{m} W^{2m-\mu_j-1/q,q}(\partial G) \ .$

The eigenvalues of $\mathcal{U}_0^*(\lambda^*)$ can be calculated with the help of the following lemma.

12.2. LEMMA. *If λ_0 is an eigenvalue of $\mathcal{U}_0(\lambda)$ then $\lambda_0^* = \overline{\lambda}_0 + i(N - 2m)$ is an eigenvalue of $\mathcal{U}_0^*(\lambda^*)$.*

P r o o f . We follow the ideas of V. G. MAZ'JA, B. A. PLAMENEVSKIĬ [5].

(i) We first show that Green's formula (12.1) in K implies Green's formula in the domain G , namely,

$$(12.6) \quad \int_G L(\omega,D_\omega,\lambda)w_1 \, \overline{w}_2 \, d\omega + \sum_{j=1}^m \int_{\partial G} M_j(\omega,D_\omega,\lambda)w_1 \, \overline{Q_j(\omega,D_\omega, \overline{\lambda} + i(N-2m)w_2} \, d\sigma$$

$$= \int_G w_1 \, \overline{L^*(\omega,D_\omega, \overline{\lambda} + i(N-2m)w_2} \, d\omega$$

$$+ \sum_{j=1}^m \int_{\partial G} N_j(\omega,D_\omega,\lambda)w_1 \, \overline{M_j^*(\omega,D_\omega, \overline{\lambda} + i(N-2m)w_2} \, d\sigma$$

for all functions $w_1 \in W^{2m,p}(\Omega)$, $w_2 \in W^{2m,q}(\Omega)$; Q_j and N_j are defined by $T_{j,0}(0,D_x) = r^{-(2m-1-m_j)} Q_j(\omega,D_\omega,rD_r)$ and $S_{j,0}(0,D_x) = r^{-(2m-1-\mu_j)} N_j(\omega,D_\omega,rD_r)$, respectively. For this purpose let us introduce a function $\eta \in C_0^\infty(R)$ with the properties

$$(12.7) \quad \eta(t) \geq 0 , \quad \int_{-\infty}^{+\infty} \eta(t) \, dt = 1 , \quad \text{supp } \eta \subset (1,\infty) .$$

We write $\eta(\log r) = \hat{\eta}(r)$ and consider a function $\kappa = \kappa(r) \in C_0^\infty(R)$ such that $\kappa(r) \equiv 1$ for $r \in \text{supp } \hat{\eta}$. Inserting into (12.1) the functions

$$u = r^{i\lambda} \eta(\log r) w_1(\omega) , \quad v = r^{i(\overline{\lambda}+i(N-2m))} \kappa(r) w_2(\omega) ,$$

where $w_1 \in W^{2m,p}(G)$, $w_2 \in W^{2m,q}(\Omega)$, we obtain for $r = e^t$

$$\int_K A_0 u \, \overline{v} \, dx = \int_0^\infty \int_G A_0 u \, \overline{v} \, r^{N-1} \, dr \, d\omega$$

$$= \int_{-\infty}^{+\infty} e^{-i\lambda t} \int_G L(\omega,D_\omega,D_t)e^{i\lambda t} \, \eta(t)w_1(\omega)\kappa(e^t)\overline{w}_2 \, dt \, d\omega$$

$$= \int_G [L(\omega,D_\omega,\lambda)w_1] \, \overline{w}_2 \, d\omega ,$$

$$\int_K u \, \overline{A_0^*v} \, dx = \int_G w_1 \, \overline{L^*(\omega,D_\omega, \overline{\lambda} + i(N-2m)w_2)} \, d\omega$$

and similarly the boundary integral terms in (12.6). Notice that we have used that $\int_{-\infty}^{+\infty} D_t^k \eta(t) \, dt = 0$ for $k > 0$ and $\int_{-\infty}^{+\infty} \eta(t) \, dt = 1$.

(ii) Let λ_0 be an eigenvalue of $\mathcal{U}_0(\lambda)$ and let $u_0 \in W^{2m,p}(G)$ be an eigenfunction corresponding to λ_0 (see Definition (6.3)). Inserting u_0 into (12.6) we obtain

$$(12.8) \qquad \int_G u_0 \overline{L^*\!\left(\omega, D_\omega, \overline{\lambda}_0 + i(N-2m)w_2\right)}\, d\omega$$

$$+ \sum_{j=1}^{m} \int_{\partial G} N_j(\omega, D_\omega, \lambda_0) u_0\, \overline{M_j^*\!\left(\omega, D_\omega, \overline{\lambda}_0 + i(N-2m)w_2\right)}\, d\sigma = 0$$

for every $w_2 \in W^{2m,q}(G)$. Assume $\lambda_0^* = \overline{\lambda}_0 + i(N-2m)$ is not an eigenvalue of $\mathcal{U}_0^*(\lambda^*)$. Since $\mathcal{U}_0^*(\lambda^*)$ has the same properties as $\mathcal{U}_0(\lambda)$ the operator $L^*(\omega, D_\omega, \lambda_0^*)$ maps the subspace $\{w_2 \in W^{2m,q}(G) : M_j^*(\omega, D_\omega, \lambda_0^*)w_2 = 0 \,,\ j = 1,..,m\}$ into the whole space $L^q(G)$. It follows from (12.8) that $u_0 = 0$. However, this is impossible because u_0 is an eigenfunction.

The following lemma can be proved in the same way as the lemma in 7.2.

<u>12.3. LEMMA.</u> *Let* λ_0 *be an eigenvalue of* $\mathcal{U}_0(\lambda)$. *The homogeneous adjoint problem*

$$(12.9) \qquad \begin{aligned} A_0^*(D_x)v &= 0 \quad in \ K\,, \\ B_{j,0}^*(D_x)v &= 0 \quad on \ \partial K\,,\ \ j = 1,\ldots,m \end{aligned}$$

has the solution

$$(12.10) \qquad v = r^{\,i(\overline{\lambda}_0 + i(N-2m))} \sum_{k=0}^{\ell} \frac{1}{k!}\, (i \log r)^k\, v_{\ell-k}(\omega)\,,$$

where $(v_0,\ldots,v_\ell)$ *is a Jordan chain of* $\mathcal{U}_0^*(\lambda^*)$ *of the length* $\ell + 1$ *with respect to the eigenvalue* $\lambda_0^* = \overline{\lambda}_0 + i(N - 2m)$.

<u>12.4. THE BIORTHONORMALITY CONDITION.</u> (i) Let us start with some notations : Let λ_μ be an eigenvalue of $\mathcal{U}_0(\lambda)$ and $I_\mu^* = \dim \ker \mathcal{U}_0^*\!\left(\overline{\lambda}_\mu + i(N-2m)\right)$. Let

$$\begin{pmatrix} v_\mu^{0,\sigma} \\ \vdots \\ v_\mu^{\kappa_{\mu_\sigma}-1,\sigma} \end{pmatrix}$$

be a Jordan chain of $\mathcal{U}_0^*(\lambda^*)$ of the length κ_{μ_σ} with respect to the eigenvalue $\overline{\lambda}_\mu + i(N-2m)$, $\sigma \in \{1,\ldots,I_\mu^*\}$. We denote

$$(12.11) \qquad v_\gamma = v_{\mu,\sigma,k}(r,\omega) = r^{\,i(\overline{\lambda}_\mu + i(N-2m))} \sum_{q=0}^{k} \frac{1}{q!}\, (i \log r)^q\, v_\mu^{k-q,\sigma}(\omega)\,,$$

$$(12.12) \qquad v_{\gamma'} = v_{\mu,\sigma,k'}(r,\omega)$$

where $k' = \kappa_{\mu_\sigma}-1-k$, $k = 0,\ldots,\kappa_{\mu_\sigma}-1$, v_γ and $v_{\gamma'}$ are solutions of (12.9).

(ii) V. G. MAZ'JA, B. A. PLAMENEVSKIĬ [2], [5], have proved that one can choose the canonical system of Jordan chains of $\mathcal{O}_0(\lambda)$ with respect to the eigenvalue λ_μ (see (7.10)) and the canonical system of Jordan chains of $\mathcal{U}_0^*(\lambda^*)$ with respect to the eigenvalues $\lambda_\mu^* = \bar{\lambda}_\mu + i(N-2m)$ in such a way that a *biorthonormality condition* is satisfieed, namely

$$(12.13) \qquad \sum_{\nu=0}^{k} \sum_{p=\nu+1}^{\tau+\nu+1} \frac{1}{p!} \left(\frac{\partial^p L(\lambda_\mu)}{\partial \lambda^p} u_\mu^{(k-\nu,\sigma)}, \; v_\mu^{\tau-p+\nu+1,\sigma'} \right)_G$$

$$+ \sum_{j=1}^{m} \sum_{\nu=0}^{k} \sum_{p=\nu+1}^{\tau+\nu+1} \frac{1}{p!} \left(\frac{\partial^p M_j(\lambda_\mu)}{\partial \lambda^p} u_\mu^{(k-\nu,\sigma)}, \; \sum_{s=0}^{\tau-p+\nu+1} \frac{1}{s!} \frac{\partial^s Q_j}{\partial \lambda^s}\left(\lambda_\mu + i(N-2m)\right) \right.$$

$$\left. \cdot \; v_\mu^{\tau-p+\nu+1,\sigma'} \right)_{\partial G} = \delta_{\sigma,\sigma'} \cdot \delta_{\kappa_{\mu_\sigma}-k-1,\tau}$$

$k = 0,\ldots,\kappa_{\mu_\sigma}-1$, $\sigma = 1,\ldots,I_\mu$, $\tau = 0,\ldots,\kappa_{\mu_\sigma}-1$, $\sigma' = 1,\ldots,I_\mu$ and

$(f,g)_G = \int_G f \, \bar{g} \, d\omega$, $(f,g)_{\partial G} = \int_{\partial G} f \, \bar{g} \, d\sigma_\omega$. We now formulate the main result of V. G. MAZ'JA, B. A. PLAMENEVSKIĬ [2], [5].

12.5. THEOREM. *Let the assumptions of Theorem 7.4 and the conditions (12.1) and (12.13) be satisfied. Then the coefficients* c_γ *in the expansion (7.14)*

$$u = \sum_{\gamma \in I} c_\gamma u_\gamma + w$$

are given by the formula

$$(12.14) \qquad c_\gamma = \int_K f \, \overline{iv_\gamma'} \, dx + \sum_{j=1}^{m} \int_{\partial K} g_j \, \overline{iT_{j,0}v_\gamma'} \, d\sigma \; .$$

§ 13 .   T h e   c o e f f i c i e n t   f o r m u l a   i n   a   b o u n d e d   d o m a i n

We use the coefficient formula (12.14) to derive the corresponding formula for bounded domains with conical points.

13.1. LEMMA. *Assume the conditions of Theorem 10.2 are satisfied and* $\{B_j\}_{j=1,\ldots,m}$ *is a normal system of boundary operators. Then the solution* u *of problem (5.3) admits the expansion (10.8)*

$$u = \sum_{\gamma \in I} c_\gamma \, \eta \, u_\gamma + w_1$$

where

$$(13.1) \qquad c_\gamma = \int_\Omega A_0(0,D_x)\eta u \, \overline{iv_\gamma'} \, dx + \sum_{j=1}^{m} \int_{\partial\Omega} B_{j,0}(0,D_x)\eta u \, \overline{iT_{j,0}v_\gamma'} \, d\sigma \; .$$

P r o o f : The conditions of Theorem 10.2 guarantee that the right hand sides F and G_j of the boundary value problem

$$A_0(0,D_x)\eta u = F \quad \text{in} \quad \Omega \, ,$$

$$B_{j,0}(0,D_x)\eta u = G_j \quad \text{on} \quad \partial\Omega \, , \quad j = 1,\ldots,m$$

are contained in $V^{\ell,p}(K,\beta) \cap V^{\ell_1,p_1}(K,\beta_1)$ and $V^{\ell+2m-m_j-1/p,p}(\partial K,\beta)$ $\cap V^{\ell_1+2m-m_j-1/p_1,p_1}(\partial K,\beta_1)$, respectively. Therefore Theorem 12.5 yields that

$$\eta u = \sum_{\gamma \in I} c_\gamma u_\gamma + w$$

where

$$c_\gamma = \int_K A_0(0,D_x)\eta u \, \overline{iv_{\gamma'}} \, dx + \sum_{j=1}^{m} \int_{\partial K} B_{j,0}(0,D_x)\eta u \, \overline{iT_{j,0}v_{\gamma'}} \, d\sigma$$

$$= \int_\Omega A_0(0,D_x)\eta u \, \overline{iv_{\gamma'}} \, dx + \sum_{j=1}^{m} \int_{\partial\Omega} B_{j,0}(0,D_x)\eta u \, \overline{iT_{j,0}v_{\gamma'}} \, d\sigma \, .$$

<u>REMARK</u>. In formula (13.1) the unknown solution u occurs. Nevertheless this formula is meaningful if instead of u we consider a numerical solution of problem (5.3) and calculate approximately the constant c_γ , e.g. by an iterative procedure (see Section 6). In some special cases formula (13.1) has a simpler form :

<u>13.2. LEMMA</u>. *Let be* $A_0(0,D_x) = A(x,D_x)$ *and* $B_j(x,D_x) = \dfrac{\partial^{j-1}}{\partial n^{j-1}}$, $j = 1,\ldots,m$. *Let the assumptions of Theorem 10.1 be satisfied. Then the solution* u *of the Dirichlet problem*

$$A_0(0,D_x)u = f \quad in \quad \Omega \, ,$$

$$\frac{\partial^{j-1}u}{\partial n^{j-1}} = 0 \quad on \quad \partial\Omega \setminus \{0\} \, , \quad j = 1,\ldots,m$$

admits the expansion

$$u = \sum_{\gamma \in I} c_\gamma \, \eta \, u_\gamma + w_1$$

where

$$(13.2) \qquad c_\gamma = \int_\Omega f \, \overline{i\eta v_{\gamma'}} \, dx - \int_\Omega u \, \overline{iA^*(\eta v_{\gamma'})} \, dx \, .$$

P r o o f : We have to calculate the individual terms of (13.1). We omit the details of these calculations and describe only the main steps :

$$\int_\Omega A(\eta u)\overline{iv_{\gamma'}} \, dx = \int_\Omega \sum_{|\alpha|=2m} \sum_{\alpha' \leq \alpha} a_\alpha \binom{\alpha}{\alpha'} D_x^{\alpha'} u D_x^{\alpha-\alpha'} \eta \overline{iv_{\gamma'}} \, dx =$$

$$= \int_\Omega \eta A u \; \overline{iv_{\gamma'}} \, dx + \int_\Omega \sum_{|\alpha|=2m} \sum_{\alpha'<\alpha} a_\alpha \binom{\alpha}{\alpha'} u(-1)^{|\alpha'|} D_x^{\alpha'} \big[D_x^{\alpha-\alpha'} \eta \overline{iv_{\gamma'}} \big] dx$$

$$= \int_\Omega \eta f \; \overline{iv_{\gamma'}} \, dx + \int_\Omega \sum_{|\alpha|=2m} \sum_{\alpha'<\alpha} a_\alpha u(-1)^{|\alpha'|} \binom{\alpha}{\alpha'} \sum_{\alpha''\leq\alpha'} \big[\binom{\alpha'}{\alpha''} D_x^{\alpha-\alpha''} \eta D_x^{\alpha''} \overline{iv_{\gamma'}} \big] dx$$

$$= \int_\Omega \eta f \; \overline{iv_{\gamma'}} \, dx - \int_\Omega u \Big[\sum_{|\alpha|=2m} a_\alpha \sum_{\alpha''\leq\alpha} \binom{\alpha}{\alpha''} D_x^{\alpha-\alpha''} \eta D^{\alpha''} \overline{iv_{\gamma'}} - \eta a_\alpha D_x^\alpha \overline{iv_{\gamma'}} \Big] dx$$

$$= \int_\Omega \eta f \; \overline{iv_{\gamma'}} \, dx + \int_\Omega u \sum_{|\alpha|=2m} \eta a_\alpha D^\alpha \overline{iv_\gamma} \, dx - \int_\Omega u \sum_{|\alpha|=2m} \overline{\bar{a}_\alpha D^\alpha (\eta iv_{\gamma'})} \, dx$$

$$= \int_\Omega f \; \overline{iv_\gamma} \, dx - \int_\Omega u \; \overline{A^*(\eta iv_{\gamma'})} \, dx \; .$$

We have used the identities $\displaystyle\sum_{\alpha-\alpha'\leq\alpha-\alpha''} (-1)^{|\alpha-\alpha'|} \binom{\alpha-\alpha''}{\alpha-\alpha'} = 0$ and $A^* v_{\gamma'} = 0$ for $x \in \operatorname{supp} \eta$. Further we have

$$B_{j,0}(0,D_x)(\eta u) = \frac{\partial^{j-1}}{\partial n^{j-1}}(\eta u) = \sum_{k=0}^{j-1} \binom{j-1}{k} \frac{\partial^{j-1-k} u}{\partial n^{j-1-k}} \frac{\partial^k}{\partial n^k} \eta = 0 \quad \text{on} \quad \partial K \; .$$

Let us now consider a more general boundary value problem than the Dirichlet problem

13.3. <u>REMARK</u>. (i) Let

$$A(x,D_x) = f \quad \text{in} \quad \Omega \; ,$$

$$B_j(x,D_x) = 0 \quad \text{on} \quad \partial\Omega \; , \quad j = 1,\dots m \; .$$

Assume $A_0(0,D_x) = A_0(x,D_x)$ and $B_{j,0}(0,D_x) = B_{j,0}(x,D_x)$ and let the assumptions of Theorem 10.1 be satisfied. Then the coefficients c_γ in the expansion

$$u = \sum_{\gamma\in I} c_\gamma \, \eta \, u_\gamma + w_1$$

are given by

$$(13.3) \qquad c_\gamma = \int_\Omega \eta f \; \overline{iv_{\gamma'}} \, dx - \int_\Omega u \; \overline{A_0^*(0,D) i \eta v_{\gamma'}} \, dx + a(\eta,u,v_{\gamma'})$$

where $\displaystyle a(\eta,u,v_{\gamma'}) = - \int_\Omega \eta A_1(x,D_x) u \; \overline{iv_{\gamma'}} \, dx + b_1(\eta,u,v_{\gamma'}) + \sum_{j=1}^m b_{2,j}(\eta,u,v_{\gamma'}) \; .$

The boundary bilinear form $b_1(\eta,u,v_{\gamma'})$ is calculated by the Gauss integral formula in the following way :

$$\int_\Omega \sum_{|\alpha|=2m} a_\alpha \sum_{\alpha'<\alpha} \binom{\alpha}{\alpha'} D_x^{\alpha-\alpha'} \eta D_x^{\alpha'} u \; \overline{iv_\gamma} \, dx$$

$$= \int_\Omega \sum_{|\alpha|=2m} a_\alpha \sum_{\alpha'<\alpha} u(-1)^{|\alpha'|} \Big[\sum_{\beta\leq\alpha'} \binom{\alpha'}{\beta} D_x^{\alpha-\beta} \eta D^\beta \overline{iv_{\gamma'}} \Big] dx + b_1(\eta,u,v_{\gamma'}) =$$

$$= - \int_{\Omega} u \; \overline{iA^*(0,D_x)\eta v_{\gamma'}} \; dx + b_1(\eta,u,v_{\gamma'}) \; .$$

The boundary bilinear forms $b_{2,j}(\eta,u,v_{\gamma'})$ are defined by

$$b_{2,j}(\eta,u,v_{\gamma'}) = - \int_{\partial\Omega} \eta B_{j,1}(x,D_x)u \; \overline{T_{j,0}iv_{\gamma'}} \; d\sigma$$

$$+ \int_{\partial\Omega} \sum_{|\alpha_j|=m_j} b_{\alpha_j} \sum_{\alpha'<\alpha_j} \binom{\alpha_j}{\alpha'} D_x^{\alpha_j-\alpha'} \eta D^{\alpha'}u \; \overline{T_{j,0}iv_{\gamma'}} \; d\sigma \; .$$

(ii)   Let us present an example in order to illustrate formula (13.3). We consider the mixed boundary value problem for the Laplace operator in a plane polygonal domain Ω :

$$(13.4) \qquad \begin{aligned} &\Delta u = - f \quad \text{in} \quad \Omega \; , \\ &B_1(x,D_x) = B(x,D_x) = c_1 u + c_2 \frac{\partial u}{\partial n} = 0 \quad \text{on} \quad \partial\Omega \; , \end{aligned}$$

where $c_1 = \{ \begin{smallmatrix} 1 \\ 0 \end{smallmatrix} , \; c_2 = \{ \begin{smallmatrix} 0 \\ 1 \end{smallmatrix}$ for $\{ \begin{smallmatrix} x \in \partial\Omega_1 \\ x \in \partial\Omega_2 \end{smallmatrix}$, (see Fig. 12) and $f \in L^2(\Omega)$.

Fig. 12

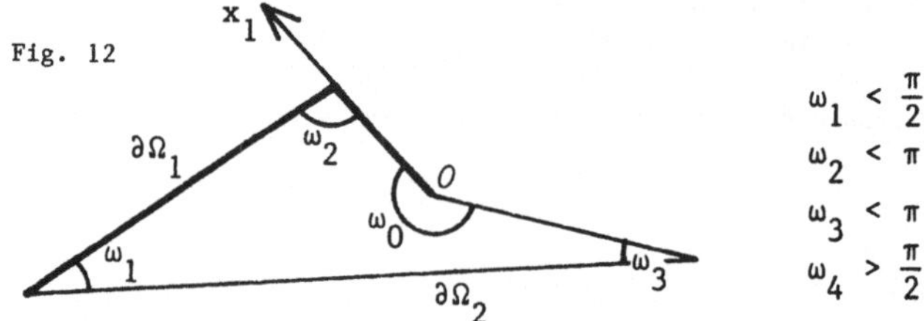

Green's formula yields

$$T_{1,0}v = Tv = c_1 \frac{\partial v}{\partial n} - c_2 v$$

Consequently,

$$(13.5) \qquad c_{\gamma} = \int_{\Omega} \Delta(\eta u)\overline{iv_{\gamma'}} \; dx + \int_{\partial\Omega} B(\eta u)\overline{Tiv_{\gamma'}} \; ds = \int_{\Omega} \eta\Delta u \overline{iv_{\gamma'}} \; dx$$

$$- \int_{\Omega} u\overline{\Delta(\eta iv_{\gamma'})} \; dx + 2 \int_{\partial\Omega} v \frac{\partial\eta}{\partial n} \; \overline{iv_{\gamma'}} \; ds + \int_{\partial\Omega} B(\eta u)\overline{Tiv_{\gamma'}} \; ds =$$

$$= \int_{\Omega} \eta\Delta u \overline{iv_{\gamma'}} \; dx - \int_{\Omega} u \; \overline{\Delta(\eta iv_{\gamma'})} \; dx + \int_{\partial\Omega_2} u \frac{\partial\eta}{\partial n} \; \overline{iv_{\gamma'}} \; ds \; .$$

§ 14. Examples

The crucial point for using the coefficient formulae is to know the functions $v_{\gamma'}$. Let us show some examples where the calculation of $v_{\gamma'}$ is easy.

14.1. __THE DIRICHLET PROBLEM FOR THE LAPLACE OPERATOR.__ (i) We consider the problem (1.1), namely

$$- \Delta u = f \quad \text{in} \quad \Omega ,$$
$$u = 0 \quad \text{on} \quad \partial\Omega ,$$

where Ω is a plane domain with the property (10.1) and $f \in L^2(\Omega)$. Assume that the angle ω_0 at O is bigger than π . Then

$$(14.1) \qquad u = c \, \eta \, r^{\pi/\omega_0} \sin \frac{\pi}{\omega_0} \omega + w$$

where $w \in W^{2,2}(\Omega)$ (see (1.26)).

(ii) For the function $u_\gamma = u_{(1,1,0)} = r^{\pi/\omega_0} \sin \frac{\pi}{\omega_0} \omega$ let us determine the corresponding function $v_{\gamma'} = v_{(1,1,0)}$. Since $\lambda_1 = - i \frac{\pi}{\omega_0}$ is the eigenvalue of $\mathcal{U}_0(\lambda)$ which occurs in (14.1), it follows from Lemma 12.3 that

$$v_{\gamma'} = r^{i(\overline{\lambda}_1 + i(N-2m))} v_1^{(0,1)}(\omega) = r^{-\pi/\omega_0} v_1^{(0,1)}(\omega) .$$

The function $v_1^{(0,1)}(\omega)$ is an eigenfunction of $\mathcal{U}_0^*(\lambda^*)$ corresponding to the eigenvalue $\lambda_1^* = i\pi/\omega_0$. Since the Dirichlet problem for the Laplace operator is selfadjoint we obtain $v_1^{(0,1)}(\omega) = c_1 \sin \frac{\pi}{\omega_0} \omega$. The constant c_1 is to be determined in such a way that the biorthonormality condition (12.13) is satisfied.

(iii) The condition (12.13) has in our case the form

$$\left(\frac{\partial L}{\partial \lambda} (- i \frac{\pi}{\omega_0}) \sin \frac{\pi}{\omega_0} \omega, \ c_1 \sin \frac{\pi}{\omega_0} \omega\right)_\Omega + 0 = 1 .$$

Since $L = - \lambda^2 + \frac{\partial^2}{\partial\omega^2}$ (cf. (1.7)), we get $\overline{c_1} \displaystyle\int_0^{\omega_0} - 2(- i \frac{\pi}{\omega_0}) \sin^2 \frac{\pi}{\omega_0} \omega \, d\omega = 1$,

i.e. $\overline{c_1} = 1/i\pi$ and $v_{\gamma'} = - \frac{1}{i\pi} r^{\pi/\omega_0} \sin \frac{\pi}{\omega_0} \omega$.

Formula (13.2) yields

$$(14.2) \qquad c = \frac{1}{\pi} \int_\Omega f\eta r^{-\pi/\omega_0} \sin \frac{\pi}{\omega_0} \omega \, dx + \frac{1}{\pi} \int_\Omega u\Delta(\eta \sin \frac{\pi}{\omega_0} \omega \, r^{-\pi/\omega_0}) \, dx .$$

14.2. __A MIXED BOUNDARY VALUE PROBLEM FOR THE LAPLACE OPERATOR.__ (i) Let us consider the problem (13.4). Formula (2.11) implies that the solution u of (13.4) admits the expansion

$$(14.3) \qquad u = c \, \eta \, r^{\pi/2\omega_0} \sin \frac{\pi}{2\omega_0} \omega + w ,$$

where $w \in W^{2,2}(\Omega)$, provided $\frac{\pi}{2} < \omega_0 < \frac{3}{2}\pi$.

(ii) For the eigenfunction $u_\gamma = u_{(1,1,0)} = r^{\pi/2\omega_0} \sin \frac{\pi}{2\omega_0} \omega$ and the corresponding eigenvalue $\lambda_1 = - i \dfrac{\pi}{2\omega_0}$ we determine the function $v_{\gamma'}$ $= v_{(1,1,0)}$. Lemma 12.3 implies that $v_{\gamma'} = r^{i(\overline{\lambda}_1+i(N-2m))} v_1^{(0,1)}(\omega)$ $= r^{-\pi/2\omega_0} v_1^{(0,1)}(\omega)$. Since the problem (13.4) is selfadjoint we obtain $v_1^{(0,1)}(\omega) = c_1 \sin \dfrac{\pi}{2\omega_0} \omega$, where the constant c_1 is given by the condition (12.13). Analogously as in Example 14.1 we obtain that $\overline{c}_1 = \dfrac{2}{i\pi}$. Therefore we have

$$v_{\gamma'} = - \frac{2}{i\pi} r^{-\pi/2\omega_0} \sin \frac{\pi}{2\omega_0} \omega \ .$$

(iii) We use formula (13.5) for the calculation of the coefficient c in (14.3) :

$$(14.4) \qquad c = \frac{2}{\pi} \Big[\int_\Omega \eta f r^{-\pi/2\omega_0} \sin \frac{\pi}{2\omega_0} \omega \ dx + \int_\Omega u \Delta (\eta r^{-\pi/2\omega_0} \sin \frac{\pi}{2\omega_0} \omega) \ dx$$

$$- \int_{\partial\Omega_2} u \frac{\partial \eta}{\partial n} r^{-\pi/2\omega_0} \sin \frac{\pi}{2\omega_0} \omega \ ds \Big].$$

14.3. <u>THE DIRICHLET PROBLEM FOR THE BIHARMONIC OPERATOR</u>. (i) Let Ω be a bounded plane domain with the property (10.1). We consider the problem (3.1), namely

$$\Delta^2 u = f \quad \text{in} \quad \Omega \ ,$$
$$u = \frac{\partial u}{\partial n} = 0 \quad \text{on} \quad \partial\Omega \setminus \{0\} \ ,$$

where $f \in L^2(\Omega)$. We want to calculate the coefficients in the expansion (3.16). Let us restrict ourselves to the most frequent case that $N_1 = N_3$, which means we have such an angle ω_0 that the zeros of $D(\lambda)$ $= \sin^2(i\lambda - 1)\omega_0 - (i\lambda - 1)^2 \sin^2\omega_0 = 0$ are simple. Then

$$(14.5) \qquad u = \sum_{\gamma \in I} c_\gamma \eta u_\gamma + w = \sum_{-3 < \operatorname{Im} \lambda_\mu < -1} c_{(\mu,1,0)} r^{i\lambda_\mu} u_\mu^{(0,1)}(\omega) + w \ ,$$

where $w \in W^{4,2}(\Omega)$, and

$$(14.6) \qquad u_\mu^{(0,1)}(\omega) = \Big[\sin i\lambda_\mu \omega_0 - \frac{i\lambda_\mu}{i\lambda_\mu - 2} \sin(i\lambda_\mu - 2)\omega_0 \Big] \Big[\cos i\lambda_\mu \omega - \cos(i\lambda_\mu - 2)\omega \Big]$$

$$- \Big[\cos i\lambda_\mu \omega_0 - \cos(i\lambda_\mu - 2)\omega_0 \Big] \Big[\sin i\lambda_\mu \omega - \frac{i\lambda_\mu}{i\lambda_\mu - 2} \sin(i\lambda_\mu - 2)\omega \Big] \ .$$

(See (3.4) and P. GRISVARD [1].)

(ii) It follows from Lemma 12.3 that

$$v_{\gamma'} = v_{(\mu,1,0)} = r^{i(\overline{\lambda}_\mu+i(N-2m))} v_\mu^{(0,1)}(\omega) = r^{i(\overline{\lambda}_\mu-2i)} v_\mu^{(0,1)}(\omega) \ .$$

Since $\alpha_0^*(\overline{\lambda}_\mu - 2i) = \overline{\alpha_0(\lambda_\mu)}$ we obtain $v_\mu^{(0,1)}(\omega) = k_{(\mu,0,1)} \overline{u_\mu^{(0,1)}(\omega)}$, where $k_{(\mu,0,1)}$ is a constant, determined by formula (12.13), that is

$$\overline{k_{(\mu,0,1)}} \int_0^{\omega_0} \frac{\partial L}{\partial \lambda}(\lambda_\mu)\left[u_\mu^{(0,1)}(\omega)\right]^2 d\omega = 1 \ .$$

Here, $\quad L(\lambda) = \dfrac{\partial^4}{\partial\omega^4} + 2(-\lambda^2-2i\lambda+2)\dfrac{\partial^2}{\partial\omega^2} + (-\lambda^2)(i\lambda-2)^2 \quad$ (see (3.3)).

(iii) Formula (13.2) yields

$$(14.7) \qquad c_{(\mu,1,0)} = \int_\Omega f \ \overline{i\eta v_{(\mu,1,0)}} \ dx - \int_\Omega u \ \overline{i\Delta^2(\eta v_{(\mu,1,0)})} \ dx$$

where $v_{(\mu,1,0)} = k_{(\mu,0,1)} r^{i(\overline{\lambda}_\mu-2i)} \ \overline{u_\mu^{(0,1)}(\omega)} \ .$

Chapter II

FINITE ELEMENT METHODS

Boundary value problems are difficult to treat numerically when they are
defined in domains with nonsmooth boundaries. In this situation standard techniques lose accuracy near the resulting singularities, and global pollution
takes place. The reason for the appearance of this effect is the lower regularity of the solutions of such problems, in comparison with those having smooth
boundaries. The regularity of solutions of boundary value problems in domains
with conical points was studied in detail in Chapter I, Section 3.

We use these regularity results to demonstrate the forms of the errors
for standard finite element solutions in L^p-spaces, $2 \le p \le \infty$, to these
problems, and to propose a modification to the standard methods. This modification is the so-called Dual Singular Function Method, considered by M. DOBRO-
WOLSKI [1] and H. BLUM [1], and involves the use of the singular functions to
augment the spaces of approximating functions and the use of the "dual singular
functions". Many alternatives to the augmentation techniques exist for dealing
with singularities; for example local mesh refinement near a singularity, and
more recently the use of modified multi-grid methods. However, the augmentation
methods are good examples of the use of the analytical form of a singularity
and of the resulting regularity results.

Section 5 : *Standard finite element methods
in domains with conical points*

In this section we first give a short survey of well-known results about
weakly formulated boundary value problems. There are many books and papers on

this subject and we refer to J. NEČAS [1], S. FUČÍK, A. KUFNER [1], P.G. CIARLET
[1], where both the linear and nonlinear cases are extensively studied. Only the
most important results required here are presented. Then we consider the standard Finite Element Methods in the Sobolev spaces $W^{m,p}(\Omega)$ for solving weakly
formulated boundary value problems in domains with conical points. We investigate the link between the convergence rate with a decreasing mesh size and the
boundary singularities, and give error estimates (see also A.-M. SÄNDIG [1]).

§ 15 . W e a k   s o l u t i o n s . E x i s t e n c e   a n d
u n i q u e n e s s

__15.1.__ Let $\Omega \subset R^N$ be a bounded domain with a boundary $\partial\Omega$ containing conical
points (cf. § 5). We consider the real boundary value problem

$$\text{(15.1)} \qquad Au = \sum_{|\alpha|\leq m} (-1)^{|\alpha|} D^\alpha\big(a_\alpha(x,D_m u)\big) = f \quad \text{in} \quad \Omega ,$$

$$B_j u = \sum_{|\alpha_j|\leq m_j} b_{\alpha_j}(x) D^{\alpha_j} u = 0 \qquad \text{on } \partial\Omega , \quad j = 1,\ldots,m ,$$

where $D_m u = \{D^{\alpha'} u\}_{|\alpha'|\leq m}$.

__15.2. WEAK SOLUTION__. Let V be the subspace of $W^{m,p}(\Omega)$ which is characte-
rized by the vanishing of the essential boundary conditions of (15.1) in the
sense of traces.

__DEFINITION__. The element $u \in V$ is a _weak solution_ of the boundary value problem (15.1) if

$$\text{(15.2)} \qquad a(u,v) = \sum_{|\alpha|\leq m} \int_\Omega a_\alpha\big(x,D_m u(x)\big) D^\alpha v(x) \ dx = <f,v>$$

for every $v \in V$, where f is an element from the dual space V^* of V .
Equation (15.2) is the weak formulation of (15.1).

__15.3. REMARK__. It is assumed that for (15.1) the corresponding weak solution satisfies the non-essential boundary conditions. It is of course possible to consider problem (15.2) at the beginning, where V is an appropriate Banach space,
without reference to (15.1). Weak solutions in non smooth domains are also investigated in A.-M. SÄNDIG [2], [3].

__15.4. GALERKIN SOLUTION__.

__DEFINITION__. Let S be a finite dimensional subspace of V . Then $u_G \in S$ is
called a _Galerkin solution_ of (15.2) if

$$\text{(15.3)} \qquad a(u_G,v) = <f,v> \quad \text{for every} \quad v \in S .$$

Let us consider the existence and uniqueness theorems for solutions of
problem (15.2). We start with the _linear case_, i.e. with $a_\alpha(x;D_m)$ in (15.1)

being of the form $\displaystyle\sum_{|\beta|\leq m} a_{\alpha\beta}(x)D^{\beta}u$.

<u>15.5.</u> <u>THEOREM OF LAX-MILGRAM</u> (see e.g. J. NEČAS [1]). *Let* H *be a real Hilbert space. Let* f *be a given element of the dual space* H^{*} *of* H . *Assume* a(u,v) *is a real bilinear form on* H × H *with the following properties:*

(15.4) $|a(u,v)| \leq c_{1}\|u;H\|\cdot\|v;H\|$ *for every* u, v ∈ H ,

(15.5) $|a(u,u)| \geq c_{2}\|u;H\|^{2}$ *for every* u ∈ H .

Then there exists a unique u* ∈ H *such that*
$$a(u*,v) = <f,v> \;\; for\;every \;\; v \in H$$
and we have the estimate
$$\|u*;H\| \leq \frac{1}{c_{2}}\,\|f;H^{*}\| \;.$$

This theorem guarantees the existence and uniqueness of a solution of (15.2) if H = V and (15.4) and (15.5) are satisfied. In the same way we obtain the uniqueness and existence of the Galerkin solution u_{G} , taking S instead of H in the Lax-Milgram theorem.

<u>15.6.</u> <u>CÉA's LEMMA</u> (see e.g. P. G. CIARLET [1]). *Assume that the assumptions of Theorem 15.5 are satisfied for the bilinear form given in* (15.2) *and for* V = H . *Then the difference between the solution* u *of* (15.2) *and the Galerkin solution* u_{G} *of* (15.3) *can be estimated as follows:*

(15.6) $\|u - u_{G};\,H\| \leq \dfrac{c_{1}}{c_{2}}\,\inf_{w\in S}\,\|u - w;\,H\| \;.$

<u>15.7.</u> <u>EXISTENCE AND UNIQUENESS FOR THE NONLINEAR CASE</u>. Let us introduce an operator T , which maps the real reflexive Banach space V into its dual space V^{*} , by the relation
$$(Tu,v) = a(u,v) - <f,v> \;\; for\;every \;\; v \in V ,$$
where a(u,v) is given by (15.2). The following theorems are generalizations of Theorem 15.5 and of Lemma 15.6.

<u>15.8.</u> <u>THEOREM</u> (see e.g.   A. LOUIS [1]). *Assume that* T *has the following properties:*

(i) T : V → V^{*} *is finitely continuous.* (That means: Let S be a finite dimensional subspace of V , (u_{n}) a sequence of elements u_{n} ∈ S which converges to an element u ∈ S . Then the sequence (Tu_{n},v) converges to (Tu,v) for every v ∈ V .)

(ii) T *is strictly monotone.* (That means:
$$(Tu - Tv,\, u - v) > 0$$
for every u, v ∈ V , u ≠ v .)

73

(iii) T *is coercive.* (That means: there is a real continuous monotone function $K(t)$ defined on $[0,\infty)$ with $\lim_{t\to 0} K(t) = \infty$ and $K(0) = 0$, such that

$$(Tu,u) \geq K\big(\|u;V\|\big)\cdot\|u;V\| \quad \text{for every} \quad u \in V \;.)$$

Then there exist a unique $u* \in V$ *with* $(Tu*,v) = 0$ *for every* $v \in V$.

15.9. THEOREM (P. G. CIARLET, M. H. SCHULTZ, R. S. VARGA [1]). *Assume*

(j) $T : V \to V^*$ *is finitely continuous;*

(jj) T *is uniformly monotone* (that means
$$(Tu - Tv, u - v) \geq c\|u - v; V\|^2 \;, \quad c > 0$$
for every $u, v \in V$);

(jjj) T *is bounded Lipschitz continuous* (that means
$$\|Tu - Tv; V^*\| \leq c(M)\|u - v; V\|$$
for every u , v with $\|u;V\| \leq M$, $\|v;V\| \leq M$).

Then the difference between the solution u *of* (15.2) *and the Galerkin solution* u_G *of* (15.3) *can be estimated as follows:*

$$(15.7) \qquad \|u - u_G; V\| \leq c \inf_{w \in S} \|u - w; V\| \;.$$

REMARK. The condition (jj) in Theorem 15.9 imply the conditions (ii) and (iii) of Theorem 15.8.

§ 16 . F i n i t e   e l e m e n t   s p a c e s

16.1. The estimates (15.6) and (15.7) are very useful for obtaining estimates of the error $u - u_G$ if we can solve the following problem: *For every* $u \in V$, *find an element of* S *which approximates* u *closely enough.* To this end we need more information about the spaces S . As spaces S we take the so-called "Finite Element" spaces for which an interpolation theory is well developed. (See P. G. CIARLET [1], P. G. CIARLET, P. A. RAVIART [1], [2], J. T. ODEN, G.F. CAREY [1].)

Let us briefly describe the finite element spaces S which we use in the following, together with the corresponding interpolation theorems.

16.2. PARTITION OF $\overline{\Omega}$. We consider a partition Π of our bounded domain $\overline{\Omega} \subset R^n$ with conical points (cf. § 5) into a finite number E of subdomains $\overline{\Omega}_e \in \Pi$ such that

(i) every element $\overline{\Omega}_e$ is closed and consists of a nonempty interior Ω_e and a Lipschitzian boundary $\partial\Omega_e$;

(ii) $\overline{\Omega} = \bigcup_{e=1}^{E} \overline{\Omega}_e$;

(iii) $\Omega_e \cap \Omega_f = \emptyset$ for arbitrary distinct elements $\overline{\Omega}_e$, $\overline{\Omega}_f \in \Pi$.

<u>16.3. LOCAL APPROXIMATION</u>. For each $\overline{\Omega}_e \in \Pi$ we introduce a finite dimensional space S_e spanned by linearly independent local interpolation functions. We approximate the restriction $u|_{\overline{\Omega}_e}$ of an element $u \in W^{m,p}(\Omega)$ by a linear combination of these interpolation functions. The coefficients of such linear combinations are usually taken to be the values of u and the values of various partial derivatives of u up to the order $s \geq 0$ at the set of the nodal points of $\overline{\Omega}_e$. In this case it makes sense to require

$$W^{m,p}(\Omega_e) \subset C^s(\overline{\Omega}_e) .$$

<u>16.4. GLOBAL APPROXIMATION</u>. Global approximations are obtained by fitting together the local approximations in such a way that the supports of the global interpolation functions are only contained in one of the sets

$$\overline{\Omega}_e \cup \text{(the elements of } \Pi \text{ adjacent to } \overline{\Omega}_e \text{)}, \quad e \in E .$$

The global interpolation functions provide a basis for our finite dimensional subspace $S \subset W^{\ell,q}(\Omega)$, and

$$S_e = \left\{ s|_{\overline{\Omega}_e} , \quad s \in S \right\} .$$

In many cases S consists of piecewise polynomial shape functions of a certain degree on each $\Omega_e \in \Pi$.

<u>16.5. FAMILIES OF FINITE ELEMENT SPACES</u>. We introduce a discretization parameter h (the *mesh size*), h approaching zero, and consider a family of the above described partitions Π_h of $\overline{\Omega}$ with the following properties:

(i) Each $\Omega_{e,h} \in \Pi_h$ contains a ball with the radius c h and is contained in a ball with the radius h . c > 0 is independent of h (strong regularity).

(ii) For every $O_i \in M \subset \partial\Omega$ (O_i is a conical point or a point where the boundary conditions change their type if N = 2 , cf. 5.2 (ii)) there is a fixed number N_0 , independent of h and i , such that

$$(16.1) \quad \bigcup_{e \in I_i} \overline{\Omega}_{e,h} = \overline{U}_i(h) \cap \overline{\Omega}$$

where $U_i(h)$ is a neighborhood of O_i and the index set I_i contains at most N_0 elements. (See Fig. 13.)   Fig. 13 shows that the property (ii) is satisfied if we take e.g. a radial partition near O_i . Such partitions were used also by J. R. WHITEMAN [1]. The spaces $S_{e,h}$ and S_h are defined for every partition Π_h in the same way as S_e and S above.

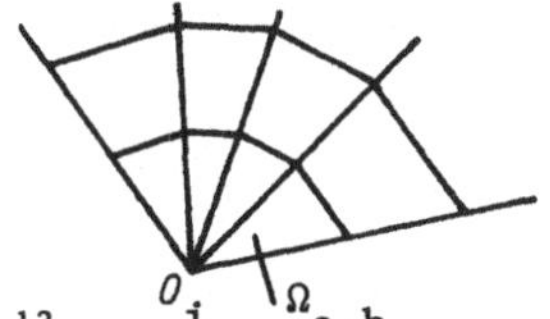

Fig. 13

16.6. <u>INTERPOLATION OPERATORS</u>. We introduce the global interpolation operator

$$(16.2) \qquad I_h : W^{m,p}(\Omega) \to S_h \subset W^{\ell,q}(\Omega) \ ,$$

where $I_h u = \sum_{i=1}^{L} \ell_i(u) e_i$ and e_i are the global basis functions of S_h generated by the finite element method as indicated in 16.4.

For the local interpolation operator we have

$$(16.3) \qquad I_h\big|_{W^{m,p}(\Omega_{e,h})} = I_{e,h} : W^{m,p}(\Omega_{e,h}) \to S_{e,h} \subset W^{\ell,q}(\Omega_{e,h}) \ .$$

We are interested in the quality of the approximation $I_h u$ of u . The key to this question lies in the character of the local interpolation operator $I_{e,h}$. Assume

(j) The family of the finite element spaces $\{S_h\}$ has the following local approximation property :

$$(16.4) \qquad \|u - I_h u; W^{\ell,q}(\Omega_{e,h})\| \le c\, h^{m-\ell-N(\frac{1}{p}-\frac{1}{q})} \|\nabla_m u; L^p(\Omega_{e,h})\|$$

for every $u \in W^{m,p}(\Omega)$ and $\Omega_{e,h} \in \Pi_h$, provided $W^{m,p}(\Omega) \subset W^{\ell,q}(\Omega)$, $0 \le \ell \le m$, and $1 \le p \le \infty$, $1 \le q \le \infty$.

(jj) The following inverse inequalities hold for all $u_h \in S_h \subset W^{\ell,q}(\Omega)$:

$$(16.5) \qquad \|\nabla_\ell u_h; L^q(\Omega_{e,h})\| \le c\, h^{N(\frac{1}{q}-\frac{1}{q_1})} \|\nabla_\ell u_h; L^{q_1}(\Omega_{e,h})\|$$

for every $\Omega_{e,h} \in \Pi_h$, $1 \le q_1 \le q \le \infty$;

here ∇_k , $k = 0,1,\dots$, denotes the field of all derivatives of the order k , $\Omega \subset R^N$.

<u>REMARK</u>. Let S_h consist of piecewise polynomial shape functions of a degree $\le k-1$ and let I_h preserve polynomial shape functions of the degrees $\le k-1$. ($k > \frac{n}{p} + s$, s was defined in 16.3.) In this case we restrict the property (j) to $m \le k$. Then the condition 16.5 (i) usually implies the conditions (j) and (jj) (it depends mainly on the shape of the domain Ω). In this connection we refer to P. G. CIARLET [1]. The conditions (j) and (jj) allow us to derive global estimates in the domain Ω .

16.7. <u>THEOREM</u>. *Assume that the conditions* (j) *and* (jj) *are satisfied. Then for* $p = q$,

$$(16.6) \qquad \|u - I_h u; W^{\ell,p}(\Omega)\| \le c\, h^{m-\ell} \|\nabla_m u; L^p(\Omega)\| \ ,$$

while for $q = \infty$,

$$(16.7) \qquad \|u - I_h u; W^{\ell,\infty}(\Omega)\| \le c\, h^{m-\ell-\frac{N}{p}} \|\nabla_m u; L^p(\Omega)\| \ .$$

Further,

$$(16.8) \qquad \|\nabla_\ell u_h; L^q(\Omega)\| \le c\, h^{N(\frac{1}{q}-\frac{1}{q_1})} \|\nabla_\ell u_h; L^{q_1}(\Omega)\|$$

for $q_1 \leq q$ *and for every* $u_h \in S_h$.

P r o o f . Let $q = p$, then (16.4) implies that

$$\|u - I_h u; W^{\ell,p}(\Omega)\|^p = \sum_{e=1}^{E_h} \|u - I_h u; W^{\ell,p}(\Omega_{e,h})\|^p$$

$$\leq (c\, h^{m-\ell})^p \sum_{e=1}^{E_h} \|\nabla_m u; L^p(\Omega_{e,h})\|^p = (c\, h^{m-\ell})^p \, \|\nabla_m u; L^p(\Omega)\|^p .$$

For $q = \infty$ we obtain from (16.4)

$$\|u - I_h u; W^{\ell,\infty}(\Omega)\| = \max_{0 \leq |\alpha| \leq \ell} \|D^\alpha(u - I_h u); L^\infty(\Omega)\|$$

$$= \|D^{\alpha_0}(u - I_h u); L^\infty(\Omega)\| = \operatorname*{ess\,sup}_{x \in \Omega} \; |D^{\alpha_0}(u(x) - I_h u(x))| =$$

$$= \max_{\Omega_{e,h} \in \Pi_h} \|D^{\alpha_0}(u - I_h u); L^\infty(\Omega_{e,h})\| = \|D^{\alpha_0}(u - I_h u); L^\infty(\Omega_{e_0,h})\|$$

$$\leq c\, h^{m-\ell-N/p} \|\nabla_m u; L^p(\Omega_{e_0,h})\| \leq c\, h^{m-\ell-N/p} \|\nabla_m u; L^p(\Omega)\| .$$

The inverse inequality (16.8) follows from (16.5), since

$$\|\nabla_\ell u_h; L^q(\Omega)\| = \left[\sum_{e=1}^{E_h} \|\nabla_\ell u_h; L^q(\Omega_{e,h})\|^q \right]^{1/q}$$

$$\leq c\, h^{N(\frac{1}{q} - \frac{1}{q_1})} \left[\sum_{e=1}^{E_h} \|\nabla_\ell u_h; L^{q_1}(\Omega_{e,h})\|^q \right]^{1/q}$$

$$\leq c\, h^{N(\frac{1}{q} - \frac{1}{q_1})} \left[\sum_{e=1}^{E_h} \|\nabla_\ell u_h; L^{q_1}(\Omega_{e,h})\|^{q_1} \right]^{1/q_1} = c\, h^{N(\frac{1}{q} - \frac{1}{q_1})} \|\nabla_\ell u_h; L^{q_1}(\Omega)\| .$$

Here we have used the Jensen inequality (see e.g. P. G. CIARLET [1]).

§ 17 . E r r o r e s t i m a t e s i n $W^{m,2}(\Omega)$

<u>17.1.</u> We restrict ourselves to linear boundary value problems as given in the
weak formulation (15.2), namely: Find a solution $u \in V \subset W^{m,2}(\Omega)$ such that

$$(17.1) \qquad a(u,v) = \sum_{\substack{|\alpha| \leq m \\ |\beta| \leq m}} \int_\Omega a_{\alpha\beta}(x) D^\alpha u D^\beta v \, dx = \int_\Omega f\, v\, dx$$

for every $v \in V$. Here $a(u,v)$ is a real bilinear form, $f \in L^p(\Omega)$, Ω is
a bounded domain with conical points or boundary points, where the boundary
conditions change their type. Assume that the conditions (15.4) and (15.5) of
the Lax-Milgram theorem are fulfilled. Without difficulty we can also deal with
nonlinear problems which satisfy the conditions of Theorem 15.9 and for which
regularity theorems in weighted Sobolev spaces are known. (As an example see
the Navier-Stokes equation in § 4 or the results of H. BLUM [2], M. DOBROWOLSKI

[2], P. TOLKSDORF [1].) Let $u \in V \subset W^{m,2}(\Omega)$ be the solution of the boundary value problem (17.1). We denote by $u_G = P_h u$ the Galerkin solution of (17.1) in the finite element space S_h , that means,

(17.2) $a(P_h u, v) = \langle f, v \rangle$ for every $v \in S_h$.

We say that $P_h u$ is the finite element solution of (17.1). We will estimate the error $u - P_h u$ in $W^{m,2}(\Omega)$. Regularity results and the assumption (ii) of 16.5 will play an important role.

17.2. <u>REGULARITY RESULTS</u>. We want to use the regularity results of § 9 and § 11 for the error estimates. To this end we require that the solution $u \in V$ of 17.1 satisfy

(17.3) $u \in \tilde{V}^{m,2}(\Omega, \vec{\beta})$,

where $\tilde{V}^{m,2}(\Omega, \vec{\beta})$ is the weighted Roĭtberg-Berezanskiĭ space, defined by (0.19) and (0.22) for the set $M = \{O_j\}_{j=1,\ldots,s} \subset \partial\Omega$, where O_j is a conical point or, in the plane case, O_j is a point where the boundary conditions change their type, $\vec{\beta} = (\beta_1, \ldots, \beta_s)$, $\beta_j \geq 0$ for $j = 1,\ldots,s$. The regularity results of § 9 and § 11, formulated for the points $O_j \in M$ and the corresponding operators $\mathcal{U}_0^j(\lambda)$ (see (9.2)), now read as follows :

(i) If the strips $\frac{N}{2} - 2m \leq \operatorname{Im} \lambda \leq \frac{N}{2} + \beta_j - m$ are free of eigenvalues of $\mathcal{U}_0^j(\lambda)$, $j = 1,\ldots,s$, then the solution $u \in V$ of (17.1) is contained in $V^{2m,2}(\Omega, \vec{0}) \subset W^{2m,2}(\Omega)$.

(ii) If the strips $\frac{N}{p} + \gamma_j - 2m + k \leq \operatorname{Im} \lambda \leq \frac{N}{2} - m + \beta_j$ are free of eigenvalues of $\mathcal{U}_0^j(\lambda)$, $j = 1,\ldots,s$, then the solution $u \in V$ of (17.1) is contained in $V^{2m-k,p}(\Omega, \vec{\gamma})$, where k is an appropriate integer, $1 \leq k \leq m$.

The assertion (i) is a special case of (ii).

Let us introduce two examples illustrating the assumption (17.3) and the assertions (i) and (ii).

17.3. <u>EXAMPLES</u>. (i) Let us consider the Dirichlet problem for the Laplace operator in a plane polygonal domain with corner points O_j and with the corresponding angles ω_j . There is a uniquely determined solution $u \in V = W_0^{1,2}(\Omega)$ of

$$a(u,v) = \int_\Omega \left(\frac{\partial u}{\partial x_1} \frac{\partial v}{\partial x_1} + \frac{\partial u}{\partial x_2} \frac{\partial v}{\partial x_2} \right) dx = \int_\Omega f\, v\, dx \quad \text{for every } v \in W_0^{1,2}(\Omega)$$

for $f \in L^p(\Omega)$, $p \geq 2$. The property (17.3) is satisfied when choosing $\beta_j = 0$ for $j = 1,\ldots,s$, since obviously

$$W_0^{1,2}(\Omega) \subset \tilde{V}^{1,2}(\Omega, \vec{0}) .$$

Let us consider the eigenvalues of $\mathcal{U}_0^j(\lambda)$, namely $-\frac{i\pi}{\omega_j}$ (see 1.3 (iii)) $j = 1,\ldots,s$. From 17.2 (i) we obtain

78

(17.4)   If $-\frac{\pi}{\omega_j} < -1$, then the strip $-1 \le \mathrm{Im}\,\lambda \le 0$ is free of eigenvalues of $\mathfrak{A}_0^j(\lambda)$ and $u \in W^{2,2}(\Omega)$

From 17.2 (ii) we obtain

(17.5)   If $-\frac{\pi}{\omega_j} < \frac{2}{p} - 1$, then $u \in V^{1,p}(\Omega,\vec{0}) \subset W^{1,p}(\Omega)$.

(ii) Consider the mixed (Dirichlet-Neumann) problem for the Laplace operator in a polygonal domain Ω
(see Fig. 14). The space V is
the closure with respect to the
norm in $W^{1,2}(\Omega)$ of all functions from $C^\infty(\overline{\Omega})$ vanishing on
$\Gamma_1 \cup \Gamma_2$. We verify that condition (17.3) holds for M =
$\{O_1,O_2,O_3\}$ and $\vec{\beta} = (\varepsilon,\varepsilon,\varepsilon)$,

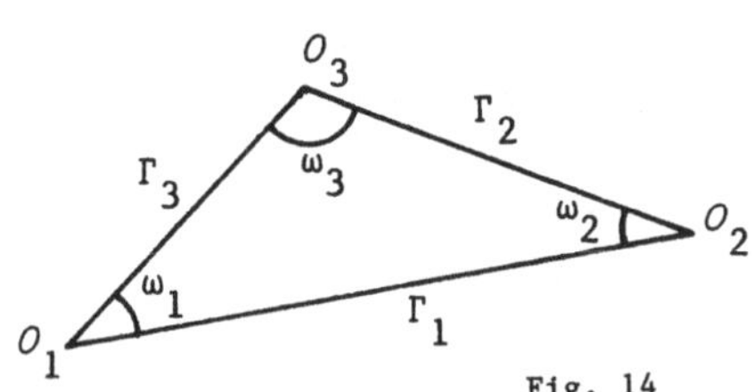

Fig. 14

where ε is a small positive real number. We have

$$V \subset W^{1,2}(\Omega) = W^{1,2}(\Omega,d_M,0) \subset W^{1,2}(\Omega,d_M,\varepsilon) = V^{1,2}(\Omega,\vec{\beta})$$

(see 0.11, 0.12 and D. E. EDMUNDS, A. KUFNER, J. RÁKOSNÍK [1]). Further, the
norm of $\tilde{V}^{1,2}(\Omega,\vec{\beta})$ is equivalent to the norm

$$\|u;\ V^{1,2}(\Omega,\vec{\beta})\| + \|\Delta u;\ V^{-1,2}(\Omega,\vec{\beta})\| .$$

This norm is finite, since $L^p(\Omega) \subset V^{-1,2}(\Omega,\vec{\beta})$, $\vec{\beta} = (\varepsilon,\varepsilon,\varepsilon) = \vec{\varepsilon}$. Therefore u
$\in \tilde{V}^{1,2}(\Omega,\vec{\varepsilon})$.

Let us consider the eigenvalues of $\mathfrak{A}_0^j(\lambda)$ which are calculated in § 2.
From 17.2 (i) we obtain

(17.6)   If $-\frac{\pi}{2\omega_1} < -1$ and $-\frac{\pi}{2\omega_3} < -1$ and $-\frac{\pi}{\omega_2} < -1$,

then $u \in W^{2,2}(\Omega)$.

From 17.2 (ii) we obtain

(17.7)   If $-\frac{\pi}{2\omega_1} < \frac{2}{p} - 1$ and $-\frac{\pi}{2\omega_3} < \frac{2}{p} - 1$ and $-\frac{\pi}{\omega_2} < \frac{2}{p} - 1$,

then $u \in W^{1,p}(\Omega)$.

<u>17.4.   CONSIDERABLE SINGULARITIES.</u> The assertion (i) of 17.2 provides condi-
tions under which the solution u of (17.1) has a regularity "similar" to that
appearing for smooth domains, provided $f \in L^2(\Omega)$. Let us denote by λ_-^j one
of the eigenvalues of $\mathfrak{A}_0^j(\lambda)$ for which the strip $\mathrm{Im}\,\lambda_-^j < \mathrm{Im}\,\lambda \le \frac{N}{2} + \beta_j - m$
is free of eigenvalues of $\mathfrak{A}_0^j(\lambda)$, j = 1,...,s . Assertion (i) of 17.2 can be
formulated in the following form :

If $\mathrm{Im}\,\lambda_-^j < \frac{N}{2} - 2m$ then $u \in W^{2m,2}(\Omega)$.

This motivates the following definition :

(17.8) DEFINITION. The point $O_j \in M$ is a *considerable singularity* if
$$\frac{N}{2} - 2m < \operatorname{Im} \lambda_-^j \; .$$

We have in the example (i) of 17.3 : If $\omega_j > \pi$, then O_j is a considerable singularity. In the example (ii) of 17.3 we have : If $\omega_j > \frac{\pi}{2}$, $j = 1$ or 3 , then O_1 or O_3 is a considerable singularity, if $\omega_2 > \pi$, then O_2 is a considerable singularity.

Assume for simplicity that only one considerable singularity O is contained in M . We write λ_- instead of λ_-^j and use the spaces $V^{\ell,p}(\Omega,\gamma)$ $= H_{\{0\}}^{\ell,p}(\Omega,d_{\{0\}},p\gamma)$ (see 0.4) instead of $V^{\ell,p}(\Omega,\vec{\gamma})$. We determine numbers $p > 2$ and γ for which the solution u of (17.1) is contained in $V^{2m,2}(\Omega,\gamma)$ $\cap W^{m+k,p}(\Omega)$, where k is an appropriate integer, $0 \le k \le m-1$. It follows from (ii) of 7.2 (instead of k insert the number $m - k$) that

$$\text{if } \operatorname{Im} \lambda_- < \frac{N}{2} + \gamma - 2m \text{ and } \operatorname{Im} \lambda_- < \frac{N}{p} - m - k$$
$$\text{then } u \in V^{2m,2}(\Omega,\gamma) \cap W^{m+k,p}(\Omega) \; .$$

Therefore we take a small real positive number ε_0 and set

(17.9) $\gamma_0 = \operatorname{Im} \lambda_- - \frac{N}{2} + 2m + \varepsilon_0$, $\frac{N}{p_0} = \operatorname{Im} \lambda_- + m + k + \varepsilon_0$

so that

(17.10) $p_0 = \dfrac{N}{\operatorname{Im} \lambda_- + m + k + \varepsilon_0} > 2 \; .$

The condition $p_0 > 2$ gives a bound for k . Hence we conclude : *If O is a considerable singularity, then the solution u of (17.1) is contained in* $V^{2m,2}(\Omega,\gamma_0) \cap W^{m+k,p_0}(\Omega)$.

We now consider such global interpolation operators (see 16.2) that

(17.11) $I_h : W^{m+k,p_0}(\Omega) \to S_h \subset W^{m,2}(\Omega) \; .$

17.5. THEOREM. *Let $a(u,v)$ be a real bilinear form which satisfies the conditions of the Lax-Milgram theorem 15.5 for $H = V \subset W^{m,2}(\Omega)$. Assume there is only one considerable singularity on $\partial\Omega$. Let $\{S_h\}$ be a family of finite element spaces with the properties (i) and (ii) of 16.5 and let I_h be the global interpolation operator (17.11) which has the properties (j) and (jj) of 16.6. Assume the solution $u \in V$ of (17.1) satisfies (17.3) and (17.10) is meaningful. Then the finite element solution $P_h u \in S_h$ approximates u in the following way :*

(17.12) $\| u - P_h u; W^{m,2}(\Omega) \| \le c \, h^{-\operatorname{Im} \lambda_- - m + \frac{N}{2} - \varepsilon} \| f; L^{p_0}(\Omega) \|$

where $c = c(N_0)$, N_0 is given by (16.1), $f \in L^{p_0}(\Omega)$.

P r o o f . We denote the neighborhood of O occuring in (16.1) by $U(h)$ and the corresponding index set by I . Then the relation (16.1) has the form

$$\bigcup_{e \in I} \overline{\Omega}_{e,h} = \overline{U}(h) \cap \overline{\Omega} \; .$$

Theorems 15.6 and 16.7 and the relations (16.1) and (16.4) yield

$$\|u - P_h u; \; W^{m,2}(\Omega)\|^2 \le c^2 \|u - I_h u; \; W^{m,2}(\Omega)\|^2$$

$$\le c^2 \Big[\sum_{e \in I} \|u - I_h u; \; W^{m,2}(\Omega_{e,h})\|^2 + \|u - I_h u; \; W^{m,2}(\Omega \setminus U(h))\|^2 \Big]$$

$$\le c_1^2 \Big[\sum_{e \in I} h^{(-(\frac{N}{P_0} - \frac{N}{2})+k)2} \|\nabla_{m+k} u; \; L^{P_0}(\Omega_{e,h})\|^2 + h^{2m} \|\nabla_{2m} u; \; L^2(\Omega \setminus U(h))\|^2 \Big]$$

$$\le c_1^2 \Big[N_0^{P_0/(P_0-2)} h^{2(-\frac{N}{P_0} + \frac{N}{2} +k)} \|\nabla_{m+k} u; \; L^{P_0}(U(h) \cap \Omega)\|^2$$

$$+ h^{2m} \|\nabla_{2m} u; \; L^2(\Omega \setminus U(h))\|^2 \Big] \; .$$

It follows that

$$\|u - P_h u; \; W^{m,2}(\Omega)\| \le c_2 h^{-\frac{N}{P_0} + \frac{N}{2} +k} \Big[\|\nabla_{m+k} u; \; L^{P_0}(U(h) \cap \Omega)\|$$

$$+ h^{-k+m+\frac{N}{P_0} - \frac{N}{2}} \|\nabla_{2m} u; \; L^2(\Omega \setminus U(h))\| \Big]$$

$$= c_2 h^{-\mathrm{Im}\, \lambda_- - m + \frac{N}{2} - \varepsilon_0} \Big[\|\nabla_{m+k} u; \; L^{P_0}(U(h) \cap \Omega)\| + h^{\gamma_0} \|\nabla_{2m} u; \; L^2(\Omega \setminus U(h))\| \Big]$$

$$\le c_2 h^{-\mathrm{Im}\, \lambda_- - m + \frac{N}{2} - \varepsilon_0} \Big[\|u; \; V^{2m,P_0}(\Omega, m-k)\| + \|u; \; V^{2m,2}(\Omega, \gamma_0)\| \Big]$$

$$\le c_3 h^{-\mathrm{Im}\, \lambda_- - m + \frac{N}{2} - \varepsilon_0} \|f; \; L^{P_0}(\Omega)\| \; .$$

In the last estimate we have used the inclusion $u \in V^{2m,P_0}(\Omega, m-k) \subset$ $\subset V^{m+k,P_0}(\Omega,0)$ and the inequalities $\|u; \; V^{2m,P_0}(\Omega, m-k)\| \le c\|f; \; L^{P_0}(\Omega, m-k)\|$ $\le c\|f; \; L^{P_0}(\Omega)\|$ and $\|u; \; V^{2m,2}(\Omega, \gamma_0)\| \le c\|f; \; L^2(\Omega, \gamma_0)\| \le c\|f; \; L^{P_0}(\Omega)\|$. These last two estimates suggest that Theorem 17.5 is valid, provided we only require that the right hand side satisfies $f \in L^2(\Omega, \gamma)$, where γ is to be determined.

17.6. THEOREM. *Assume that the assumptions of Theorem 17.5 are satisfied and* $f \in L^2(\Omega, \gamma_0 - \varepsilon_0/2)$ *. Assume there is an integer* k *,* $0 \le k \le m-1$ *, such that* $P_0 > 2$ *for* $N \ge 2$ *and* $\mathrm{Im}\, \lambda_- + \varepsilon_0 + m - N/2 + k + 1 \ge 0$ *for* $N > 2$ *. Then the finite element solution* $P_h u \in S_h$ *approximates the solution* u *of* (17.1) *in the following way :*

$$(17.13) \quad \|u - P_h u; \; W^{m,2}(\Omega)\| \le c h^{-\mathrm{Im}\, \lambda_- - m + N/2 - \varepsilon_0} \|f; \; L^2(\Omega, \gamma_0 - \varepsilon_0/2)\| \; ,$$

where ε_0 *is an arbitrary small positive real number.*

P r o o f : We show that $V^{2m,2}(\Omega, \gamma_0 - \varepsilon_0/2) \subset V^{k+m,P_0}(\Omega,0)$. We have (see (0.13))

$$(17.14) \qquad V^{2m,2}(\Omega, \gamma_0 - \varepsilon_0/2) \subset V^{2m-1,p}(\Omega, \gamma_0 - \varepsilon_0/2)$$

for $p \in [1,\infty)$ if $N = 2$ and for $p \in [1, \frac{2N}{N-2}]$ if $N > 2$. Further (cf. (0.28)),

$$(17.15) \qquad V^{2m-1,p}(\Omega, \gamma_0 - \varepsilon_0/2) \subset V^{m+k,p}(\Omega, \gamma_0 - \varepsilon_0/2 - m+k+1) \subsetneq V^{m+k,p_0}(\Omega, 0)$$

if $p \geq p_0$ and $\gamma_0 - \varepsilon_0/2 - m + k + 1 + \frac{N}{p} < \frac{N}{p_0}$. If $N = 2$ we choose p so large that $2/p < \varepsilon_0/4$. The imbeddings (17.14) and (17.15) hold. If $N > 2$ we choose $p = 2N/(N-2)$, that means $N/p = N/2 - 1$. The imbedding (17.15) follows from the assumptions of our theorem. The rest of the proof is the same as for Theorem 17.5.

(17.16)   <u>REMARK</u>. If there are some considerable singularities $O_j \in M$, $j = 1,\ldots,J$, we have to set the exponent $\underset{j=1,\ldots,J}{Min} \ (- Im \ \lambda_-^j)$ instead of $(- Im \ \lambda_-)$ and the number

$$\underset{j=1,\ldots,J}{Min} \ \left\{ \frac{N}{Im \ \lambda_-^j + m + k + \varepsilon_0} \right\} \text{ instead of } p_0 \text{ in (17.12).}$$

<u>17.7. EXAMPLES</u>. (i) We consider the Dirichlet problem for the Laplace operator in a plane polygonal domain Ω with only one corner point O with the interior angle $\omega_0 > \pi$. (See 17.3 (i).) We obtain

$$(17.17) \qquad \| u - P_h u; \ W^{1,2}(\Omega) \| \leq c \ h^{\pi/\omega_0 - \varepsilon_0} \| f; \ L^{p_0}(\Omega) \| \ ,$$

where $p_0 = 2/(1 - \pi/\omega_0 + \varepsilon_0)$. This estimate is of the same kind as that obtained by M. DOBROWOLSKI [1]. There it is shown that the estimate (17.7) is optimal.

(ii) We consider the Dirichlet problem for the biharmonic operator in the weak formulation in a plane polygonal domain Ω with only one considerable singularity O (that means $N/2 - 2m = - 3 < Im \ \lambda_-$ and the corresponding angle $\omega_0 > 126°$, see Fig. 6). We get the error estimate

$$(17.18) \qquad \| u - P_h u; \ W^{2,2}(\Omega) \| \leq c \ h^{-Im \ \lambda_- - 1 - \varepsilon_0} \| f; \ L^{p_0}(\Omega) \| \ .$$

We refer to H. MELZER, R. RANNACHER [1] for the calculation of $Im \ \lambda_- = Im \ \lambda_-(\omega_0)$. Such an estimate can be found in H. BLUM [1] as well.

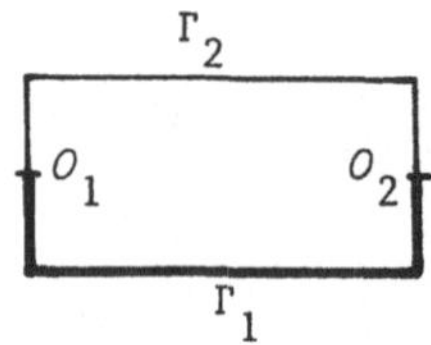

Fig. 15

(iii) Let Ω be a rectangle and $\partial\Omega = \Gamma_1 \cup \Gamma_2$, see Fig. 15. We consider the mixed boundary value problem

$$a(u,v) = \int_{\Omega} \left(\frac{\partial u}{\partial x_1} \frac{\partial v}{\partial x_1} + \frac{\partial u}{\partial x_2} \frac{\partial v}{\partial x_2} \right) dx = \int_{\Omega} f \, v \, dx$$

for every $v \in V$, where V is the closure with respect to the norm of $W^{1,2}(\Omega)$ of all smooth functions which vanish on Γ_1 . The points O_1 and O_2 are considerable singularities. Since $\text{Min} \, (-\text{Im} \, \lambda_-^1, \, -\text{Im} \, \lambda_-^2) = \frac{1}{2}$ we have the error estimate

$$(17.19) \quad \|u - P_h u; \, W^{1,2}(\Omega)\| \leq c \, h^{1/2 - \varepsilon_0} \, \|f; \, L^{P_0}(\Omega)\|$$

where $p_0 = 2/(1/2 + \varepsilon_0)$.

§ 18. Error estimates in $L^p(\Omega)$, $2 \leq p \leq \infty$

Error estimates in which derivatives of orders less than m *appear are often sufficient for the purpose of numerical calculations. Again we study this problem only for the linear case. The main idea is to use the Aubin-Nitsche method and estimates like those in § 17.*

18.1 THE AUBIN-NITSCHE METHOD. This technique was proposed independently by J. P. AUBIN [1] and J. NITSCHE [1]. The Aubin-Nitsche method involves the construction of an auxiliary problem which makes it possible to estimate $u - P_h u$ in $L^p(\Omega)$, $2 \leq p \leq \infty$. Here u is the solution of the linear problem (17.1) that means

$$a(u,v) = \langle f,v \rangle \quad \text{for every } v \in V , \quad f \in L^p(\Omega) \, ;$$

$P_h u$ is the finite element solution (see (17.2)); $a(u,v)$ satisfies the assumptions of the Lax-Milgram theorem 15.5.

For the real bilinear form $a(u,v)$ we consider the *quasi-adjoint problem*: Find a solution $u_g \in V$ such that for an element $g \in V^*$,

$$(18.1) \quad a^*(u_g,v) = a(v,u_g) = \langle g,v \rangle \quad \text{for every } v \in V$$

(V^* denotes the dual space). The assumptions of the Lax-Milgram theorem 15.5 are valid also for $a^*(u,v)$. We assume that the corresponding operator $\mathcal{A}_0^*(\lambda)$ has the same eigenvalue λ_- as $\mathcal{A}_0(\lambda)$.

Let X be a Banach space with $V \subset X$. Then $X^* \subset V^*$. The *auxiliary problem* is the following : Find a solution $u_g \in V$ such that for any $g \in X^*$,

$$(18.2) \quad a(v,u_g) = \langle g,v \rangle \quad \text{for every } v \in V .$$

We obtain the following error estimate of $u - P_h u$ in the space X :

$$\|u - P_h u; \, X\| = \|u - P_h u; \, X^{**}\| = \sup_{\|g;X^*\|=1} |\langle g, \, u - P_h u \rangle|$$

$$= \sup_{\substack{g \in X^* \\ g \neq 0}} \frac{|\langle g, \, u - P_h u \rangle|}{\|g; \, X^*\|} = \sup_{\substack{g \in X^* \\ g \neq 0}} \frac{|a(u - P_h u, \, u_g)|}{\|g; \, X^*\|} =$$

$$= \sup_{\substack{g \in X^* \\ g \neq 0}} \frac{|a(u - P_h u, u_g - I_h u_g)|}{\|g; X^*\|} \leq c \|u - P_h u; X_1\| \sup_{\substack{g \in X^* \\ g \neq 0}} \frac{\|u_g - I_h u_g; X_2\|}{\|g; X^*\|} ,$$

or shortly

$$(18.3) \qquad \|u - P_h u; X\| \leq c \|u - P_h u; X_1\| \sup_{\substack{g \in X^* \\ g \neq 0}} \frac{\|u_g - I_h u_g; X_2\|}{\|g; X^*\|}$$

where X_1 and X_2 are suitable Banach spaces, e.g. $X_1 = X_2 = W^{m,2}(\Omega)$, or $X_1 = W^{m,p}(\Omega)$, $X_2 = W^{m,q}(\Omega)$, $\frac{1}{p} + \frac{1}{q} = 1$.

To demonstrate the Aubin-Nitsche method we first derive an estimate of $u - P_h u$ in $L^2(\Omega)$.

<u>18.2.</u> <u>THEOREM</u> (error estimate in $L^2(\Omega)$). *Assume the assumptions of Theorem 17.6 are valid,* $f \in L^2(\Omega)$ *. Then*

$$(18.4) \qquad \|u - P_h u; L^2(\Omega)\| \leq c \, h^{2(-\mathrm{Im}\, \lambda_- -m+N/2 -\varepsilon_0)} \|f; L^2(\Omega)\| ,$$

where $\varepsilon_0 > 0$ *is any small real number.*

P r o o f : In the estimate (18.3) let $X = L^2(\Omega)$, $X_1 = X_2 = W^{m,2}(\Omega)$. We have $V \subset L^2(\Omega)$. Therefore the estimate (18.3) is applicable. Using the estimate (17.13) we obtain

$$\|u - P_h u; L^2(\Omega)\| \leq c \|u - P_h u; W^{m,2}(\Omega)\| \sup_{\substack{g \in X^* \\ g \neq 0}} \frac{\|u_g - I_h u_g; W^{m,2}(\Omega)\|}{\|g; L^2(\Omega)\|}$$

$$\leq c \, h^{-\mathrm{Im}\, \lambda_- -m+N/2 -\varepsilon_0} \|f; L^2(\Omega, \gamma_0 - \tfrac{\varepsilon_0}{2})\| \, h^{-\mathrm{Im}\, \lambda_- -m+N/2 -\varepsilon_0}$$

$$\cdot \sup_{\substack{g \in X^* \\ g \neq 0}} \frac{\|g; L^2(\Omega, \gamma_0 - \varepsilon_0/2)\|}{\|g; L^2(\Omega)\|}$$

$$\leq c \, h^{2(-\mathrm{Im}\, \lambda_- -m+N/2 -\varepsilon_0)} \|f; L^2(\Omega, \gamma_0 - \varepsilon_0/2)\|$$

$$\leq c \, h^{2(-\mathrm{Im}\, \lambda_- -m+N/2 -\varepsilon_0)} \|f; L^2(\Omega)\| .$$

<u>18.3.</u> <u>THEOREM</u> (error estimates in $L^p(\Omega)$, $2 < p \leq \infty$). *Assume the assumptions of Theorem 17.6 (or Theorem 17.5) are valid,* $f \in L^p(\Omega)$; $2 < p \leq \infty$ *. Let* $S_h \subset L^p(\Omega)$ *. Then*

$$(18.5) \qquad \|u - P_h u; L^p(\Omega)\| \leq c \, h^{2(-\mathrm{Im}\, \lambda_- -m-\varepsilon_0+N/2) + N/p - N/2} \|f; L^p(\Omega)\| .$$

The estimate (18.5) can be improved in the following cases:

(i) *If* $S_h \subset W^{m,p_0}(\Omega)$ *and* $\gamma_p' = \mathrm{Im}\, \lambda_- + 2m - N + N/p < 0$ *then*

$$(18.6) \qquad \|u - P_h u; L^p(\Omega)\| \leq c \, h^{N/p -\mathrm{Im}\, \lambda_- -\varepsilon_0} \|f; L^p(\Omega)\| .$$

(ii) *If* $S_h \subset W^{m,p_0}(\Omega)$ *and* $\beta_p = \text{Im } \lambda_- + m + k + 1 - N + N/p \geq 0$ *then*

(18.7) $\quad \| u - P_h u, L^p(\Omega) \|$

$$\leq \begin{cases} c\, h^{2(-\text{Im } \lambda_- - m - \varepsilon_0 + N/2)} \| f; L^p(\Omega) \| & \textit{for } \ p \leq p_0 \\[2mm] c\, h^{\text{Min}\left[2(-\text{Im } \lambda_- - m - \varepsilon_0 + N/2) + N/p - N/p_0, \ N/p - \text{Im } \lambda_- - \varepsilon_0\right]} \| f; L^p(\Omega) \| & \\[1mm] & \textit{for } \ p > p_0 \end{cases}$$

with p_0 *and* k *defined by* (17.10).

P r o o f . First we prove the estimate (18.5). Let us consider the whole set $M = \{0_i\}_{i=1,\ldots,s} \subset \partial\Omega$ and the corresponding neighborhoods $U^i(h)$ defined by (16.1).

$$\| u - P_h u; L^p(\Omega) \| \leq \| u - I_h u; L^p(\Omega) \| + \| I_h u - P_h u; L^p(\Omega) \|$$

$$\leq c\left[\left\| u - I_h u; L^p\left(\bigcup_{i=1}^{s} U^i(h) \cap \Omega\right) \right\| + \left\| u - I_h u; L^p\left(\Omega \setminus \bigcup_{i=1}^{s} U^i(h)\right) \right\| \right]$$

$$+ \| I_h u - P_h u; L^p(\Omega) \| .$$

We estimate separately the last three terms : For the first term we have

(18.8) $\quad \left\| u - I_h u; L^p\left(\bigcup_{i=1}^{s} U^i(h) \cap \Omega\right) \right\| \leq c\, h^{m+k-N/p_0+N/p} \| \nabla_{m+k} u; L^{p_0}(\Omega) \|$

$$\leq c\, h^{m+k-N/p_0+N/p} \| f; L^2(\Omega) \| ,$$

since $W^{m+k,p_0}(\Omega) \subset L^p(\Omega)$. We have used the estimate (16.4), the regularity results of 17.2 and (17.14) and (17.15). For the second term we have

(18.9) $\quad \left\| u - I_h u; L^p\left(\Omega \setminus \bigcup_{i=1}^{s} U^i(h)\right) \right\| \leq c\, h^{2m} \left\| \nabla_{2m} u; L^p\left(\Omega \setminus \bigcup_{i=1}^{s} U^i(h)\right) \right\|$

$$\leq c\, h^{2m-\gamma_p} \| u; V^{2m,p}(\Omega,\gamma_p) \| \leq c\, h^{2m-\gamma_p} \| f; L^p(\Omega,\gamma_p) \| .$$

We have used the smoothness of u in $\Omega \setminus \bigcup_{i=1}^{s} U^i(h)$ and the relations $u \in$ $\in V^{2m,p}(\Omega,\gamma_p)$, $\gamma_p = \text{Im } \lambda_- + 2m - N/p - \varepsilon_0$ (cf. § 9). For the third term the estimate (16.8) yields

(18.10) $\quad \| I_h u - P_h u; L^p(\Omega) \| \leq c\, h^{N/p-N/2} \| I_h u - P_h u; L^2(\Omega) \|$

$$\leq c\, h^{N/p-N/2} \| I_h u - u; L^2(\Omega) \| + c\, h^{N/p-N/2} \| P_h u - u; L^2(\Omega) \|$$

$$\leq c\, h^{N/p-N/2+m+k+N/2-N/p_0} \| \nabla_{m+k} u; L^{p_0}(\Omega) \|$$

$$+ c\, h^{N/p-N/2+2m-\gamma_0} \| f; L^2(\Omega) \| + c\, h^{N/p-N/2} \| P_h u - u; L^2(\Omega) \| .$$

Here we have used the same facts as in (18.8) and (18.9) for the estimate of $\| I_h u - u; L^2(\Omega) \|$. Using (18.4) for $\| P_h u - u; L^2(\Omega) \|$ we finally obtain

$$(18.11) \quad \|I_h u - P_h u; L^p(\Omega)\| \le c\, h^{N/p - \mathrm{Im}\,\lambda_- - \varepsilon_0} \|f; L^2(\Omega)\|$$

$$+ c\, h^{N/p - N/2 + 2(-\mathrm{Im}\,\lambda_- - m - \varepsilon_0 + N/2)} \|f; L^2(\Omega)\| .$$

From (18.8), (18.9) and (18.11) we conclude

$$\|u - P_h u; L^p(\Omega)\| \le c\, h^{N/p - \mathrm{Im}\,\lambda_- - \varepsilon_0} \|f; L^p(\Omega)\|$$

$$+ c\, h^{N/p - N/2 + 2(-\mathrm{Im}\,\lambda_- - m - \varepsilon_0 + N/2)} \|f; L^2(\Omega)\| .$$

Since $\dfrac{N}{p} - \mathrm{Im}\,\lambda_- - \varepsilon_0 \ge \dfrac{N}{p} - \dfrac{N}{2} + 2(-\,\mathrm{Im}\,\lambda_- - m - \varepsilon_0 + \dfrac{N}{2})$, we obtain (18.5).

We now prove the estimates (18.6) and (18.7) in two steps; first for $2 < p \le p_0$ and then for $p_0 < p \le \infty$:

<u>First step</u> : Let p be a real number with $2 < p \le p_0$. Taking into account that $W^{m,2}(\Omega) \subset L^p(\Omega)$ we can use the Aubin-Nitsche-method, inserting $X = L^p(\Omega)$, $X_1 = W^{m,p}(\Omega)$, $X_2 = W^{m,q}(\Omega)$ with $\dfrac{1}{p} + \dfrac{1}{q} = 1$ in (18.3). We obtain

$$(18.12) \quad \|u - P_h u; L^p(\Omega)\| \le c\|u - P_h u; W^{m,p}(\Omega)\| \sup_{\substack{g \in L^q(\Omega) \\ g \ne 0}} \frac{\|u_g - I_h u_g; W^{m,q}(\Omega)\|}{\|g; L^q(\Omega)\|}$$

$$\le c\Big[\|u - I_h u; W^{m,p}(\Omega)\|$$

$$+ \|I_h u - P_h u; W^{m,p}(\Omega)\| \Big] \sup_{\substack{g \in L^q(\Omega) \\ g \ne 0}} \frac{\|u_g - I_h u_g; W^{m,q}(\Omega)\|}{\|g; L^q(\Omega)\|} .$$

Let us estimate the first term of the first factor. We use the inclusion $W^{m+k,p_0}(\Omega) \subset W^{m,p}(\Omega)$ and proceed analogously to (18.8) and (18.9) :

$$(18.13) \quad \|u - I_h u; W^{m,p}(\Omega)\| \le c\Big[\|u - I_h u; W^{m,p}\big(\bigcup_{i=1}^{s} U^i(h) \cap \Omega\big)\|$$

$$+ \|u - I_h u; W^{m,p}\big(\Omega \setminus \bigcup_{i=1}^{s} U^i(h)\big) \|\Big] \le c\, h^{k - N/p_0 + N/p} \|\nabla_{m+k} u; L^{p_0}(\Omega)\|$$

$$+ c\, h^m \|\nabla_{2m} u; L^p\big(\Omega \setminus \bigcup_{i=1}^{s} U^i(h)\big)\| \le c\, h^{k - N/p_0 + N/p} \|f; L^2(\Omega)\|$$

$$+ c\, h^{m - \gamma_p} \|f; L^p(\Omega; \gamma_p)\| \le c\, h^{-\mathrm{Im}\,\lambda_- - m - \varepsilon_0 + N/p} \|f; L^p(\Omega)\| .$$

Further, it follows from (16.8) and (17.13) that

$$(18.14) \quad \|I_h u - P_h u; W^{m,p}(\Omega)\| \le c\, h^{N/p - N/2} \|I_h u - P_h u; W^{m,2}(\Omega)\|$$

$$\le c\, h^{N/p - \mathrm{Im}\,\lambda_- - m - \varepsilon_0} \|f; L^2(\Omega)\| .$$

We now estimate the second factor of (18.12). Since $g \in L^q(\Omega)$ it follows from the regularity results of § 9 that $u_g \in V^{2m,q}(\Omega; \gamma_p' + \varepsilon_0)$ where $\gamma_p' = \gamma_q = \mathrm{Im}\,\lambda_+ + 2m - N + \dfrac{N}{p} = \mathrm{Im}\,\lambda_- + 2m - \dfrac{N}{q}$. If the assumption (i) of our theorem is valid, then we choose ε_0 so small that $\gamma_p' + \varepsilon_0 \leq 0$, which implies $u_g \in V^{2m,q}(\Omega; \gamma_p' + \varepsilon_0) \subset V^{2m,q}(\Omega,0)$. We obtain

$$(18.15) \qquad \| u_g - I_h u_g; W^{m,q}(\Omega) \| \leq c\, h^m \| \nabla_{2m} u_g; L^q(\Omega) \| \leq c\, h^m \| g; L^q(\Omega) \| \ .$$

If the assumption (ii) of our theorem is valid then $u_g \in V^{2m,q}(\Omega, \gamma_p' + \varepsilon_0/2)$ $\subset V^{2m-1,p_1}(\Omega, \gamma_p' + \varepsilon_0/2) \subset V^{m+k,p_0}(\Omega,0)$, where $p_1 = \dfrac{Nq}{N-q}$ (cf. the proof of Theorem 17.6). Therefore,

$$(18.16) \qquad \| u_g - I_h u_g; W^{m,q}(\Omega) \| \leq c \Big[\| u_g - I_h u_g; W^{m,q}(U(h) \cap \Omega) \|$$
$$+ \| u_g - I_h u_g; W^{m,q}(\Omega \setminus U(h)) \| \Big] \leq c\, h^{k - N/p_0 + N/q} \| \nabla_{m+k} u_g; L^{p_0}(\Omega) \|$$
$$+ c\, h^m \| \nabla_{2m} u_g; L^q(\Omega \setminus U(h)) \| \leq c\, h^{k - N/p_0 + N/q} \| g; L^q(\Omega) \|$$
$$+ c\, h^{m - \gamma_p' - \varepsilon_0} \| g; L^q(\Omega) \| \leq c\, h^{-m - \operatorname{Im}\lambda_- - \varepsilon_0 + N/q} \| g; L^q(\Omega) \| \ .$$

Inserting (18.13), (18.14) and (18.15) or (18.16) into (18.12) and assuming (i) is valid we obtain

$$(18.17) \qquad \| u - P_h u; L^p(\Omega) \| \leq c\, h^{N/p - \operatorname{Im}\lambda_- - m - \varepsilon_0 + m} \| f; L^p(\Omega) \|$$
$$= c\, h^{N/p - \operatorname{Im}\lambda_- - \varepsilon_0} \| f; L^p(\Omega) \| \ ,$$

while

$$(18.18) \qquad \| u - P_h u; L^p(\Omega) \| \leq c\, h^{N/p - \operatorname{Im}\lambda_- - m - \varepsilon_0 - m - \operatorname{Im}\lambda_- - \varepsilon_0 + N/q} \| f; L^p(\Omega) \|$$
$$= c\, h^{2(-\operatorname{Im}\lambda_- - m - \varepsilon_0 + N/2)} \| f; L^p(\Omega) \|$$

provided (ii) is valid.

$\underline{\text{Second step}}$: Let p be a real number, $p_0 < p \leq \infty$. We use the same ideas as in the first part of our proof. First we obtain

$$\| u - P_h u; L^p(\Omega) \| \leq c\, h^{N/p - \operatorname{Im}\lambda_- - \varepsilon_0} \| f; L^p(\Omega) \| + \| I_h u - P_h u; L^p(\Omega) \| \ .$$

Then we use again (16.8) but for $L^{p_0}(\Omega)$ instead of $L^2(\Omega)$:

$$\| I_h u - P_h u; L^p(\Omega) \| \leq c\, h^{N/p - N/p_0} \| I_h u - P_h u; L^{p_0}(\Omega) \|$$
$$\leq c\, h^{N/p - N/p_0} \Big[\| I_h u - u; L^{p_0}(\Omega) \| + \| u - P_h u; L^{p_0}(\Omega) \| \Big]$$
$$\leq c\, h^{N/p - N/p_0 + m + k} \| \nabla_{m+k} u; L^{p_0}(\Omega) \| + c\, h^{N/p - N/p_0} \| u - P_h u; L^{p_0}(\Omega) \|$$
$$\leq c\, h^{N/p - \operatorname{Im}\lambda_- - \varepsilon_0} \| f; L^p(\Omega) \| + c\, h^{N/p - N/p_0} \| u - P_h u; L^{p_0}(\Omega) \| \ .$$

If (i) is valid then (18.17) and (18.6) imply

$$\| u - P_h u; L^p(\Omega) \|$$
$$\leq c\, h^{N/p - \operatorname{Im}\lambda_- - \varepsilon_0} \| f; L^p(\Omega) \| + c\, h^{N/p - N/p_0 + N/p_0 - \operatorname{Im}\lambda_- - \varepsilon_0} \| f; L^p(\Omega) \| \ .$$

If (ii) is valid then (18.18) implies the second part of formula (18.7).

<u>18.4. REMARKS CONCERNING THEOREM 18.3</u>. (i) The error estimates from our theorem are not very nice. It should be possible to improve (18.5) in a uniform manner. However, to this aim we should need some theoretical results concerning the solvability of $a(u,v) = \langle f,v \rangle$ for a pair of weighted spaces of the form $V^{m,p}(\Omega,\alpha_p) \times V^{m,q}(\Omega,\alpha_q)$, where $\frac{1}{p} + \frac{1}{q} = 1$.

(ii) If $m = 1$ we have $\gamma_p' = \beta_p$ and (18.6) and (18.7) are better than (18.5).

<u>18.5. EXAMPLES</u>. (i) We consider the Dirichlet problem for the Laplace operator in a plane polygonal domain Ω with only one corner point 0 with the interior angle $\omega_0 > \pi$ (see 17.3 (i)). Theorem 18.2 yields :

$$(18.19) \qquad \|u - P_h u; L^2(\Omega)\| = O\!\left(h^{2\pi/\omega_0 - \varepsilon}\right) , \qquad \varepsilon > 0 , \text{ any real number.}$$

Since $\gamma_\infty' = -\dfrac{\pi}{\omega_0} < 0$, Theorem 18.3 gives

$$(18.20) \qquad \|u - P_h u; L^\infty(\Omega)\| = O\!\left(h^{\pi/\omega_0 - \varepsilon}\right) .$$

(ii) We consider the Dirichlet problem for the biharmonic operator in the weak formulation in a plane polygonal domain Ω with only one considerable singularity 0 with the angle ω_0 . We have

$$(18.21) \qquad \|u - P_h u; L^2(\Omega)\| = O\!\left(h^{2(-\operatorname{Im}\lambda_- - 1 - \varepsilon_0)}\right) .$$

If $\omega_0 \sim 126°$, that means $-\operatorname{Im}\lambda_- = 3$, we get

$$\|u - P_h u; L^2(\Omega)\| = O(h^{4-\varepsilon}) .$$

If $\omega_0 = 2\pi$, that means $-\operatorname{Im}\lambda_- = 1{,}5$, we get

$$\|u - P_h u; L^2(\Omega)\| = O(h^{1-\varepsilon}) .$$

We now give an estimate in $L^\infty(\Omega)$. For $126° < \omega_0 < \pi$ we have $\gamma_\infty' < 0$ and therefore

$$(18.22) \qquad \|u - P_h u; L^\infty(\Omega)\| = O\!\left(h^{-\operatorname{Im}\lambda_- - \varepsilon}\right) ,$$

while

$$\|u - P_h u; L^\infty(\Omega)\| = O(h^{3-\varepsilon}) \quad \text{for} \quad \omega_0 \sim 126° ,$$

$$\|u - P_h u; L^\infty(\Omega)\| = O(h^{2-\varepsilon}) \quad \text{for} \quad \omega_0 \sim \pi .$$

If $\pi < \omega_0 < 2\pi$ we have $\gamma_\infty' \gtrless 0$ and $\beta_\infty < 0$. The estimate (18.5) is meaningless and we use the imbedding $L^\infty(\Omega) \supset W^{2,2}(\Omega)$ and the estimate (17.13)

$$(18.23) \qquad \|u - P_h u; L^\infty(\Omega)\| = O\!\left(h^{-\operatorname{Im}\lambda_- - 1 - \varepsilon}\right) ,$$

$$\|u - P_h u; L^\infty(\Omega)\| = O\!\left(h^{1/2 - \varepsilon}\right) \quad \text{for} \quad \omega_0 = 2\pi .$$

(iii) Consider the mixed boundary value problem (iii) from 17.7. We have

$$(18.24) \quad \|u - P_h u; \ L^2(\Omega)\| = O(h^{1-\varepsilon})$$

$$(18.25) \quad \|u - P_h u; \ L^\infty(\Omega)\| = O(h^{1/2 - \varepsilon}) \ .$$

Section 6 : *A Modified Finite Element Method in domains with conical points*

As we have seen in the previous section the accuracy of the error estimate depends on the angle ω_0 of the conical point and is not so good as for smooth domains. Thus our task now is to find modified Finite Element Methods which would improve the convergence properties.

In order to achieve this goal we will use our knowledge of the asymptotic expansion of the solution near a conical point. H. BLUM [1] and M. DOBROWOLSKI [1] have proposed an iterative process to improve the finite element solution $P_h u$, in which knowledge of the singular functions and their coefficients is fundamental. This iterative process in a natural way leads to the so-called "Dual Singular Function Method". This method consists in augmenting the finite element spaces S_h with n singular functions to produce S_h^n , and with n so-called dual singular functions to produce S_h^{-n} . The bilinear form $a(u,v)$ will be extended on $S_h^n \times S_h^{-n}$ to $a'(u,v)$ and we have to find a solution $u \in S_h^n$ of $a'(u,v) = \langle f,v \rangle$ for all $v \in S_h^{-n}$.

We present the iterative method together with the Dual Singular Function Method, both proposed by H. BLUM [1] and M. DOBROWOLSKI [1] for some special cases, and give error estimates based on the results of § 17 and § 18.

§ 19 . An iterative method

<u>19.1,</u> Let Ω be a bounded domain with only one boundary point O with a considerable singularity (that means, the corresponding angle ω_0 is so large that the solution u of $a(u,v) = \langle f,v \rangle$, $f \in L^2(\Omega)$, is not contained in $W^{2m,2}(\Omega)$, see 17.4). If the assumptions of Theorem 10.2 are satisfied u admits an expansion (see (10.8))

$$(19.1) \qquad u = \eta \sum_{\gamma \in I} c_\gamma u_\gamma + w \ ,$$

where $w \in W^{m_1,p}(G)$, $m \leq m_1 \leq 2m$, $p \geq 2$. The coefficients c_γ are given by formula (13.1) or, in special cases, by (13.2) and (13.3). We restrict ourselves here to boundary value problems of the type 13.3 (i) and require that, moreover,

$$(19.2) \qquad c_\gamma = \int_\Omega \eta f \ \overline{iv_{\gamma'}} dx - \int_\Omega u \ \overline{A_0^*(0,D) i\eta v_{\gamma'}} \ dx + a(\eta; u, v_{\gamma'}) \ ,$$

where $a(\eta; u, v_{\gamma'})$ is defined for functions $u \in W^{m,2}(\Omega)$ and

(19.3) $|a(\eta;u,v_{\gamma,})| \leq k_{\gamma}\|u; W^{\ell,2}(\Omega)\|$

for an integer ℓ with $0 \leq \ell \leq m$, $k_{\gamma} > 0$ being a constant, independent of $u \in W^{m,2}(\Omega)$. The condition (19.3) is satisfied e.g. if $A_0(0,D) = A(x,D)$ and $B_j(x,D) = \dfrac{\partial^{j-1}}{\partial n^{j-1}}$, $j = 1,\ldots,m$, or if the problem 13.4 is considered. It follows from (19.3) that

(19.4) $\hat{a}(\eta;u,v_{\gamma,}) = - \displaystyle\int_{\Omega} u\, \overline{A_0^*(0,D)i\eta v_{\gamma,}}\, dx + a(\eta,u,v_{\gamma,})$

satisfies the inequality

(19.5) $|\hat{a}(\eta;u,v_{\gamma,})| \leq \ell_{\gamma}\|u; W^{\ell,2}(\Omega)\|$,

where $\ell_{\gamma} > 0$ is a constant independent of $u \in W^{m,2}(\Omega)$.

We now introduce the iterative procedure.

<u>19.2.</u> <u>ITERATIVE PROCEDURE.</u>

1°   Set $c_{\gamma}^0 = 0$ for $\gamma \in I$ and $P_h^1 u = P_h u$.

2°   Find c_{γ}^1 from
$$c_{\gamma}^1 = (f,\eta i v_{\gamma,}) + \hat{a}(\eta;P_h^1 u,v_{\gamma,})$$
and determine $P_h \eta u_{\gamma}$ for $\gamma \in I$.

3°   Set $P_h^2 u = P_h u + \displaystyle\sum_{\gamma \in I} c_{\gamma}^1(\eta u_{\gamma} - P_h \eta u_{\gamma})$.

4°   Determine c_{γ}^j from
$$c_{\gamma}^j = (f,\eta i v_{\gamma,}) + \hat{a}(\eta;P_h^j u,v_{\gamma,})$$
and set

(19.6) $P_h^{j+1} u = P_h u + \displaystyle\sum_{\gamma \in I} c_{\gamma}^j(\eta u_{\gamma} - P_h \eta u_{\gamma})$ for $j \geq 2$.

We want to estimate the error $u - P_h^j u$ in the norm of the spaces X , $X = W^{m,2}(\Omega)$ or $X = L^p(\Omega)$, $2 \leq p \leq \infty$. We start with the estimate of $|c_{\gamma} - c_{\gamma}^{j-1}|$ which plays an important role.

<u>19.3.</u> <u>LEMMA</u> (error estimate of the coefficients). *Assume that Theorem* 10.2 *holds and that the coefficients in formula* (19.1) *are given by* (19.2) *where* (19.5) *is valid for* $\ell = 0$ *or* $\ell = m$. *Further assume that Theorem* 18.2 *holds if* $\ell = 0$ *and Theorem* 17.5 *or Theorem* 17.6 *are satisfied if* $\ell = m$. *Then*

(19.7) $|c_{\gamma} - c_{\gamma}^1| \leq c\, h^{2(-\mathrm{Im}\,\lambda_{-}-m+N/2-\varepsilon)}$ *if* $\ell = 0$,

(19.8) $|c_{\gamma} - c_{\gamma}^1| \leq c\, h^{(-\mathrm{Im}\,\lambda_{-}-m+N/2-\varepsilon)}$ *if* $\ell = m$.

P r o o f : If $\ell = 0$ we have

$$|c_\gamma - c_\gamma^1| = |\hat{a}(\eta; u - P_h u, v_\gamma,)| \leq \ell_\gamma \|u - P_h u; L^2(\Omega)\|$$
$$\leq c\, h^{2(-\text{Im } \lambda_- - m + N/2 - \varepsilon)} \|f; L^2(\Omega)\| \quad \text{for} \quad \gamma \in I .$$

If $\ell = m$ we get

$$|c_\gamma - c_\gamma^1| \leq \ell_\gamma \|u - P_h u; W^{m,2}(\Omega)\| \leq c\, h^{-\text{Im } \lambda_- - m + N/2 - \varepsilon} \quad \text{for} \quad \gamma \in I .$$

<u>19.4.</u> <u>NOTATION</u>. Let us introduce the following notation:

$$- \text{Im } \lambda_- - m + \frac{N}{2} - \varepsilon = a_\varepsilon .$$

Further, we denote the exponent of the convergence rate in the space X for the smooth remainder $w \in W^{m_1, p}(\Omega)$ in the expansion (19.1) by $e(w,X)$, that means

$$\|w - P_h w; X\| \leq c\, h^{e(w,X)} ;$$

and analogously

$$\|\eta u_\gamma - P_h \eta u_\gamma; X\| \leq c\, h^{e(\eta u_\gamma, X)} .$$

Let

$$(19.9) \qquad a_j = \text{Min}\left[e(w, W^{m,2}(\Omega)), \quad a_\varepsilon + (j - 1)\,\text{Min}_\gamma\left(e(\eta u_\gamma, W^{m,2}(\Omega))\right)\right] ,$$

$$(19.10) \qquad b_j = \text{Min}\left[e(w, L^2(\Omega)), \quad 2a_\varepsilon + (j - 1)\,\text{Min}_\gamma\left(e(\eta u_\gamma, L^2(\Omega))\right)\right] , \quad j = 1, 2, \ldots$$

<u>19.5.</u> <u>THEOREM</u> (error estimate). *Assume that the assumptions of Lemma* 19.3 *are satisfied. If $\ell = 0$, then for $j \geq 2$ we have*

$$(19.11) \qquad \|u - P_h^j u; L^2(\Omega)\| \leq c\, h^{b_j}$$

and

$$(19.12) \qquad \|u - P_h^j u; X\| \leq c\, h^{\text{Min}_\gamma\left[e(w,X),\, b_{j-1} + e(\eta u_\gamma, X)\right]} ,$$

where $X = L^p(\Omega)$, $2 < p \leq \infty$ or $X = W^{m,2}(\Omega)$, and

$$(19.13) \qquad \sum_{\gamma \in I} |c_\gamma - c_\gamma^j| \leq c\, h^{b_j} .$$

If $\ell = m$, then for $j \geq 2$ we have

$$(19.14) \qquad \|u - P_h^j u; W^{m,2}(\Omega)\| \leq c\, h^{a_j}$$

and

$$(19.15) \qquad \|u - P_h^j u; L^p(\Omega)\| \leq c\, h^{\text{Min}_\gamma\left(e(w, L^p(\Omega)),\, a_{j-1} + e(\eta u_\gamma, L^p(\Omega))\right)} ,$$

where $2 \leq p \leq \infty$, and

$$(19.16) \qquad \sum_{\gamma \in I} |c_\gamma - c_\gamma^j| \leq c\, h^{a_j} .$$

P r o o f : Since

$$u = \eta \sum_{\gamma \in I} c_\gamma u_\gamma + w$$

we get

$$P_h u = \sum_{\gamma \in I} c_\gamma P_h \eta u_\gamma + P_h w$$

and

$$P_h^j u = P_h u + \sum_{\gamma \in I} c_\gamma^{j-1} (\eta u_\gamma - P_h \eta u_\gamma)$$

$$= P_h w + \sum_{\gamma \in I} \left[c_\gamma P_h \eta u_\gamma + c_\gamma^{j-1} (\eta u_\gamma - P_h \eta u_\gamma) \right] .$$

Therefore

$$u - P_h^j u = w - P_h w + \sum_{\gamma \in I} (c_\gamma - c_\gamma^{j-1})(\eta u_\gamma - P_h \eta u_\gamma)$$

and

$$(19.17) \qquad \|u - P_h^j u; X\| \leq \|w - P_h w; X\| + \sum_{\gamma \in I} |c_\gamma - c_\gamma^{j-1}| \, \|\eta u_\gamma - P_h \eta u_\gamma; X\|$$

for $j \geq 2$. Let $\ell = 0$. It follows from (19.17) and (19.7) that

$$\|u - P_h^2 u; X\| \leq \|w - P_h w; X\| + c \, h^{2a_\epsilon} \sum_{\gamma \in I} \|\eta u_\gamma - P_h \eta u_\gamma; X\|$$

$$\leq c \, h^{e(w,X)} + c \, h^{2a_\epsilon + \min_\gamma e(\eta u_\gamma; X)}$$

$$\leq c \, h^{\min\left[e(w,X), \, 2a_\epsilon + \min_\gamma e(\eta u_\gamma, X)\right]} .$$

If $X = L^2(\Omega)$, we have

$$\|u - P_h^2 u; L^2(\Omega)\| \leq c \, h^{b_2} .$$

Since $|c_\gamma - c_\gamma^j| \leq c \|u - P_h^j u; L^2(\Omega)\|$, we get (19.11) and (19.12) from (19.17) by induction and (19.13) immediately follows. Let $\ell = m$. Then (19.8) and (19.17) yield

$$\|u - P_h^2 u; X\| \leq \|w - P_h w; X\| + c \, h^{a_\epsilon} \sum_{\gamma \in I} \|\eta u_\gamma - P_h \eta u_\gamma; X\|$$

$$\leq c \, h^{\min\left[e(w,X), \, a_\epsilon + \min_\gamma e(\eta u_\gamma, X)\right]} .$$

If $X = W^{m,2}(\Omega)$, we have

$$\|u - P_h^2 u; W^{m,2}(\Omega)\| \leq c \, h^{a_2} .$$

Since $|c_\gamma - c_\gamma^j| \leq c \|u - P_h^j u; W^{m,2}(\Omega)\|$ we obtain the estimates (19.14), (19.15) and (19.16) by induction.

19.6. __EXAMPLES__. (i) We consider again the Dirichlet problem for the Laplace operator in a plane polygonal domain Ω with only one corner point O with the interior angle $\omega_0 > \pi$ (see 17.3 (i) and 18.5 (i)). The solution $u \in$ $\in W_0^{1,2}(\Omega)$ of $a(u,v) = \int_\Omega \left(\frac{\partial u}{\partial x} \frac{\partial v}{\partial x} + \frac{\partial u}{\partial y} \frac{\partial v}{\partial y} \right) dx\,dy = (f,v)$, where $v \in W_0^{1,2}(\Omega)$ and $f \in L^2(\Omega)$, admits the expansion (see (1.26)) $u = c_1 \eta \, r^{\pi/\omega_0} \sin \frac{\pi}{\omega_0} + w$

where $w \in W^{2,2}(\Omega)$. The conditions of Theorem 19.5 are satisfied for $\ell = 0$. Since $a_\varepsilon = \pi/\omega_0 - \varepsilon$, we obtain the following error estimates :

$$\| u - P_h^2 u; \ L^2(\Omega) \| = O\big(h^{\text{Min}(2,\ 4\pi/\omega_0\ -\varepsilon)}\big) \ ,$$

$$|c_1 - c_1^2| = O\big(h^{\text{Min}(2,\ 4\pi/\omega_0\ -\varepsilon)}\big) \ ,$$

$$\| u - P_h^2 u, \ W^{1,2}(\Omega) \| = O(h^1) \ .$$

These estimates are of the same quality as those for smooth domains.

(ii) Let us consider the Dirichlet problem for the biharmonic operator in the weak formulation in a plane polygonal domain Ω with only one considerable singularity O with the angle ω_0 (cf. 18.5 ·(ii), 17.7 (ii)). The solution $u \in W_0^{2,2}(\Omega)$ admits the expansion (see (3.16))

$$u = \eta \sum_{\gamma \in I} c_\gamma u_\gamma + w \ ,$$

where $w \in W^{4,2}(\Omega)$. The conditions of Theorem 11.5 are satisfied for $\ell = 0$. Since $a_\varepsilon = - \text{Im} \, \lambda_- - 1 - \varepsilon$ we get

$$\| u - P_h^2 u; \ L^2(\Omega) \| = O\big(h^{\text{Min}(4,\ 4(-\text{Im} \, \lambda_- -1-\varepsilon))}\big) \ ,$$

$$\sum_{\gamma \in I} |c_\gamma - c_\gamma^2| = O\big(h^{\text{Min}(4,\ 4(-\text{Im} \, \lambda_- -1-\varepsilon))}\big) \ ,$$

$$\| u - P_h^2 u; \ W^{2,2}(\Omega) \| = O\big(h^{\text{Min}(2,\ 3(-\text{Im} \, \lambda_- -1-\varepsilon))}\big) \ .$$

That means : If $126° < \omega_0 < \pi$ then

$$\| u - P_h^2 u; \ L^2(\Omega) \| = O(h^4) \ ,$$

$$\sum_{\gamma \in I} |c_\gamma - c_\gamma^2| = O(h^4) \ ,$$

$$\| u - P_h^2 u; \ W^{2,2}(\Omega) \| = O(h^2) \ .$$

If $\omega_0 = 2\pi$, then

$$\| u - P_h^2 u; \ L^2(\Omega) \| = O(h^{2-\varepsilon}) \ ,$$

$$\sum_{\gamma \in I} |c_\gamma - c_\gamma^2| = O(h^{2-\varepsilon}) \ ,$$

$$\| u - P_h^2 u; \ W^{2,2}(\Omega) \| = O(h^{3/2 \ -\varepsilon}) \ .$$

(iii) We consider the mixed boundary value problem (iii) of 17.7. Formula (2.11) yields the expansion

$$u = \eta_1 c_1 \ r_1^{1/2} \cos \tfrac{1}{2} \overset{1}{\omega} + \eta_2 c_2 \ r_2^{1/2} \cos \tfrac{1}{2} \overset{2}{\omega} + w(x) \ ,$$

where $r_i = |x - O_i|$, $\overset{i}{\omega}$ are the polar angles with respect to O_i , $i = 1,2$, $0 < \overset{i}{\omega} \leq \pi$, and $w(x) \in W^{2,2}(\Omega)$. It follows from (13.5) that $\ell = 0$ and that the assumptions of Theorem 19.5 are satisfied. Since $a_\varepsilon = \tfrac{1}{2} - \varepsilon$, we obtain

$$\|u - P_h^2 u; \ L^2(\Omega)\| = O(h^{2-\varepsilon}) \ ,$$

$$|c_1 - c_1^2| = O(h^{2-\varepsilon}) \ ,$$

$$|c_2 - c_2^2| = O(h^{2-\varepsilon}) \ ,$$

$$\|u - P_h^2 u; \ W^{1,2}(\Omega)\| = O(h^1) \ .$$

§ 20. Dual Singular Function Method

20.1. <u>TEST AND TRIAL SPACES</u>. The Dual Singular Function Method, shortly DSFM, was proposed for the first time by H. BLUM [1] and M. DOBROWOLSKI [2]. This method is a finite element method which employs different test and trial spaces. These are defined in the following way :

For simplicity, let Ω be a bounded domain with only one boundary point O with a considerable singularity. We consider a boundary value problem of the type 13.3 (i) given in the weak formulation (17.1), that means,

$$a(u,v) = \ <f,v> \quad \text{for every} \quad v \in V \ ,$$

$f \in L^2(\Omega)$. Assume the assumptions of Theorem 10.2 are satisfied and u admits an expansion (19.1)

$$u = \sum_{\gamma \in I} c_\gamma \eta u_\gamma + w \ .$$

Observe that the bilinear form

$$(20.1) \qquad a(\eta u_\gamma, v) = (A\eta u_\gamma; v)$$

is defined and ηu_γ has the trivial expansion

$$(20.2) \qquad \eta u_\gamma = 1 \cdot \eta u_\gamma + 0 \ .$$

Let n be the number of the indices of I . The test and trial spaces are

$$(20.3) \qquad S_h^n = S_h \oplus_{\gamma \in I} \eta u_\gamma \ ,$$

$$(20.4) \qquad S_h^{-n} = S_h \oplus_{\gamma' \in I'} \eta v_{\gamma'} \ ,$$

where $v_{\gamma'}$ is defined by (12.12), η is a cut-off function defined by (1.2) and $I' = \{\gamma'\}_{\gamma \in I}$.

20.2. <u>DSFM-SOLUTIONS</u>. We assume that the coefficients c_γ are given by (19.2) where (19.5) is satisfied. Let us extend the bilinear form $a(u,v)$ to a bilinear form $a'(u,v)$ over $S_h^n \times S_h^{-n}$ by the definitions

$$(20.5) \qquad a'(u,v_h) = a(u,v_h) \ , \qquad\qquad \text{for} \quad u \in S_h^n \ , \quad v_h \in S_h$$

$$(20.6) \qquad a'(u_h, \eta v_{\gamma'}) = - \ \hat{a}(\eta; u_h, v_{\gamma'}) \ , \quad \text{for} \quad u_h \in S_h \ , \quad \gamma' \in I'$$

$$(20.7) \qquad a'(\eta u_\gamma, v_{\mu'}, \eta) = (A\eta u_\gamma, i\eta v_{\mu'}) \ , \quad \text{for} \quad \gamma \in I \ , \quad \mu' \in I' \ .$$

Let us write

(20.7) in a more suitable form. Since

$$c_\gamma = c_\gamma(f) = (f, \eta i v_{\gamma'}) + \hat{a}(\eta; u, v_{\gamma'}) \ ,$$

it follows from (20.2) that

$$c_\gamma(A\eta u_\gamma) = 1 = (A\eta u_\gamma, \eta i v_{\gamma'}) + \hat{a}(\eta; u_\gamma, v_{\gamma'})$$

and

$$c_\mu(A\eta u_\gamma) = 0 = (A\eta u_\gamma, \eta i v_{\mu'}) + \hat{a}(\eta; u_\gamma, v_{\mu'}) \quad \text{for} \ \mu \neq \gamma \ , \quad \mu \in I \ .$$

Therefore we conclude

(20.8) $\qquad a'(\eta u_\gamma, v_{\mu'}, \eta) = \delta_{\gamma\mu} - \hat{a}(\eta, u_\gamma, v_{\mu'}) \ .$

We now define the DSFM-solution $P_h^n u \in S_h^n$ of $a(u,v) = <f,v>$.

<u>DEFINITION</u>. $P_h^n u \in S_h^n$ is the DSFM-*solution* of $a(u,v) = <f,v>$ if

(20.9) $\qquad a'(P_h^n u, v) = (f,v) \quad$ for every $\ v \in S_h^{-n}$.

<u>20.3. THEOREM</u>. *Assume that the assumptions of Lemma 19.3 are satisfied. Then
the DSFM-solution* $P_h^n u$ *of* $a(u,v) = <f,v>$ *is uniquely determined and the
following error estimates hold: If* $\ell = 0$ *then*

(20.10) $\qquad \|u - P_h^n u; \ X\| = O(h^{e(w,X)})$

where $X = W^{m,2}(\Omega)$, $X = L^p(\Omega)$, $2 \leq p \leq \infty$. *If* $\ell = m$ *then*

(20.11) $\qquad \|u - P_h^n u; \ W^{m,2}(\Omega)\| = O\left(h^{e(w,W^{m,2}(\Omega))}\right)$.

(See 19.4 for the meaning of $e(w,X)$.)

Further, we have : *If* $\lim\limits_{j \to \infty} P_h^j u$ *exists, then* $\lim\limits_{j \to \infty} P_h^j u = P_h^n u$, *where* $P_h^j u$
is the iterative solution defined by (19.6).

P r o o f : (i) First we show that $P_h^n u$ is uniquely determined. Since
$P_h^n u \in S_h^n$ and

$$a'(P_h^n u, v_h) = a(P_h^n u, v_h) = a(P_h u, v_h) = (f, v_h) \quad \text{for every} \ v_h \in S_h$$

we obtain

(20.12) $\qquad P_h^n u = P_h u + \sum\limits_{\gamma \in I} c_\gamma^h (\eta u_\gamma - P_h \eta u_\gamma)$.

We calculate the coefficients c_γ^h using (20.6) and (20.8) :

$$a'(P_h^n u, v_\mu, \eta) = a'(P_h u, v_\mu, \eta) + a'\Big(\sum\limits_{\gamma \in I} c_\gamma^h(\eta u_\gamma - P_h \eta u_\gamma, \ v_\mu, \eta)$$

(20.13)

$$= -\hat{a}\ (\eta; P_h u, v_{\mu'}) + \sum\limits_{\gamma \in I} c_\gamma^h \big[\delta_{\gamma\mu} + \hat{a}(\eta, P_h \eta u_\gamma - \eta u_\gamma, \ v_{\mu'})\big] = (f, v_\mu, \eta) \ .$$

Therefore the coefficients c_γ^h , $\gamma \in I$, satisfy the $n \times n$ linear system

(20.14) $\qquad \sum\limits_{\gamma \in I} c_\gamma^h \big[\delta_{\gamma\mu} + \hat{a}(\eta, P_h \eta u_\gamma - \eta u_\gamma, \ v_{\mu'})\big] = (f, v_\mu, \eta) + \hat{a}(\eta, P_h u, v_{\mu'})$

for $\mu' \in I'$. Since $\hat{a}(\eta, P_h u_\gamma - u_\gamma, \ v_{\mu'})$ is small provided h is sufficiently small, the system (20.14) is uniquely solvable. Consequently, $P_h^n u$ exists

and is unique.

(ii) We now prove that

$$(20.15) \qquad c_\gamma^h = (f, \eta v_\gamma) + \hat{a}(\eta; P_h^n u, v_\gamma) \qquad \text{for } \gamma \in I .$$

The numbers c_γ^h are well defined by (20.15). Let us insert them in (20.13). We obtain

$$a'(P_h^n u, v_\mu, \eta) = -a'(\eta; P_h u, v_\mu) + c_\mu^h + \sum_{\gamma \in I} c_\gamma^h \hat{a}(\eta, P_h \eta u_\gamma - \eta u_\gamma, v_\mu)$$

$$= \hat{a}(\eta, P_h^n u - P_h u, v_\mu) + (f, \eta v_\mu) + \sum_{\gamma \in I} c_\gamma^h \hat{a}(\eta, P_h u_\gamma - \eta u_\gamma, v_\mu)$$

$$= (f, \eta v_\mu) .$$

Consequently,

$$P_h^n u = P_h u + \sum_{\gamma \in I} c_\gamma^h (\eta u_\gamma - P_h \eta u_\gamma) ,$$

where $c_\gamma^h = (f, \eta v_\gamma) + \hat{a}(\eta, P_h^n u, v_\gamma)$. Comparing this result with (19.6) we conclude : If $\lim_{j \to \infty} P_h^j u$ exists, then $\lim_{j \to \infty} P_h^j u = P_h^n u$.

(iii) Finally, we prove the estimates (20.10) and (20.11). Since

$$u = \sum_{\gamma \in I} c_\gamma \eta u_\gamma + w , \quad P_h u = \sum_{\gamma \in I} c_\gamma P_h \eta u_\gamma + P_h w$$

and

$$P_h^n u = P_h u + \sum_{\gamma \in I} c_\gamma^h (\eta u_\gamma - P_h \eta u_\gamma)$$

$$= P_h w + \sum_{\gamma \in I} \left[c_\gamma^h \eta u_\gamma + (c_\gamma - c_\gamma^h) P_h \eta u_\gamma \right] ,$$

we have

$$u - P_h^n u = w - P_h w + \sum_{\gamma \in I} (c_\gamma - c_\gamma^h)(\eta u_\gamma - P_h \eta u_\gamma) .$$

Therefore

$$(20.16) \qquad \| u - P_h^n u; X \| \le \| w - P_h w; X \| + \sum_{\gamma \in I} | c_\gamma - c_\gamma^h | \, \| \eta u_\gamma - P_h \eta u_\gamma; X \| .$$

Let $\ell = 0$. (20.15), (19.2) and (19.5) yields the inequality

$$| c_\gamma - c_\gamma^h | = | \hat{a}(\eta, u - P_h^n u, v_\gamma) | \le c \| u - P_h^n u, L^2(\Omega) \| \le c \| u - P_h^n u; X \| ,$$

where $X = W^{m,2}(\Omega)$ or $L^p(\Omega)$, $2 \le p \le \infty$. If h is sufficiently small we get

$$\sum_{\gamma \in I} | c_\gamma - c_\gamma^h | \, \| \eta u_\gamma - P_h \eta u_\gamma; X \| \le \delta \| u - P_h^n u; X \| ,$$

where $\delta < 1$. Since (20.16) implies

$$\| u - P_h^n u; X \| \le c \| w - P_h w; X \| ,$$

we have proved the estimate (20.10). $-$ If $\ell = m$, then

$$| c_\gamma - c_\gamma^h | = | \hat{a}(\eta, u - P_h^n u, v_\gamma) | \le c \| u - P_h^n u; W^{m,2}(\Omega) \|$$

and we analogously obtain the estimate (20.11).

Chapter III

ELLIPTIC BOUNDARY VALUE PROBLEMS IN DOMAINS WITH EDGES

In this chapter we study the solvability of elliptic boundary value problems in domains with edges in the weighted function spaces $V^{\ell,p}(\Omega,\beta)$ (see (0.18)), the behavior of the solutions near the edge by means of an asymptotic expansion, and the calculation of the coefficients appearing in this expansion. The key step for these investigations is to reduce these problems locally to a special boundary value problem in a dihedral angle.

Section 7 : *A s p e c i a l b o u n d a r y v a l u e p r o b l e m i n a d i h e d r a l a n g l e*

§ 21. A n i n t r o d u c i n g e x a m p l e

First we illustrate and motivate the above program with the help of a relatively simple example, namely, the Dirichlet problem for the Poisson equation in a threedimensional domain with edges.

<u>21.1. FORMULATION OF THE PROBLEM.</u> Let Ω be a polyhedral domain as in Fig.16.

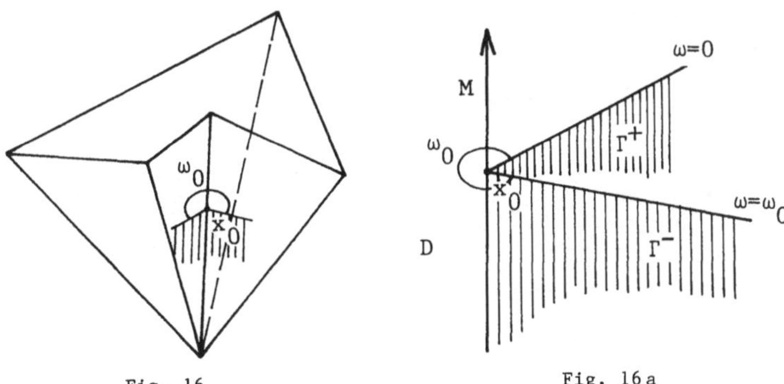

Fig. 16   Fig. 16a

The problem is the following : For a given $f \in L^2(\Omega)$ investigate the smoothness of the weak solution $u \in W_0^{1,2}(\Omega)$ of the threedimensional Dirichlet problem

$$(21.1) \quad \begin{aligned} -\Delta u(x_1,x_2,x_3) &= f(x_1,x_2,x_3) && \text{for } x = (x_1,x_2,x_3) \in \Omega \ , \\ u(x_1,x_2,x_3) &= 0 && \text{for } x \in \partial\Omega \end{aligned}$$

near a point x_0 of an edge. (We assume that x_0 lies on a single edge, that is, corner points are excluded.)

<u>21.2 THE BOUNDARY VALUE PROBLEM IN A DIHEDRAL ANGLE.</u> Let ω_0 be the angle

of the edge at which x_0 is situated (see Fig. 16). We choose x_0 as the origin and consider the dihedral angle

(21.2) $D = K \times M$,

where $K = \{(x_1,x_2) = (r \cos \omega, r \sin \omega) \in R^2: 0 < r < \infty, 0 < \omega < \omega_0\}$ and $M = \{x_3: -\infty < x_3 < \infty\}$ (see Fig. 16a). The sides of K are $\gamma^{\pm} = \{(x_1,x_2) \in R^2: \omega = 0 \text{ or } \omega = \omega_0\}$ and the faces of D are $\Gamma^{\pm} = \gamma^{\pm} \times M$. There is a ball-neighborhood $U(x_0)$ of x_0 with a radius r_0 such that $D \cap U(x_0) = \Omega \cap U(x_0)$. We again consider a cut-off function $\eta \in C^{\infty}(R^3)$, $0 \leq \eta(x) \leq 1$ with

(21.3) $\eta(x) \equiv \begin{cases} 1 & \text{for } |x| < \dfrac{r_0}{2}, \\ 0 & \text{for } |x| > r_0. \end{cases}$

Multiplying the weak solution $u \in W_0^{1,2}(\Omega)$ of the problem (21.1) by η we get a boundary value problem for $u_1 = \eta u$ in the dihedral angle D :

$$-\frac{\partial^2 u_1}{\partial x_1^2} - \frac{\partial^2 u_1}{\partial x_2^2} - \frac{\partial^2 u_1}{\partial x_3^2} = f_1(x) \quad \text{in } D ,$$

(21.4)

$$u_1(x) = 0 \qquad \text{on } \Gamma^{\pm} ,$$

where $f_1 = -\eta \Delta u - u \Delta \eta - 2\eta_{x_1} u_{x_1} - 2\eta_{x_2} u_{x_2} - 2\eta_{x_3} u_{x_3}$. One could conjecture that the boundary regularity of the solution u_1 of (21.4) will be determined by a plane Dirichlet problem in the cone K . This indeed is the case.

21.3. A PLANE DIRICHLET PROBLEM. We write (21.4) in the form

$$-\frac{\partial^2 u_1}{\partial x_1^2} - \frac{\partial^2 u_1}{\partial x_2^2} = f_1(x) + \frac{\partial^2 u_1}{\partial x_3^2} = F(x) \quad \text{in } D ,$$

(21.5)

$$u_1(x) = 0 \qquad \text{on } \Gamma^{\pm} .$$

In order to get a plane problem we fix a point $x_3 = z_0 \in M$ and consider the *plane cone* $K_{z_0} = K \times \{z_0\}$ with the sides $\gamma_{z_0}^{\pm} = \gamma^{\pm} \times \{z_0\}$. From (21.5) we get the plane Dirichlet problem

$$-\Delta_y u_1(y,z_0) = F(y,z_0) \quad \text{in } K_{z_0} ,$$

(21.6)

$$u_1(y,z_0) = 0 \qquad \text{on } \gamma_{z_0}^{\pm} ,$$

where $y = (x_1,x_2)$.

If we know whether the right hand side $F(y,z_0)$ belongs to a weighted Sobolev space $V^{\ell,2}(K_{z_0},\beta)$ then we can describe the regularity of $u_1(y,z_0)$ near the corner point $(0,0,z_0)$ of K_{z_0} . We will show later in § 30 that $F(x) \in L^2(D)$ and therefore $F(y,z_0) \in L^2(K_{z_0})$ for a.e. $z_0 \in M$. Using the results from § 1 or the more general ones from § 7 we conclude : If $\omega_0 > \pi$

then $u_1(y,z_0)$ has the form

$$(21.7) \qquad u_1(y,z_0) = c(z_0)r^{\pi/\omega_0} \sin \frac{\pi}{\omega_0}\omega + w_1(y,z_0)$$

where $w_1(y,z_0) \in W^{2,2}(K_{z_0})$; if $\omega_0 \leqq \pi$ then $u_1(y,z_0) \in W^{2,2}(K_{z_0})$.

21.4. <u>REGULARITY AND THE EXPANSION NEAR AN EDGE</u>. The following questions arise in connection with the expansion (21.7) : *Is* $w_1(y,z)$ *an element of* $W^{2,2}(D)$ *if* $\omega_0 > \pi$ *and does* $u_1(y,z) \in W^{2,2}(D)$ *hold if* $\omega_0 \leqq \pi$ *? Is the coefficient* $c(z)$ *sufficiently smooth and how can we calculate it ?*

The answers, which we will give in § 30 and § 33, are as follows :

(i) If $\dfrac{\partial f}{\partial z}$ and $\dfrac{\partial^2 f}{\partial z^2}$ are elements of $L^2(\Omega)$ for $z = x_3$, then $w_1(y,z) \in W^{2,2}(D)$ provided $\omega_0 > \pi$, and $u_1(y,z) \in W^{2,2}(D)$ provided $\omega_0 \leqq \pi$. The coefficient $c(z)$ is contained in $W^{2,\ -\pi/\omega_0 +3}(M)$ for $\omega_0 > \pi$. (We refer to A. KUFNER, O. JOHN, S. FUČÍK [1] for the definition of the Sobolev spaces $W^{k,p}(M)$ with k noninteger.) The coefficient $c(z)$ can be calculated from

$$(21.8) \qquad c(z) = \frac{\Gamma\left(\frac{1}{2} + \frac{\pi}{\omega_0}\right)}{(\sqrt{\pi})^3 \Gamma\left(\frac{\pi}{\omega_0}\right)} \int_D f_1(x)|x - z|^{-\pi/\omega_0 -1} (\sin\Theta)^{\pi/\omega_0} \sin\frac{\pi\omega}{\omega_0}\, dx$$

where $\Gamma(x)$ denotes the Gamma function and $\sin\Theta = \dfrac{\sqrt{x_1^2 + x_2^2}}{|x - z|}$, $x - z =$ $= (x_1,x_2,x_3-z)$. Therefore in this case we get the expansion (near the edge point x_0)

$$(21.9) \qquad \eta u(x) = u_1(x) = u_1(y,z) = c(z)r^{\pi/\omega_0} \sin\frac{\pi}{\omega_0}\omega + w_1(y,z)$$

where $c(z)$ is given by (21.8).

(ii) If the right hand side $f(x) = f(y,z)$ is not sufficiently smooth with respect to z we get an expansion similar to (21.9) where instead of $c(z)$ a coefficient-function $\hat{c}(x) \in \bigcap\limits_{q\geqq 1} W^{q,2}(D, d_M, 2(\frac{\pi}{\omega_0} - 2 + q))$ occurs (see (0.11) for the definition of $W^{q,2}(D, d_M, 2(\frac{\pi}{\omega_0} - 2 + q))$).

§ 22 . F o r m u l a t i o n   o f   s o m e   b o u n d a r y   v a l u e
         p r o b l e m s

22.1. <u>THE DOMAINS</u>. (i) Let $\Omega \subset R^N$ be a domain with a compact closure $\overline{\Omega}$, bounded by an (N-1)-dimensional manifold $\partial\Omega$. Assume there is a closed smooth (N-2)-dimensional subset $M = M_1 \cup \ldots \cup M_{T-1} \subset \partial\Omega$, the set of nonintersecting edges. M divides $\partial\Omega$ in smooth disjoint connected components $\Gamma_1,\ldots,\Gamma_T$ such that $\partial\Omega = M \cup \Gamma_1 \cup \ldots \cup \Gamma_T$ (see Fig. 17). Assume that in a neighborhood of each point of M the domain Ω is diffeomorphic to an N-dimensional dihedral angle D .

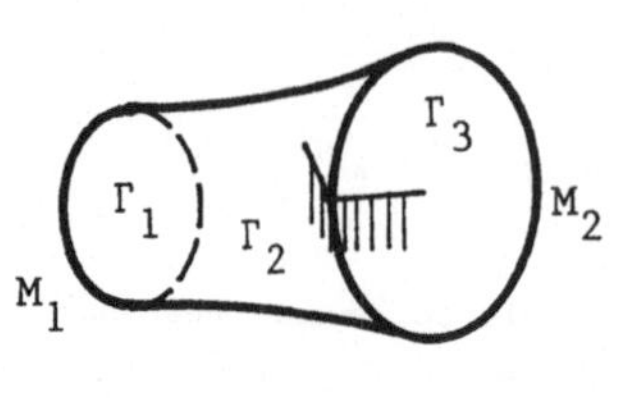

Fig. 17

(ii) A dihedral angle D is defined as $D = K \times R^{N-2}$, where $K = \{y = (y_1,y_2) = (r \cos \omega, r \sin \omega) \in R^2$, $0 < r < \infty$, $0 < \omega < \omega_0\}$ is an infinite cone with the sides $\gamma^+ = \{y \in R^2, \omega = 0\}$ and $\gamma^- = \{y \in R^2, \omega = \omega_0\}$. The faces of D are $\Gamma^\pm = \gamma^\pm \times R^{N-2}$ and the edge of D is $M_D = (0,0) \times R^{N-2}$ (see Fig. 16a).

22.2. THE DIFFERENTIAL OPERATORS. We consider the linear differential operators

(22.1) $\qquad A(x,D_x) = \sum_{|\alpha| \leq 2m} a_\alpha(x)D_x^\alpha$ defined for $x \in \bar\Omega$,

(22.2) $\qquad B_j^{(q)}(x,D_x) = \sum_{|\alpha| \leq m_{q_j}} b_{j,\alpha}^{(q)}(x)D_x^\alpha$ defined for $x \in \Gamma_q$, $1 \leq q \leq T$,

$\qquad\qquad\qquad\qquad\qquad\qquad 1 \leq j \leq m$, $m_{q_j} \leq 2m-1$,

using the notation

(22.3) $\qquad D_x^\alpha = (-i)^{|\alpha|} \dfrac{\partial^{|\alpha|}}{\partial x^\alpha} = (-i)^{|\alpha|} \dfrac{\partial^{\alpha_1 + \ldots + \alpha_N}}{\partial x_1^{\alpha_1} \ldots \partial x_N^{\alpha_N}}$.

Assume that A is elliptic in Ω and $\{B_1^{(q)},\ldots,B_m^{(q)}\}$, $q = 1,\ldots T$, is a normal systems on Γ_q which covers A . (For more details see J. WLOKA [1].) Suppose that all coefficients are sufficiently smooth in $\bar\Omega$ or $\bar\Gamma_q$, respectively. We denote by

(22.4) $\qquad \mathcal{U}(x,D_x) = \{A(x,D_x); B_1^{(1)}(x,D_x), B_2^{(1)}(x,D_x),\ldots, B_m^{T}(x,D_x)\}$

$\qquad\qquad\qquad = \{A(x,D_x); B_j^{(q)}(x,D_x)\}_{\substack{j=1,\ldots,m \\ q=1,\ldots,T}}$

the operator defined by (22.1) and (22.2). We have

(22.5) $\qquad \mathcal{U}(x,D_x): V^{2m+\ell,p}(\Omega,\kappa(\cdot)) \to V^{\ell,p}(\Omega,\kappa(\cdot)) \times$

$\qquad\qquad\qquad\qquad\qquad\qquad\qquad \times \sum_{q=1}^{T} \prod_{j=1}^{m} V^{\ell+2m-m_{q_j} - \frac{1}{p}, p}(\Gamma_q,\kappa(\cdot))$

where $\kappa = \kappa(\zeta)$ is a smooth function defined for every $\zeta \in M$. (Cf. 0.6 (iii) for the definition of these weighted spaces.)

22.3. BOUNDARY VALUE PROBLEMS IN Ω AND D . (i) Now the boundary value problem in Ω is : *Investigate the solvability of*

(22.6) $\qquad \begin{aligned} & A(x,D_x)u = f(x) && in\ \Omega\ , \\ & B_j^{(q)}(x,D_x)u = g_j^{(q)}(x) && on\ \Gamma_q\ ,\ q = 1,\ldots,T\ ,\ j = 1,\ldots,m\ , \end{aligned}$

in $V^{p,k}(\Omega,\kappa(\cdot))$ *and study the behavior of the solutions* u *near the edge*

set M . Another formulation is : *Investigate the properties of the operator (22.5), for instance : Under what conditions is* $\mathcal{U}(x,D_x)$ *a Fredholm operator?*

In order to solve this problem we proceed in the following natural way : We consider a sufficiently fine covering of Ω and a partition of unity subordinate to this covering. From the assumption that in a neighborhood of each point of M the set Ω is diffeomorphic to an N-dimensional dihedral angle it follows that (22.6) can be locally transformed near an edge point by a diffeomorphism into a boundary value problem in a dihedral angle. By the well-known scheme of fitting together the local results we obtain the result for the domain Ω , too.

(ii) Let us now formulate the boundary value problems in a dihedral angle. The problem (22.6) will be transformed by the above described diffeomorphic mapping into a boundary value problem of the following type (for simplicity we use the same variables and similar notation):

$$(22.7) \qquad A(x,D_x)u = \sum_{|\alpha|\leq 2m} a_\alpha(x)D^\alpha u(x) = f(x) \quad \text{in} \quad D ,$$

$$B_j^\pm(x,D_x)u = \sum_{|\alpha|\leq m_j^\pm} b_{j,\alpha}^\pm(x)D^\alpha u(x) = g_j^\pm(x) \quad \text{on} \quad \Gamma^\pm , \quad j = 1,2,\ldots,m ,$$

where D is a dihedral angle defined by 22.1 (ii). We denote the corresponding operator by $\mathcal{U}_D(x,D_x) = \{A(x,D_x), B_1^+(x,D_x),\ldots, B_m^+(x,D_x), B_1^-(x,D_x),\ldots, B_m^-(x,D_x)\}$,

$$(22.8) \qquad \mathcal{U}_D(x,D_x):V^{\ell+2m,P}(D,\kappa(\cdot)) \to$$
$$\to V^{\ell,P}(D,\kappa(\cdot)) \times \sum_\pm \prod_{j=1}^m V^{\ell+2m-m_j^\pm-\frac{1}{P},P}(\Gamma^\pm,\kappa(\cdot))$$

where $\kappa = \kappa(z)$ is a smooth function defined for every $z \in M_D$, and $x = (y,z)$, $y \in K$, $z \in R^{N-2}$.

The properties of $\mathcal{U}_D(x,D_x)$ are determined by the properties of the operators which are given by the principal parts of $A(x,D_x)$ and $B_j^\pm(x,D_x)$, $j = 1,\ldots,m$, with frozen coefficients. Therefore we first consider special boundary value problems of the form

$$(22.9) \qquad A_0(D_x)u = \sum_{|\alpha|=2m} a_\alpha D_x^\alpha u = f \quad \text{in} \quad D ,$$

$$B_{j,0}^\pm(D_x)u = \sum_{|\alpha|=m_j^\pm} b_{j,\alpha}^\pm D_x^\alpha u = g_j^\pm \quad \text{on} \quad \Gamma^\pm , \quad j = 1,\ldots,m , \quad m_j^\pm \leq 2m-1 .$$

A_0 is an elliptic operator with constant coefficients, $\{B_{j,0}^\pm\}_{j=1,\ldots,m}$ are normal systems on $\Gamma^\pm$ which cover A . The corresponding operator $\mathcal{U}_0(D_x) = \{A_0(D_x),B_{1,0}^\pm(D_x),\ldots,B_{m,0}^\pm(D_x)\}$ is a continuous mapping on the following spaces (β is a real number, $\ell \geq 0$ is an integer) :

$$(22.10) \qquad \mathcal{U}_0(D_x) : V^{\ell+2m,P}(D,\beta) \to V^{\ell,P}(D,\beta) \times \sum_\pm \prod_{j=1}^m V^{2m+\ell-m_j^\pm-\frac{1}{P},P}(\Gamma^\pm,\beta) .$$

22.4. PLANE BOUNDARY VALUE PROBLEMS IN K . The crucial idea for the investigation of the problem (22.9) is to reduce it to a plane problem in the infinite cone K . Following the paper of V. G. MAZ'JA, B. A. PLAMENEVSKIĬ [6] we use the real Fourier transform with respect to the variable $z = (z_1,\ldots,z_{n-2}) \in M_D$. First we introduce some function spaces which appear in this connection.

(i) Some function spaces. Let K be an infinite cone with the vertex O ; that is,

$$K = \{y = (y_1,y_2) = (r \cos \omega, r \sin \omega) \in R^2 : 0 < r < \infty, 0 < \omega < \omega_0\} \ .$$

We consider the set

$$C^\infty_{\{O\}}(K) = \{u \in C^\infty(\bar{K}), \ \text{supp } u \ \text{bounded, supp } u \cap \{O\} = \emptyset\}$$

(cf. 0.15) and the norm

$$(22.11) \qquad \left[\int_K |y|^{P\beta} \sum_{|\alpha| \leq \ell} \left(|y|^{P(|\alpha|-\ell)} + 1 \right) |(D^\alpha_y u)(y)|^P \ dy \right]^{1/P} \ .$$

We define : The space

$(22.12) \quad E^{\ell,P}(K,\beta)$ is the closure of $C^\infty_{\{O\}}(K)$ with respect to the norm (22.11)

(β is a real number, $\ell \geq 0$ is an integer). We now introduce the trace spaces of $E^{\ell,P}(K,\beta)$. Let $\gamma^\pm$ be the sides of K . We denote by $E_0^{\ell,P}(K,\gamma^\pm,\beta)$ the closure of $C^\infty_{\bar{\gamma}^\pm}(K)$ with respect to the norm (22.11). [For the definition of $C^\infty_{\bar{\gamma}^\pm}(K)$, see (0.15).] The *trace spaces* are defined as the factor spaces

$$(22.13) \qquad E^{\ell-1/P,P}(\gamma^\pm,\beta) = E^{\ell,P}(K,\beta)/E_0^{\ell,P}(K,\gamma^\pm,\beta)$$

equipped with the corresponding factor norms.

(22.14)   REMARK. For functions u with support in $\{y : |y| < 1\}$ the norms in $V^{\ell,P}(K,\beta)$ and $E^{\ell,P}(K,\beta)$ are equivalent.

(ii) TRANSFORMS. We consider the real Fourier transform

$$(22.15) \qquad \tilde{u}(\eta) = \int_{R^{N-2}} e^{-iz\eta} u(z) \ dz \ ,$$

where $z = (z_1,z_2,\ldots,z_{N-2}) \in R^{N-2}$, $\eta = (\eta_1,\ldots,\eta_{N-2}) \in R^{N-2}$ and $z\eta = \sum_{i=1}^{N-2} z_i \eta_i$. We want to apply (22.15) to the problem (22.9) and therefore we write D^α_x in the form $D^\alpha_x = D^{\alpha'}_y D^{\alpha''}_z$, where $\alpha' = (\alpha_1,\alpha_2)$, $\alpha'' = (\alpha_3,\ldots,\alpha_N)$, $\alpha = (\alpha', \alpha'')$ and $D^{\alpha''}_z = (-i)^{|\alpha''|} \dfrac{\partial^{|\alpha''|}}{\partial z^{\alpha''}}$. Applying (22.15) to (22.9) we get

$$(22.16) \qquad A_0(D_y,\eta) \ \tilde{u}(y,\eta) = \sum_{\substack{|\alpha|=2m \\ \alpha=(\alpha',\alpha'')}} a_{(\alpha',\alpha'')} D^{\alpha'}_y \eta^{\alpha''} \tilde{u}(y,\eta) = \tilde{f}(y,\eta) \ ,$$

$$(22.16) \quad B_{j,0}^{\pm}(D_y,\eta)\tilde{u}(y,\eta) = \sum_{\substack{|\alpha|=m_j^{\pm} \\ \alpha=(\alpha',\alpha'')}} b_{j,(\alpha',\alpha'')}^{\pm} D_y^{\alpha'}\eta^{\alpha''} \tilde{u}(y,\eta) = \tilde{g}_j^{\pm}(y,\eta) \ .$$

We now introduce the variables Y and θ for $\eta \neq (0,\ldots,0)$

$$(22.17) \quad Y = y|\eta| \ , \quad \theta = \frac{\eta}{|\eta|} \ .$$

Then we can write (22.16) in the form

$$A_0(D_Y,\theta)\tilde{u}(\tfrac{Y}{|\eta|},\eta) = \sum_{\substack{|\alpha|=2m \\ \alpha=(\alpha',\alpha'')}} a_{(\alpha',\alpha'')} D_Y^{\alpha'}\theta^{\alpha''} \tilde{u}(\tfrac{Y}{|\eta|},\eta) = \frac{1}{|\eta|^{2m}} \tilde{f}(\tfrac{Y}{|\eta|},\eta),$$

$$(22.18)$$

$$B_{j,0}^{\pm}(D_Y,\theta)\tilde{u}(\tfrac{Y}{|\eta|},\eta) = \sum_{\substack{|\alpha|=m_j^{\pm} \\ \alpha=(\alpha',\alpha'')}} b_{j,(\alpha',\alpha'')}^{\pm} D_y^{\alpha'}\theta^{\alpha''} \tilde{u}(\tfrac{Y}{|\eta|},\eta)$$

$$= \frac{1}{|\eta|^{m_j^{\pm}}} \tilde{g}_j^{\pm}(\tfrac{Y}{|\eta|},\eta) \ .$$

For every fixed η (or θ) (22.18) describes a plane boundary value problem in the infinite cone K. Let us find out in which spaces the corresponding operator $\mathcal{O}(\theta) = \mathcal{O}(D_Y,\theta) = \{A_0(D_Y,\theta), B_{j,0}^{\pm}(D_Y,\theta)\}_{j=1,\ldots,m}$ acts. There are constants $c_1 > 0$ and $c_2 > 0$ such that, for $k = 0,1,2,\ldots$

$$(22.19) \quad c_1 \int_{R^{N-2}} |\eta|^{-2(\beta-k+1)} \|\tilde{u}; \, E^{k,2}(K,\beta)\| d\eta \leq \|u; \, V^{k,2}(D,\beta)\|^2$$

$$\leq c_2 \int_{R^{N-2}} |\eta|^{-2(\beta-k+1)} \|\tilde{u}; \, E^{k,2}(K,\beta)\|^2 d\eta \ ,$$

where $\tilde{u} = \tilde{u}(y,\eta) = \tilde{u}(\tfrac{Y}{|\eta|},\eta)$. Therefore we conclude : If $u \in V^{2m,2}(D,\beta)$, then $\tilde{u}(\tfrac{Y}{|\eta|},\eta) \in E^{2m,2}(K,\beta)$ for a.e. $\eta \in R^{N-2}$ and

$$(22.20) \quad \mathcal{O}(\theta) : E^{2m,2}(K,\beta) \to E^{0,2}(K,\beta) \times \sum_{\pm} \prod_{j=1}^{m} E^{2m-m_j^{\pm}-\frac{1}{2},2}(\gamma^{\pm},\beta)$$

for a.e. $\theta \in S^{N-3}$ (S^{N-3} is the unit sphere in R^{N-2}).

§ 23. Solvability of the special problem in $V^{\ell+2m,p}(D,\beta)$

In this section we study some properties of the operator (22.10), namely, the solvability of the problem (22.9) in the spaces $V^{\ell+2m,p}(D,\beta)$. If we are able to describe the properties of the operators (22.20) $\mathcal{O}(\theta)$ for $\theta \in S^{N-3}$, then by the inverse transforms of 23.4 (ii) we get solvability results for the problem (22.9), too.

23.1. **SOLVABILITY OF THE PLANE PROBLEM.** (i) The results of § 6 about the solvability of boundary value problems in an infinite cone play a crucial role in the following problem (compare Remark (22.14)). The operator $\mathcal{O}(\theta)$ defined

by (22.20) is an inhomogeneous operator for every $\theta \in S^{N-3}$. Its principal
part is given by

$$(23.1) \qquad \mathcal{U}(0) = \left\{ A_0(D_Y,0), \ B_{j,0}^{\pm}(D_Y,0) \right\}_{j=1,\ldots,m} .$$

As in § 6 we introduce polar coordinates and obtain

$$(23.2) \qquad
\begin{aligned}
A_0(D_Y,0) &= r^{-2m} \, L(\omega,D_\omega,rD_r) \ . \\
B_{j,0}^{\pm}(D_Y,0) &= r^{-m_j^{\pm}} \cdot M_j^{\pm}(\omega,D_\omega,rD_r) \ , \quad j = 1,2,\ldots,m \ .
\end{aligned}$$

Let (see again § 6)

$$\mathcal{U}_0(\lambda) = \left\{ L(\omega,D_\omega,\lambda), \ M_j^{\pm}(\omega,D_\omega,\lambda) \right\}_{j=1,\ldots,m} ,$$

where

$$(23.3) \qquad \mathcal{U}_0(\lambda) : W^{2m,2}(G) \to L^2(G) \times \mathbb{C}^m \times \mathbb{C}^m$$

and $G = \left\{ \omega: 0 < \omega < \omega_0 \right\}$. The following theorem was proved by V. G. MAZ'JA,
B. A. PLAMENEVSKIĬ [6].

(ii) <u>THEOREM</u>.

(23.4)   *If the straight line* $\operatorname{Im} \lambda = \beta + 1 - 2m$ *contains no eigenvalue of
the operator* $\mathcal{U}_0(\lambda)$ *defined by (23.3), then the operator* $\mathcal{U}(\theta)$
defined by (22.20) is a Fredholm operator for all $\theta \in S^{N-3}$.

(23.5)   *If for some* $\theta \in S^{N-3}$ *the operator (22.20) is a Fredholm operator,
then the line* $\operatorname{Im} \lambda = \beta + 1 - 2m$ *contains no eigenvalue of* $\mathcal{U}_0(\lambda)$.

   (iii) <u>REMARK</u>. The operator $\mathcal{U}(\theta)$ defined by (22.20) is an isomorphism
if and only if

(23.6)   the line $\operatorname{Im} \lambda = \beta + 1 - 2m$ contains no eigenvalue of $\mathcal{U}_0(\lambda)$;

(23.7) $\ker \mathcal{U}(\theta)$ and $\operatorname{coker} \mathcal{U}(\theta)$ are trivial for all $\theta \in S^{N-3}$.

   (iv) <u>REMARK</u>. The properties (23.6) and (23.7) will play an important
role in what follows. Verification of condition (23.7) is generally a difficult
problem. For equations of the second order the problem was studied e.g. by
A. I. KOMEČ [1] and V. G. MAZ'JA, B. A. PLAMENEVSKIĬ [3].

23.2. <u>SOLVABILITY OF THE SPECIAL PROBLEM IN THE DIHEDRAL ANGLE</u>. Let us first
investigate solvability of the special problem (22.9) in the space $V^{2m,2}(D,\beta)$.

   (i) <u>THEOREM</u>. *The operator*

$$(23.8) \qquad \mathcal{U}_0(D_x): V^{2m,2}(D,\beta) \to V^{0,2}(D,\beta) \times \sum_{\pm} \prod_{j=1}^{m} V^{2m-m_j^{\pm}-\frac{1}{2}, \, 2}(\Gamma^{\pm},\beta)$$

*is an isomorphism if and only if the conditions (23.6) and (23.7) are satis-
fied.*

   (ii) <u>REMARKS</u> to the p r o o f : We only show that the conditions (23.6)

and (23.7) are sufficient. The proof of necessity can be found in V. G. MAZ'JA,
B. A. PLAMENEVSKIĬ [6]. We first note that the problem (22.9) is uniquely solvable in $V^{2m,2}(D,\beta)$ provided the right hand sides f and $g_j^{\pm}$ are from
$V^{0,2}(D,\beta)$ and $V^{2m-m_j^{\pm}-1/2,\,2}(\Gamma^{\pm},\beta)$, if and only if the problem

$$(23.9) \qquad \begin{aligned} A_0(D_x)u &= f \quad \text{in } D , \\ B_{j,0}^{\pm}(D_x)u &= 0 \quad \text{on } \Gamma^{\pm} , \quad j = 1,\ldots,m \end{aligned}$$

with vanishing boundary values is uniquely solvable in $V^{2m,2}(D,\beta)$.

We now transform problem (23.9) into a plane problem in the same way as
in 22.4 (ii). For a.e. $\Theta \in S^{N-3}$ we get

$$(23.10) \qquad \begin{aligned} A_0(D_Y,\Theta)\tilde{u} &= \frac{1}{|\eta|^{2m}} \, \tilde{f}\left(\frac{Y}{|\eta|},\eta\right) \quad \text{in } K , \\ B_{j,0}^{\pm}(D_Y,\Theta)\tilde{u} &= 0 \quad \text{on } \gamma^{\pm} , \quad j = 1,\ldots,m . \end{aligned}$$

Since $\dfrac{1}{|\eta|^{2m}} \, \tilde{f}\left(\dfrac{Y}{|\eta|},\eta\right) \in E^{0,2}(K,\beta)$ for a.e. $\eta \in R^{N-2}$ it follows from the
conditions (23.6) and (23.7) that there exists a uniquely determined solution
$\tilde{u}\left(\dfrac{Y}{|\eta|},\eta\right) \in E^{2m,2}(K,\beta)$ of (23.10) and

$$(23.11) \qquad \left\| \tilde{u};\ E^{2m,2}(K,\beta) \right\|^2 \leq c \left\| \frac{1}{|\eta|^{2m}} \, \tilde{f};\ E^{0,2}(K,\beta) \right\|^2 .$$

Multiplying (23.11) by $|\eta|^{-2(\beta-2m+1)}$, then integrating with respect to η
$\in R^{N-2}$ and using (22.19) we get

$$\left\| u;\ V^{2m,2}(D,\beta) \right\|^2 \leq c_1 \int_{R^{N-2}} |\eta|^{-2(\beta-2m+1)} \left\| \tilde{u};\ E^{2m,2}(K,\beta) \right\|^2 d\eta$$

$$\leq c_2 \int_{R^{N-2}} |\eta|^{-2(\beta+1)} \left\| \tilde{f};\ E^{0,2}(K,\beta) \right\|^2 d\eta \leq c_3 \left\| f;\ V^{0,2}(D,\beta) \right\|^2 .$$

Here the function u is defined by the inverse transform of $\tilde{u}$ and is a solution of (23.9).

We are now proceeding to the case $p \neq 2$, $\ell \geq 0$, that is, we consider
the spaces $V^{2m+\ell,p}(D,\beta)$. The following result was proved by V. G. MAZ'JA,
B. A. PLAMENEVSKIĬ [6] with the aid of a theorem on operator-valued multipliers
and other results on operators in Banach spaces.

(iii) <u>THEOREM</u>. *Conditions (23.6) and (23.7) are necessary and sufficient
for the operator*

$$\mathcal{A}_0(D_x):\ V^{\ell+2m,p}\left(D,\ \beta+1-\tfrac{2}{p}+\ell\right) \to V^{\ell,p}\left(D,\ \beta+1-\tfrac{2}{p}+\ell\right) \times$$

$$\times \sum_{\pm} \prod_{j=1}^{m} V^{\ell+2m-m_j^{\pm}-\frac{1}{p},\,p}\left(\Gamma^{\pm},\ \beta+1-\tfrac{2}{p}+\ell\right)$$

to induce an isomorphism for any $p \in (1, \infty)$ *and* $\ell = 0, 1, 2, \ldots$.

   (iv)   <u>REMARK</u>.   The value of the power of the weight $\beta' = \beta + 1 - \frac{2}{p} + \ell$
is natural; indeed, recall that $\operatorname{Im} \lambda = \beta + 1 - 2m = \beta' + \frac{2}{p} - \ell - 2m$ (cf. § 6).
Theorem (iii) can be formulated also as follows : *Conditions*

(23.6)'   *the line* $\operatorname{Im} \lambda = \beta + \frac{2}{p} - \ell - 2m$ *contains no eigenvalue of* $\mathscr{A}_0(\lambda)$;

(23.7)'   *ker* $\mathscr{A}(\theta)$ *and*   *coker* $\mathscr{A}(\theta)$ *are trivial for all* $\theta \in S^{N-3}$, *where*

$$\mathscr{A}(\theta): E^{2m,2}(K, \beta - 1 + \tfrac{2}{p} + \ell) \to E^{0,2}(K, \beta - 1 + \tfrac{2}{p} - \ell)$$

$$\times \sum_{\pm} \prod_{j=1}^{m} E^{2m - m_j^{\pm} - \frac{1}{2},\, 2}(\gamma^{\pm}, \beta - 1 + \tfrac{2}{p} + \ell)$$

are sufficient   and necessary for

$$\mathscr{A}_0(D_x): V^{\ell + 2m, p}(D, \beta) \to V^{\ell, p}(D, \beta) \times \sum_{\pm} \prod_{j=1}^{m} V^{\ell + 2m - m_j^{\pm} - \frac{1}{p},\, p}(\Gamma^{\pm}, \beta)$$

to induce an isomorphism.

§ 24 .   R e g u l a r i t y   o f   t h e   s p e c i a l   p r o b l e m
         i n   a   d i h e d r a l   a n g l e

   In § 23 we have seen the important role which the conditions (23.6) and
(23.7) play for the unique solvability of problem (22.9) in the spaces
$V^{2m+\ell, p}(D, \beta + 1 - \frac{2}{p} + \ell)$. We now consider the following question :   *If the
right hand sides are sufficiently smooth and (23.6) and (23.7) are satisfied
how "regular" is the solution ? It* is clear that the answer depends on the
location of the eigenvalues of $\mathscr{A}_0(\lambda)$ near the line $\operatorname{Im} \lambda = \beta + 1 - 2m$.

(24.1)   Let λ_- and λ_+ be such complex numbers that the strip $\operatorname{Im} \lambda_- <$
         $< \beta + 1 - 2m < \operatorname{Im} \lambda_+$ contains no eigenvalue of the operator $\mathscr{A}_0(\lambda)$.

<u>24.1. LEMMA</u>.   *Assume the conditions (23.6) and (23.7) are satisfied for some*
β . *Then (23.7) also holds for each β' with $\operatorname{Im} \lambda_- < \beta' + 1 - 2m < \operatorname{Im} \lambda_+$.*

In the   p r o o f   of Lemma 24.1 the imbedding theorems and norm estimates for
$|y| < 1$ and $|y| \geq 1$ are used. Lemma 24.1 implies Theorem 24.2, which was
proved by V. G. MAZ'JA, B. A. PLAMENEVSKIĬ [6].

<u>24.2. THEOREM</u>.   *Assume that the conditions (23.6) and (23.7) are satisfied
for some β and that the right hand sides of problem (22.9) satisfy*

$$f \in V^{\ell_1, p_1}(D, \beta_1 + 1 - \tfrac{2}{p_1} + \ell_1) \quad V^{\ell_2, p_2}(D, \beta_2 + 1 - \tfrac{2}{p_2} + \ell_2) \quad and$$

$$g_j^{\pm} \in V^{\ell_1 + 2m - m_j^{\pm} - 1/p_1,\, p_1}(\Gamma^{\pm}, \beta_1 + 1 - \tfrac{2}{p_1} + \ell_1) \cap$$

$$\cap V^{\ell_2 + 2m - m_j^{\pm} - 1/p_2,\, p_2}(\Gamma^{\pm}, \beta_2 + 1 - 2/p_2 + \ell_2) ,$$

where ℓ_1 and ℓ_2 are nonnegative integers; p_1, $p_2 \in (1,\infty)$ and

$$\operatorname{Im} \lambda_- < \beta_i + 1 - 2m < \operatorname{Im} \lambda_+ \ , \quad i = 1,2 \ .$$

Then the solution $u \in V^{\ell_1+2m,p_1}(D, \beta_1 + 1 - \frac{2}{p_1} + \ell_1)$ of problem (22.9) is

contained in $V^{\ell_2+2m,p_2}(D, \beta_2 + 1 - \frac{2}{p_2} + \ell_2)$, too.

§ 25 . General boundary value problem in D

25.1. <u>NOTATION</u>. We now study the more general boundary value problem (22.7) and the corresponding operator $\mathcal{A}_D(x,D_x)$ defined by (22.8). (Remember that A is elliptic in D , the systems $\{B_j^\pm(x,D_x)\}_{j=1,\ldots,m}$ are normal and cover A on $\Gamma^\pm$ and the coefficients of A and $B_j^\pm$ are smooth.) For every point $\zeta \in M_D$ we consider a boundary value problem of the type (22.9), namely

$$(25.1) \qquad A_0(\zeta,D_x)u = \sum_{|\alpha|=2m} a_\alpha(\zeta)\, D_x^\alpha\, u = f_\zeta(x) \quad \text{in } D \ ,$$

$$B_{j,0}^\pm(\zeta,D_x)u = \sum_{|\alpha|=m_j^\pm} b_{j,\alpha}^\pm(\zeta)\, D_x^\alpha\, u = g_{j,\zeta}^\pm(x) \quad \text{on } \Gamma^\pm , \quad j = 1,2,\ldots,m \ ,$$

and the corresponding operator

$$(25.2) \qquad \mathcal{A}_0(\zeta,D_x) : V^{\ell+2m,p}\big(D,\kappa(\zeta)\big) \to V^{\ell,p}\big(D,\kappa(\zeta)\big) \times$$
$$\times \sum_\pm \prod_{j=1}^m V^{\ell+2m-m_j^\pm-\frac{1}{p},\,p}\big(\Gamma^\pm,\kappa(\zeta)\big) \ .$$

Further, analogously to (22.20) we introduce the operator

$$(25.3) \qquad \mathcal{A}(\zeta,\theta) : E^{2m,2}\big(K,\beta(\zeta)\big) \to E^{0,2}\big(K,\beta(\zeta)\big) \times \sum_\pm \prod_{j=1}^m E^{2m-m_j^\pm-\frac{1}{2},\,2}\big(\gamma^\pm,\beta(\zeta)\big),$$

and analogously to (23.3) the operator

$$(25.4) \qquad \mathcal{A}_0(\zeta,\lambda) : W^{2m,2}(G) \to L^2(G) \times \mathbb{C}^m \times \mathbb{C}^m \ .$$

Let us now formulate the conditions as given by (23.6) and (23.7) :

(25.5)   For all $\zeta \in M_D$ the line $\operatorname{Im} \lambda = \beta(\zeta) + 1 - 2m$ contains no eigen-
         value of $\mathcal{A}_0(\zeta,\lambda)$.

(25.6)   For all $\zeta \in M_D$ and all $\theta \in S^{N-3}$ both the kernel and the cokernel
         of $\mathcal{A}_0(\zeta,\theta)$ are trivial.

It is natural to consider the operator $\mathcal{A}_D(x,D_x)$ in the spaces

$$(25.7) \qquad \mathcal{A}_D(x,D_x) : V^{\ell+2m,p}(D,\kappa(\cdot)) \to V^{\ell,p}(D,\kappa(\cdot)) \times$$
$$\times \sum_\pm \prod_{j=1}^m V^{\ell+2m-m_j^\pm-\frac{1}{p},\,p}(\Gamma^\pm,\kappa(\cdot)) \ ,$$

where $\kappa(\cdot) = \kappa(\zeta)$ is a smooth function defined for every $\zeta \in M_D$. Finally, for every $\zeta \in M_D$ let us introduce complex numbers $\lambda_-(\zeta)$ and $\lambda_+(\zeta)$. They determine the strip $\operatorname{Im} \lambda_-(\zeta) < \beta(\zeta) + 1 - 2m < \operatorname{Im} \lambda_+(\zeta)$ which contains no eigenvalue of $\mathcal{U}_0(\zeta,\lambda)$, provided the line $\operatorname{Im}\lambda = \beta(\zeta) + 1 - 2m$ is free of eigenvalues (cf. 24.1).

25.2. A LOCAL SOLVABILITY RESULT. We want to clarify how the properties of the operator $\mathcal{U}(x,D_x)$ are conditioned by the properties of the operators $\mathcal{U}_0(\zeta,D_x)$. Let us restrict ourselves to $\zeta = 0$. The following lemma, proved by V. G. MAZ'JA, B. A. PLAMENEVSKIĬ [6], gives the answer.

(i) <u>LEMMA</u>. *Let $\varepsilon > 0$ be a sufficiently small number. Then there is an operator*

$$\mathcal{U}^{\varepsilon}(x,D_x) : V^{\ell+2m,P}\left(D,\ \beta(0)+1-\frac{2}{p}+\ell\right)$$
$$\rightarrow V^{\ell,P}\left(D,\ \beta(0)+1-\frac{2}{p}+\ell\right) \times \sum_{\pm}\ \prod_{j=1}^{m}\ V^{\ell+2m-m_j^{\pm}-1/p,\ P}\left(\Gamma^{\pm},\ \beta(0)+1-\frac{2}{p}+\ell\right)\ ,$$

$$(25.8)\qquad \mathcal{U}^{\varepsilon}(x,D_x) = \begin{cases} r^{\delta(z)}\mathcal{U}(x,D_x)r^{-\delta(z)} & \text{for}\quad |x| < \dfrac{\varepsilon}{2}\ , \\[2mm] \mathcal{U}_0(0,D_x) & \text{for}\quad |x| > \varepsilon \end{cases}$$

such that $\left\|\mathcal{U}_0(0,D_x) - \mathcal{U}^{\varepsilon}(x,D_x)\right\| < \dfrac{1}{2}\ \dfrac{1}{\left\|\mathcal{U}_0^{-1}(0,D_x)\right\|}$. *The function* $\delta(z) =$
$= \beta(z) - \beta(0)$ *is defined for* $z \in M_D$, $x = (y,z)$ *with* $y = (y_1,y_2) \in K$ *and* $z = (z_1,\ldots,z_{N-2})$.

(ii) <u>COROLLARY</u>. $\mathcal{U}^{\varepsilon}(x,D_x)$ *is an isomorphism, provided* $\mathcal{U}_0(0,D_x)$ *is an isomorphism.*

P r o o f : The operator $\left(I + \mathcal{U}_0^{-1}(0,D_x)\left(\mathcal{U}^{\varepsilon}(x,D_x) - \mathcal{U}_0(0,D_x)\right)\right)^{-1}$ exists and is bounded, since $\left\|\mathcal{U}_0^{-1}(0,D_x)\left(\mathcal{U}^{\varepsilon}(x,D_x) - \mathcal{U}_0(0,D_x)\right)\right\| < \dfrac{1}{2}$. The assertion follows from the relation

$$\left(\mathcal{U}^{\varepsilon}(x,D_x)\right)^{-1} = \left(\mathcal{U}_0(0,D_x) + \mathcal{U}^{\varepsilon}(x,D_x) - \mathcal{U}_0(0,D_x)\right)^{-1}$$
$$= \left(\mathcal{U}_0(0,D_x)\left[I + \mathcal{U}_0^{-1}(0,D_x)\left(\mathcal{U}^{\varepsilon}(x,D_x) - \mathcal{U}_0(x,D_x)\right)\right]\right)^{-1}$$
$$= \left[I + \mathcal{U}_0^{-1}(0,D_x)\left(\mathcal{U}^{\varepsilon}(x,D_x) - \mathcal{U}_0(0,D_x)\right)\right]^{-1}\mathcal{U}_0^{-1}(0,D_x)\ .$$

The following local solvability result follows from Lemma (i) and Corollary (ii).

(iii) <u>LEMMA</u>. *Assume the conditions (25.5) and (25.6) are satisfied for $\zeta = 0$. Let the right hand sides* f *and* $g_j^{\pm}$ *of (22.7) be functions with supports in the ball* $B_{\varepsilon/2} = \left\{x :\ |x| < \dfrac{\varepsilon}{2}\right\}$ *and* $f \in V^{\ell,P}\left(D,\ \beta(\cdot) + 1 - \dfrac{2}{p} + \ell\right)$, $g_j^{\pm} \in V^{\ell+2m+m_j^{\pm}-1/p,p}\left(\Gamma^{\pm},\ \beta(\cdot) + 1 - \dfrac{2}{p} + \ell\right)$. *If ε is sufficiently small, then there exists a function* $u \in V^{\ell+2m,P}\left(D,\ \beta(\cdot) + 1 - \dfrac{2}{p} + \ell\right)$ *which is a solution of the problem (22.7) in* $D \cap B_{\varepsilon/2}$ *(with boundary conditions on* $\Gamma^{\pm} \cap B_{\varepsilon/2}$ *) and*

$$
\|u; \ V^{\ell+2m,P}(D, \ \beta(\cdot) + 1 - \frac{2}{p} + \ell)\| \leq c \Big[\|f; \ V^{\ell,P}(D, \ \beta(\cdot) + 1 - \frac{2}{p} + \ell)\|
$$

$$
(25.9) \qquad + \sum_{\pm} \sum_{j=1}^{m} \|g_j^{\pm}; \ V^{\ell+2m-m_j^{\pm}-1/p, \ P}(\Gamma^{\pm}, \ \beta(\cdot) + 1 - \frac{2}{p} + \ell)\| \ .
$$

Idea of the p r o o f : The conditions (25.5) and (25.6) imply that $\mathcal{U}_0(0,D_x)$ is an isomorphism. It follows that $\mathcal{U}^\varepsilon(x,D_x)$ is an isomorphism, too. There-fore there is a solution $v \in V^{2m+\ell,P}(D, \ \beta(0)+1-\frac{2}{p}+\ell)$ of the problem

$$
\mathcal{U}^\varepsilon(x,D_x)v = \{r^{\delta(z)}f, \ r^{\delta(z)}g_j^{\pm}\}_{j=1,\ldots,m} \ .
$$

The function $u = r^{-\delta(z)}v$ satisfies the lemma.

<u>25.3. A LOCAL REGULARITY RESULT</u>. We now consider a point $\zeta \in M_D$ and the corresponding operators (25.2), (25.3) and (25.4). We formulate a regularity result for solutions of (22.7) with bounded supports.

(i) <u>LEMMA</u>. *Let* $u \in V^{\ell+2m,P}(D, \ \beta(\cdot) + 1 - \frac{2}{p} + \ell)$ *be a solution of the problem* (22.7) *and* $u \equiv 0$ *for* $|x| > R_0$. *Assume the conditions* (25.5) *and* (25.6) *are satisfied for all* $\zeta \in M_D$ *with* $|\zeta| < R_0$. *Let*
$f \in V^{\ell_1,P_1}(D, \ \beta_1(\cdot) + 1 - 2/p_1 + \ell_1)$ *and*

$g_j^{\pm} \in V^{\ell_1+2m-m_j^{\pm}-1/p_1, \ P_1}(\Gamma^{\pm}, \ \beta_1(\cdot) + 1 - 2/p_1 + \ell_1)$, $j = 1,2,\ldots,m$, *where*

$\beta_1(\cdot)$ *is a smooth function defined on* M_D *with* $\text{Im} \ \lambda_-(\zeta) < \beta_1(\zeta) + 1 - 2m <$
$< \text{Im} \ \lambda_+(\zeta)$ *for* $\zeta \in M_D$ *and* $|\zeta| < R_0$. *Then*

$u \in V^{\ell_1+2m,P_1}(D, \ \beta_1(\cdot) + 1 - 2/p_1 + \ell_1)$ *and*

$$
\|u; \ V^{\ell_1+2m,P_1}(D, \ \beta_1(\cdot) + 1 - 2/p_1 + \ell_1)\|
$$

$$
\leq c \Big[\|f; \ V^{\ell_1,P_1}(D, \ \beta_1(\cdot) + 1 - 2/p_1 + \ell_1)\|
$$

$$
(25.10) \qquad + \sum_{\pm} \sum_{j=1}^{m} \|g_j^{\pm}; \ V^{\ell_1+2m-m_j^{\pm}-1/p_1, \ P_1}(\Gamma^{\pm}, \ \beta_1(\cdot) + 1 - 2/p_1 + \ell_1)\|
$$

$$
+ \|u; \ V^{\ell+2m,P}(D, \ \beta(\cdot) + 1 - \frac{2}{p} + \ell)\| \Big] \quad \ell_1 = 0,1,\ldots \ , \quad p_1 \in (1,\infty) \ .
$$

(ii) <u>REMARK</u>. The complete proof was given by J. ROSSMANN [2]. It is ba-sed on ideas of V. G. MAZ'JA, B. A. PLAMENEVSKIĬ [6]. The partition of unity, the properties of the operator $\mathcal{U}^\varepsilon(x,D_x)$, and Theorem 24.2 are used.

Section 8 : *B o u n d a r y   v a l u e   p r o b l e m   i n*
             a   b o u n d e d   d o m a i n

§ 26. S o l v a b i l i t y   i n $V^{\ell+2m,P}(\Omega,\kappa(\cdot))$ a n d
       r e g u l a r i t y

<u>26.1.</u> <u>INTRODUCTORY REMARKS</u>. In 22.3 we have already roughly described the
connection between the problems (22.6) and (22.7). Now we are going into more
detail.

(i) Let Ω be the bounded domain introduced in 22.1, with the edge set
$M = M_1 \cup \ldots \cup M_{T-1}$. We consider a finite sufficiently fine covering $\{U_i\}_{i \in I}$
of Ω with the following properties :

(26.1)   If $U_i \cap M_k \neq \emptyset$ then $U_i \cap M_j = \emptyset$ for $i \in I$, $k \neq j$, $k, j \in \{1,\ldots,T-1\}$;

(26.2)   if $U_i \cap M = \emptyset$, then U_i is a smooth domain;

(26.3)   if $U_i \cap M \neq \emptyset$ then U_i is diffeomorphic to a part of a dihedral
          angle D_i , namely to $D_i \cap \{x: |x| < R_i\}$.

Assume that for $i = 1,2,\ldots,T_1$ the condition (26.3) is valid and for $i =$
$= T_1+1,\ldots,T_2$ the condition (26.2) is valid. Introducing local coordinate
systems in U_i , $i = 1,\ldots,T_1$, and using the diffeomorphisms we get boundary
value problems of the type (22.7) in dihedral angles D_i . Then we consider
operators (25.2), (25.3) and (25.4) for every D_i . As we know, conditions
(25.5) and (25.6) play an important role for solvability and regularity results.
We have to formulate them for all dihedral angles D_i , $i = 1,\ldots,T_1$; e.g.
for at least one point of M_{D_i} , if we want to use Lemma (iii), 25.2. However,
this is very tedious and therefore let us give a more suitable formulation.

(ii) For every point $\zeta \in M$ we introduce a dihedral angle D_ζ bounded
by the tangential planes $\Gamma^\pm(\zeta)$. Let $\Pi(\zeta)$ be the 2-dimensional plane which
is orthogonal to M_{D_ζ} at ζ . Let K_ζ be the cone from $\Pi(\zeta)$ (on the side where
Ω lies) bounded by the rays $\gamma^\pm(\zeta) = \Gamma^\pm(\zeta) \cap \Pi(\zeta)$ and let $\omega_0(\zeta)$ be the
corresponding angle (see Fig. 18).

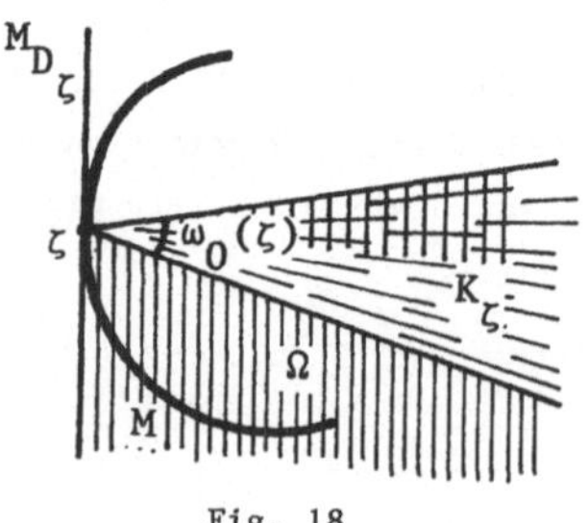

Fig. 18

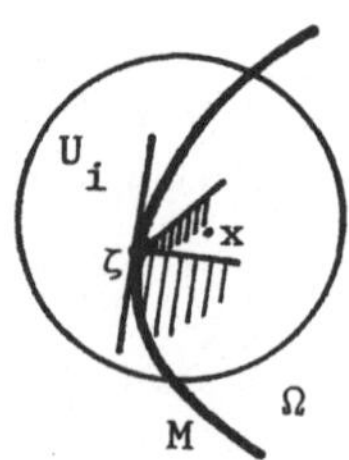

Fig. 19

110

For every U_i , $i = 1,\ldots,T_1$, we introduce the following coordinate system.
Let ζ be the point of M which has the least distance from a point $x \in U_i$.
Then we set $x = (y,\zeta)$, where $y = (y_1,y_2) \in \Pi(\zeta)$ (see Fig. 19). We write
the operator $\mathcal{O}(x,D_x)$ for $x \in U_i$, $i = 1,\ldots,T_1$, in these local coordinates and consider the operators $\mathcal{O}_{D_\zeta}(\zeta,D_x)$ defined by (22.8), $\mathcal{O}(\zeta,\theta)$ defined by (22.20), and $\mathcal{O}_0(\zeta,\lambda)$ defined by (23.3). (In these definitions we have
to set D_ζ instead of D , K_ζ instead of K , $\omega_0(\zeta)$ instead of ω_0 and
$\beta(\zeta)$ instead of β .)

Conditions (23.6) and (23.7) now assume the following form :

(26.4)   For all $\zeta \in M$ the line $\text{Im } \lambda(\zeta) = \beta(\zeta) + 1 - 2m$ contains no eigenvalue of $\mathcal{O}_0(\zeta,\lambda)$.

(26.5)   For all $\zeta \in M$ and all $\theta \in S^{N-3}$ both $\ker \mathcal{O}(\zeta,\theta)$ and
coker $\mathcal{O}(\zeta,\theta)$ are trivial.

26.2. SOLVABILITY THEOREM. *Conditions (26.4) and (26.5) are necessary and
sufficient for the operator of problem (22.6)*

(26.6) $\mathcal{O}(x,D_x) : V^{\ell+2m,P}(\Omega, \beta(\cdot) + 1 - \frac{2}{p} + \ell)$

$$\to V^{\ell,P}(\Omega, \beta(\cdot)+1- \frac{2}{p} +\ell) \times \sum_{q=1}^{T} \prod_{j=1}^{m} V^{\ell+2m-m_{q_j}-\frac{1}{p},\, P}(\Gamma_q, \beta(\cdot)+1- \frac{2}{p} +\ell)$$

to be a Fredholm operator for any $p \in (1,\infty)$, $\ell = 0,1,2\ldots$.

Sketch of the p r o o f : (i) We first show that conditions (26.4) and (26.5)
are sufficient.

- For the above mentioned covering 26.1 (i) of Ω we take a partition of
  unity $\{\phi_{U_i}\}_{i \in I}$ and consider smooth functions $\{\psi_{U_i}\}_{i \in I}$ with supports
  in U_i and $\psi_{U_i}\phi_{U_i} = \phi_{U_i}$, $i \in I$.

- Let U_{10} be an element of the covering with $U_{10} \cap M \neq \emptyset$. For simplicity
  we assume that $U_{10} \cap \Omega$ coincides with an open subset of a dihedral angle
  D_{10} , namely with $D_{10} \cap B_{\varepsilon/2}$ (otherwise we use a diffeomorphism). The
  real number $\varepsilon > 0$ is determined by the properties of the operator
  $\mathcal{O}^\varepsilon(x,D_x)$ defined by (25.8). Let $f \in V^{\ell,P}(\Omega, \beta(\cdot) + 1 - \frac{2}{p} + \ell)$ and
  $g_j^{(q)} \in V^{\ell+2m-m_{q_j}-1/p,\, P}(\Gamma_q, \beta(\cdot) + 1 - \frac{2}{p} + \ell)$ for $j = 1,\ldots,m$, $q = 1,\ldots,T$.
  Then $\psi_{10}f$ and $\psi_{10}g_j^{(q)}$ are functions with supports in $B_{\varepsilon/2}$ belonging
  to the same spaces. It follows from Lemma (iii) of 25.2 that there is a
  function $u_{10} \in V^{\ell+2m,P}(D_{10}, \beta(\cdot) + 1 - \frac{2}{p} + \ell)$ ($\beta(\cdot)$ is extended at D_{10})
  which is a solution of (22.7) in $D_{10} \cap U_{10}$ and with boundary conditions on
  $\Gamma_{10}^{\pm} \cap U_{10}$, and the estimate (25.9) holds in D_{10} . We write

$$u_{i0} = R_{i0}\{\psi_{i0}f, \psi_{i0}g_j^{(q)}\}_{\substack{j=1,\ldots,m \\ q=1,\ldots,T}}.$$

— Let U_{i_1} be an element of the covering with $U_{i_1} \cap M = \emptyset$. The existence of a right regularizer

$$R_{i_1} : \quad W^{\ell,P}(U_{i_1}) \times \prod_{j=1}^{m} W_P^{\ell+2m-m_j - \frac{1}{P}, P}(\partial U_{i_1}) \to W^{\ell+2m,P}(U_{i_1})$$

follows from the theory of elliptic boundary value problems in smooth domains.

— We put together the operators R_{i0} and R_{i1} and define

$$u = R(f,g) = \sum_{i \in I} \phi_i R_i(\psi_i f, \psi_i g).$$

g denotes shortly the right hand sides of the boundary conditions.
R is a right regularizer of the operator $\mathcal{U}(x, D_x)$ given by (26.6). We can construct a left regularizer analogously. It follows that $\mathcal{U}(x, D_x)$ is a Fredholm operator (compare also § 9).

(ii) We now proceed to the necessity of the conditions (26.4) and (26.5). Let $\zeta \in M$ be a fixed edge point. With the help of the operator $\mathcal{U}^{\varepsilon}(x, D_x)$ defined by (25.8) we can show also for the points $\zeta \neq 0$ that the operator $\mathcal{U}_0(\zeta, D_x)$ (see 22.9 and 22.10) is a Fredholm operator in the spaces

$$V^{\ell+2m,P}(D_\zeta, \; \beta(\zeta) + 1 - \frac{2}{P} + \ell)$$

$$\to V^{\ell,P}(D_\zeta, \; \beta(\zeta) + 1 - \frac{2}{P} + \ell) \times \sum_{\pm} \prod_{j=1}^{m} V^{\ell+2m-m_j^{\pm} - \frac{1}{P}, P}(\Gamma_\zeta^{\pm}, \; \beta(\zeta) + 1 - \frac{2}{P} + \ell).$$

Applying Lemma 8.1 of V. G. MAZ'JA, B. A. PLAMENEVSKIĬ [6] we obtain that $\mathcal{U}_0(\zeta, D_x)$ is an isomorphism. Theorem 23.2 (i) yields the assertion.

(iii) <u>REMARK</u>. The proof of part (i) of Theorem 26.2 is to be found in V. G. MAZ'JA, B. A. PLAMENEVSKIĬ [6]; the idea of part (ii) is due to V. G. MAZ'JA, J. ROSSMANN [1], III. Lemma 4.13.

<u>26.3. REGULARITY THEOREM.</u> *Let* $\beta = \beta(\zeta)$ *and* $\beta_1 = \beta_1(\zeta)$ *be smooth functions defined in the edge set* $M \subset \partial\Omega$ *, and let* $u \in V^{\ell+2m,P}(\Omega, \; \beta(\cdot) + 1 - \frac{2}{P} + \ell)$ *be a solution of the problem* (22.6), *where the right hand sides satisfy*

$$f \in V^{\ell_1, P_1}(\Omega, \; \beta_1(\cdot) + 1 - \frac{2}{P_1} + \ell_1), \quad g_j^{(q)} \in$$

$$\in V^{\ell_1+2m-m_{q_j}-1/P_1, P_1}(\Gamma_q, \; \beta_1(\cdot) + 1 - \frac{2}{P_1} + \ell_1), \; j = 1,2,\ldots,m, \quad q = 1,2,\ldots,T.$$

Assume that conditions (26.4) *and* (26.5) *are satisfied and* $\operatorname{Im} \lambda_-(\zeta) < \beta_1(\zeta) + 1 - 2m < \operatorname{Im} \lambda_+(\zeta)$ *for all* $\zeta \in M$ *. Then*

$$u \in V^{\ell_1+2m, P_1}(\Omega, \; \beta_1(\cdot) + 1 - \frac{2}{P_1} + \ell_1) \quad and$$

$$\| u; \ V^{\ell_1 + 2m, P_1}(\Omega, \ \beta_1(\cdot) + 1 - \frac{2}{P_1} + \ell_1) \|$$

$$\leq c \Big[\| f; \ V^{\ell_1, P_1}(\Omega, \ \beta_1(\cdot) + 1 - \frac{2}{P_1} + \ell_1) \|$$

$$\text{(26.7)} \qquad + \sum_{q=1}^{T} \sum_{j=1}^{m} \| g_j^{(q)}; \ V^{\ell_1 + 2m - m_{q_j} - 1/p_1, P_1}(\Gamma_q, \ \beta_1(\cdot) + 1 - \frac{2}{P_1} + \ell_1) \|$$

$$+ \| u; \ V^{\ell + 2m, P}(\Omega, \ \beta(\cdot) + 1 - \frac{2}{p} + \ell) \| \Big].$$

This regularity theorem follows from Lemma (i) 25.3 (compare V. G. MAZ'JA, B. A. PLAMENEVSKIĬ [6]); $\lambda_-(\zeta)$ and $\lambda_+(\zeta)$ are defined as in 25.1.

§ 27.   T h e   c a s e $\ell < 0$

We formulate some results analogously to § 11, where we have studied "weak" solutions of boundary value problems in domains with conical points.

Assume for simplicity that Ω is a bounded domain with only one edge (see 22.1). Let ℓ be an integer. We consider the spaces $\tilde{V}^{\ell+2m,2}(\Omega, \kappa(\cdot))$, defined by (0.25), and the operator $\mathcal{U}(x, D_x)$ from problem (22.6) which maps

$$\tilde{V}^{\ell+2m,2}(\Omega, \kappa(\cdot)) \quad \text{into} \quad V^{\ell,2}(\Omega, \kappa(\cdot)) \times \sum_{q=1}^{2} \prod_{j=1}^{m} V^{\ell + 2m - m_{q_j} - \frac{1}{2}, 2}(\Gamma_q, \kappa(\cdot)) \ . \ \text{The}$$

function κ is smooth and defined on the edge M .

27.1.   <u>THE SPECIAL PROBLEM IN D</u> . Let us formulate two lemmas about the solvability and regularity of the special problem (22.9) in a dihedral angle D for $\ell < 0$. They are similar to the solvability theorem 23.2 (iii) and the regularity theorem 24.2 for $\ell \geq 0$, and were proved by J. ROSSMANN [2].

(i)   <u>LEMMA</u>.   *Let ℓ be an integer. Assume the conditions (23.6) and (23.7) are satisfied. Then the operator*

$$\mathcal{U}_0(D_x) : \ \tilde{V}^{\ell+2m,2}(D, \ \beta+\ell) \to V^{\ell,2}(D, \ \beta+\ell) \times \sum_{\pm} \prod_{j=1}^{m} V^{\ell + 2m - m_j^{\pm} - \frac{1}{2}, 2}(\Gamma^{\pm}, \ \beta+\ell)$$

is an isomorphism.

(ii)   <u>LEMMA</u>.   *Assume that the conditions (23.6) and (23.7) are satisfied for some β and that the right hand sides of problem (22.9) satisfy*

$$f \in V^{\ell_1, 2}(D, \ \beta_1+\ell_1) \cap V^{\ell_2, 2}(D, \ \beta_2+\ell_2) \quad \text{and}$$

$$\substack{+ \\ j} \in V^{\ell_1 + 2m - m_j^{\pm} - \frac{1}{2}, 2}(\Gamma^{\pm}, \ \beta_1+\ell_1) \cap V^{\ell_2 + 2m - m_j^{\pm} - \frac{1}{2}, 2}(\Gamma^{\pm}, \ \beta_2+\ell_2) \ ,$$

where ℓ_1 and ℓ_2 are integers and

$$\text{Im } \lambda_- < \beta_i + 1 - 2m < \text{Im } \lambda_+ \ , \quad i = 1,2 \ .$$

113

(λ_- and λ_+ are defined by 24.1). *Then the solution* $u \in \tilde{V}^{\ell_1+2m,2}(D, \beta_1+\ell_1)$ *of problem (22.9) is contained in* $\tilde{V}^{\ell_2+2m,2}(D, \beta_2+\ell_2)$, *too.*

<u>27.2. THE GENERAL PROBLEM IN Ω</u> . The following lemma follows from 27.1, Lemma (i) :

(i) <u>LEMMA</u>. *Let ℓ be an integer. Assume the conditions (26.4) and (26.5) are satisfied. Then the operator* $\mathcal{U}(x,D_x)$ *of problem (22.6)*,

$$\mathcal{U}(x,D_x) : \tilde{V}^{\ell+2m,2}(\Omega, \beta(\cdot) + \ell)$$

$$\rightarrow V^{\ell,2}(\Omega, \beta(\cdot) + \ell) \times \sum_{q=1}^{2} \prod_{j=1}^{m} V^{\ell+2m-m_{q_j}-\frac{1}{2}, 2}(\Gamma_q, \beta(\cdot) + \ell) ,$$

is a Fredholm operator.

Further, the following regularity result holds :

(ii) <u>LEMMA</u>. *Let ℓ and ℓ_1 be integers. Assume the conditions (26.4) and (26.5) are satisfied. Let* $u \in \tilde{V}^{\ell+2m,2}(\Omega, \beta(\cdot) + \ell)$ *be a solution of problem (22.6) and let* $f \in V^{\ell_1,2}(\Omega, \beta_1(\cdot) + \ell_1)$, $g_j^{(q)} \in V^{\ell_1+2m-m_{q_j}-1/2, 2}(\Gamma_q, \beta_1(\cdot) + \ell_1)$, $q = 1,2$. *If*

$$\text{Im } \lambda_-(\zeta) < \beta_1(\zeta) + 1 - 2m < \text{Im } \lambda_+(\zeta) \quad \text{for every } \zeta \in M ,$$

then $u \in \tilde{V}^{\ell_1+2m,2}(\Omega, \beta_1(\cdot) + \ell_1)$, *too.*

§ 28 . E x a m p l e

We now give an example in which the results for $\ell < 0$ play an important role.

<u>28.1.</u> We consider the *Dirichlet problems* :

$$(28.1) \quad A(x,D_x)u = \sum_{|\alpha|,|\beta|\leq m} (-1)^{|\alpha|} D_x^{\alpha}\left(a_{\alpha\beta}(x)D_x^{\beta}u\right) = f \quad \text{in } \Omega ,$$

$$D_n^j u = 0 \quad \text{on } \partial\Omega \setminus M , \quad j = 0,1,2,\ldots,m-1 ,$$

and

$$(28.2) \quad A(x,D_x)u = \sum_{|\alpha|,|\beta|\leq m} (-1)^{|\alpha|} D_x^{\alpha}\left(a_{\alpha\beta}(x)D_x^{\beta}u\right) = f \quad \text{in } \Omega ,$$

$$D_n^j u = g_j^{(q)} \quad \text{on } \Gamma_q , \quad q = 1,2 , \quad j = 0,1,2,\ldots,m-1 ,$$

where D_n is the derivative in the direction of the normal to Γ_q , $q = 1,2$, Ω is a bounded domain in R^N with one edge M and faces Γ_1 and Γ_2 .

Assume that the Dirichlet problem (28.1) is uniquely solvable in $W_0^{m,2}(\Omega)$ for all $f \in (W_0^{m,2}(\Omega))^*$. It follows that the problem (28.2) is uniquely solvable in $\tilde{V}^{m,2}(\Omega,0)$ for $f \in V^{-m,2}(\Omega,0)$, $g_j^{(q)} \in V^{m-j-1/2,2}(\Gamma_q,0)$, $q = 1,2$.

(See J. ROSSMANN [2].) Further, one can show (see V. G. MAZ'JA, B. A. PLAMENEV-SKIĬ [6]) that the conditions (26.4) and (26.5) are satisfied for the operator $\mathfrak{A}(x,D_x) = (A(x,D_x), D_n^j)$ for $\beta(\zeta) \equiv m$. Therefore we can formulate the following result.

28.2. LEMMA. *Assume the Dirichlet problem* (28.1) *is uniquely solvable in* $W_0^{m,2}(\Omega)$ *for all* $f \in (W_0^{m,2}(\Omega))^*$. *Let* $\beta_1(\cdot)$ *be a smooth function defined on* M *and such that*

$$\operatorname{Im} \lambda_-(\zeta) < \beta_1(\zeta) - 2m + 1 < \operatorname{Im} \lambda_+(\zeta) \quad \text{for every } \zeta \in M .$$

Then the operator $\mathfrak{A}(x,D_x)$ *of problem* (28.2) *is an isomorphism of* $\tilde{V}^{\ell+2m,2}(\Omega, \beta_1(\cdot)+\ell)$ *onto* $V^{\ell,2}(\Omega, \beta_1(\cdot)+\ell) \times \sum\limits_{q=1}^{2} \prod\limits_{j=0}^{m-1} V^{\ell+2m-j-1/2,2}(\Gamma_q, \beta_1(\cdot)+\ell)$ *for* $\ell = 0, \pm 1, \pm 2, \ldots$. *If* $\ell = 0,1,2,\ldots$, *then* $\mathfrak{A}(x,D_x)$ *is an isomorphism of* $V^{\ell+2m,p}\big(\Omega, \beta_1(\cdot) + 1 - \frac{2}{p} + \ell\big)$ *onto* $V^{\ell,p}\big(\Omega, \beta_1(\cdot) + 1 - \frac{2}{p} + \ell\big)$ $\times \sum\limits_{q=1}^{2} \prod\limits_{j=0}^{m-1} V^{\ell+2m-j-1/p, p}(\Gamma_q, \beta_1(\cdot) + 1 - \frac{2}{p} + \ell)$ ($\lambda_-(\zeta)$ and $\lambda_+(\zeta)$ are defined by 24.1 for· $\beta(\zeta) \equiv m$).

P r o o f : Lemma (i) of 27.2 implies that $\mathfrak{A}(x,D_x)$ is a Fredholm operator. Using Lemma (ii) of 27.2 and the unique solvability of (28.2) for $\ell = -m$, $\beta(\zeta) \equiv 0$, $p = 2$ we obtain the assertion. For the case $\ell \geq 0$ we refer to 26.3.

28.3. COROLLARY. *Let* A *be the Laplace operator and* $\omega_0(\zeta)$ *the angle at the edge point* $\zeta \in M$. *Let* $u \in W_0^{1,2}(\Omega)$ *be the solution of* (28.1) *for a function* $f \in L^{p_1}(\Omega)$, $p_1 > 2$: *If* $\omega_0(\zeta) < \pi p_1/(p_1-2)$ *for all* $\zeta \in M$, *then* $u \in W_0^{1,p_1}(\Omega)$.

P r o o f : Lemma (ii) of 27.2 implies that $u \in V^{2,2}(\Omega,1)$. Further we have $\operatorname{Im} \lambda_-(\zeta) = - \pi/\omega_0(\zeta)$ and $\operatorname{Im} \lambda_+(\zeta) = \pi/\omega_0(\zeta)$ (see § 1). Using Theorem 26.3 with $p = 2$, $\beta(\zeta) = 1$, $\beta_1(\zeta) = 2/p_1$ and $\ell = \ell_1 = 0$ we obtain $u \in V^{2,p_1}(\Omega,1)$ and therefore $u \in W_0^{1,p_1}(\Omega)$.

Section 9 : *E x p a n s i o n s   n e a r   t h e   e d g e*

§ 29 . D e f i n i t i o n   o f   s o m e   f u n c t i o n   s p a c e s

29.1. SPACES WITH POWER WEIGHTS. Let us consider weighted spaces of the type (0.11), where the power $\varepsilon = \varepsilon(\cdot)$ is a function defined on the edge set.

(i) Let D be a dihedral angle, $D = \{x = (y,z), 0 < |y| = r < \infty$, $0 < \omega < \omega_0, z \in R^{N-2}\}$, with the edge $M = \{0\} \times R^{N-2}$ (see 22.1). We have denoted by $r = r(x) = \sqrt{y_1^2 + y_2^2}$ the distance of a point $x \in D$ from M and by ω the polar angle. Let $\beta = \beta(z)$ be a real smooth function defined for $z \in R^{N-2}$. As in (0.11) we define the weighted spaces

(29.1) $W^{k,p}(D,d_M,\beta(\cdot))$ and $W_0^{k,p}(D,d_M,\beta(\cdot))$

equipped with the norm

(29.2) $\|u; W^{k,p}(D,d_M,\beta(\cdot))\| = \Big(\sum_{|\alpha| \leq k} \int_D r^{p\beta(z)} |D^\alpha u|^p \, dx \Big)^{1/p} \; ;$

$k \geq 0$ is an integer, $1 < p < \infty$.

(ii) Let Ω be a bounded domain with the edge set M (see 22.1), and let $\beta = \beta(\cdot)$ be a smooth real function defined on M . Let $r = r(x) = $ $= \mathrm{dist}(x,M) = |x - z|$, $z \in M$ (z is locally uniquely defined). We define $W^{k,p}(\Omega,d_M,\beta(\cdot))$ analogously to (29.1) and (29.2).

29.2. <u>TRACES ON EDGES</u>. (i) We consider a dihedral angle with the edge M . We have already defined trace spaces on the faces $\Gamma^\pm$ of D in 0.7. Now, we are interested in the traces of functions from $W^{k,p}(D,d_M,\beta(\cdot))$ on the edge M , which are defined (if they exist) by

(29.3) $\lim_{r \to 0} \overline{u}(r,z) = \lim_{r \to 0} \frac{1}{\omega_0} \int_0^{\omega_0} u(r,\omega,z) \, d\omega = t(z)$.

V. G. MAZ'JA, J. ROSSMANN [1], I, have proved that for some power functions β the trace $t(z)$ is well defined and that t is an element of a BESOV space of the following kind :

(ii) <u>DEFINITION</u>. Let $\kappa = \kappa(z)$ be a real smooth function defined for $z \in R^{N-2}$, and let $k \geq 1$ be the smallest integer such that $0 < \inf \kappa(z)$ $\leq \sup \kappa(z) < k$. We denote

(29.4) $B^{\kappa(\cdot),p}(R^{N-2}) = B^{\kappa(\cdot),p}(M) = \{ f \in L^p(R^{N-2}) : \|f; B^{\kappa(\cdot),p}(R^{N-2})\|$

$= \Big(\|f; L^p(R^{N-2})\|^p + \int_{R^{N-2}} \int_{R^{N-2}} |\Delta_z^k f(\zeta)|^p \frac{d\zeta dz}{|z|^{N-2+p\kappa(\zeta)}} \Big)^{1/p} < \infty \} \; ,$

where $\Delta_z^k f(\zeta) = \sum_{\nu=0}^{k} (-1)^\nu \binom{k}{\nu} f(\zeta + \nu z)$ is the k-th difference.

(iii) <u>LEMMA</u>. *Let* s *and* k *be integers with* $0 \leq s \leq k-1$, *and let* $\beta = \beta(z)$ *be a smooth real function defined on* R^{N-2} *such that*

(29.5) $s - \frac{2}{p} < \inf \beta(z) \leq \sup \beta(z) < s + 1 - \frac{2}{p}$ *and*

$|\nabla \beta(z)| < c_0 < \infty$ *for every* $z \in M$.

Then the traces of a function $u \in W^{k,p}(D,d_M,\beta(\cdot))$ *exist for* $z \in M$ *and* $t = t(z) \in B^{k - \frac{2}{p} - \beta(\cdot),\, p}(M)$. *Moreover, the following estimates hold :*

(29.6) $\|t; B^{k - \frac{2}{p} - \beta(\cdot),\, p}(M)\| \leq c \|u; W^{k,p}(D,d_M,\beta(\cdot))\|$

(29.7) $\quad \int_D r^{p(\beta(z)-s-1)} |u(x) - t(z)|^p \, dx \leq c \|u; \, W^{k,p}(D,d_M,\beta(\cdot))\|^p$,

where the constants c *are independent of* u .

(iv) <u>REMARK</u>. V. G. MAZ'JA, J. ROSSMANN [1] have proved the following
result : *If* $\inf \beta(z) > k - \frac{2}{p}$ *or* $\sup \beta(z) < -\frac{2}{p}$ *and* u *is a function from*
$W^{k,p}(D,d_M,\beta(\cdot))$ *with bounded support, then* u *is contained in* $V^{k,p}(D,\beta(\cdot))$,
too.

(v) Let us introduce a figure summarizing these results for functions u
$W^{k,p}(D,d_M,\beta(\cdot))$ with bounded support provided (29.5) is satisfied (see Fig.
20).

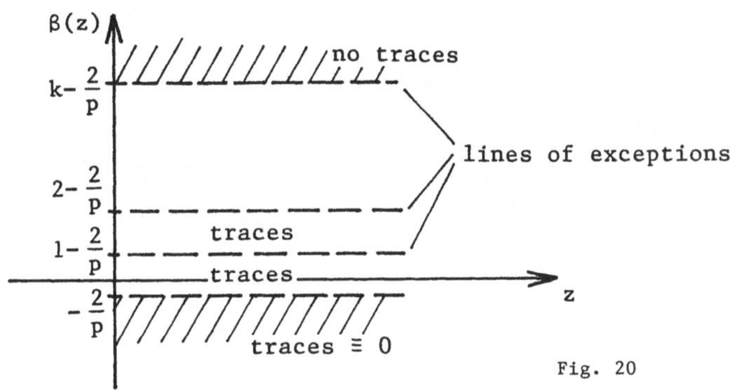

Fig. 20

(vi) <u>EXTENSIONS</u>. Conversely, we now consider functions t from
$B^{k-\frac{2}{p}-\beta(\cdot), \, p}(M)$ and construct functions v from $W^{k,p}(D,d_M,\beta(\cdot))$ such
that t is the trace of v on M . The following result can be found again
in V. G. MAZ'JA, J. ROSSMANN [1], I. : *Assume the conditions* (29.5) *are satis-*
fied for $\beta = \beta(z)$. *Then for any function* $t \in B^{k-(2/p)-\beta(\cdot), \, p}(M)$ *there is*
an extension $v \in W^{k,p}(D,d_M,\beta(\cdot))$ *such that* $v|_M = t$ *and for* $|\alpha| \leq k - s - 1$
we have

(29.8) $\quad \int_D r^{p(\beta(z)-k+|\alpha|)} |D_z^\alpha v(y,z) - D_z^\alpha t(z)|^p \, dx \leq c \|t; \, B^{k-\frac{2}{p}-\beta(\cdot), \, p}(M)\|^p$

and

(29.9) $\quad \int_D r^{-2+\varepsilon} |D_z^\alpha v(y,z)|^p \, dx \leq c \|t; \, B^{k-\frac{2}{p}-\beta(\cdot), \, p}(M)\|^p$

(ε is a small positive real number). *Further, for* $|\alpha| \geq k - s$ *or* $|\gamma| \geq 1$
we have

(29.10) $\quad \int_D r^{p(\beta(z)-k+|\alpha|+|\gamma|)} |D_y^\gamma D_z^\alpha v(y,z)|^p \, dx \leq c \|t; \, B^{k-\frac{2}{p}-\beta(\cdot), \, p}(M)\|^p$.

§ 30. **Expansions in a dihedral angle with and without tangential smoothness conditions**

We will derive expansions of the solutions of the general boundary value problem (22.7) in a dihedral angle. The smoothness in the tangential direction of the right hand sides will play an important role in this connection. These results are fundamental for expansions near the edge of the solutions of boundary value problems in a bounded domain.

30.1. <u>SMOOTHNESS OF THE SOLUTION IN THE TANGENTIAL DIRECTION.</u> (i) Let us write the derivatives of the boundary value problem (22.7) separately in the y-direction and in the z-direction :

$$A(x,D_x)u = A(y,z,D_y,D_z)u = \sum_{|\alpha|\leq 2m} a_\alpha(y,z)D_y^{\alpha_1} D_z^{\alpha_2} u = f \quad \text{in} \quad D \ ,$$

$$B_j^\pm(x,D_x)u = B_j^\pm(y,z,D_y,D_z)u = \sum_{|\alpha|\leq m_j^\pm} b_{j,\alpha}^\pm(y,z)D_y^{\alpha_1} D_z^{\alpha_2} u = g_j^\pm \quad \text{on} \quad \Gamma^\pm,$$

$j = 1,2,\ldots,m$, where $D_x^\alpha = D_y^{\alpha_1} D_z^{\alpha_2}$, $\alpha = (\alpha_1,\alpha_2)$. In order to be able to use the results for an infinite cone K in the plane, we now write (22.7) as follows :

$$(30.1) \qquad A_0(0,z,D_y,0)u = \sum_{|\alpha_1|=2m} a_{(\alpha_1,0)}(0,z) D_y^{\alpha_1} u$$

$$= f(y,z) - \sum_{|\alpha_1|=2m} [a_{(\alpha_1,0)}(y,z) - a_{(\alpha_1,0)}(0,z)] D_y^{\alpha_1} u$$

$$- \sum_{\substack{|\alpha|\leq 2m \\ |\alpha_1|<2m}} a_\alpha(y,z) D_y^{\alpha_1} D_z^{\alpha_2} u = F(y,z)$$

$$(30.2) \qquad B_{j,0}^\pm(0,z,D_y,0)u = \sum_{|\alpha_1|=m_j^\pm} b_{j,(\alpha_1,0)}^\pm(0,z) D_y^{\alpha_1} u = g_j^\pm(y,z)$$

$$- \sum_{|\alpha_1|=m_j^\pm} [b_{j,(\alpha_1,0)}^\pm(y,z) - b_{j,(\alpha_1,0)}^\pm(0,z)] D_y^{\alpha_1} u$$

$$- \sum_{\substack{|\alpha|\leq m_j^\pm \\ |\alpha_1|<m_j^\pm}} b_{j,\alpha}^\pm D_y^{\alpha_1} D_z^{\alpha_2} u = G_j^\pm(y,z) \ , \quad j = 1,2,\ldots,m \ .$$

Let $u \in V^{\ell+2m,p}(D,\kappa(\cdot))$ be a solution of problem (22.7) with bounded support in $B_{R_0} = \{x : |x| < R_0\}$, and let $f \in V^{\ell_1,p_1}(D,\kappa_1(\cdot))$, $g_j^\pm \in$

$\in V^{\ell_1+2m-m_j^\pm-1/p_1,\, p_1}(\Gamma^\pm,\kappa_1(\cdot))$, where $\kappa(\cdot) = \beta(\cdot) + 1 - \frac{2}{p} + \ell$ and $\kappa_1(\cdot)$

$= \beta_1(\cdot) + 1 - \frac{2}{p_1} + \ell_1$ are smooth functions defined for $z \in M = R^{N-2}$, $\ell \geq 0$,

$\ell_1 \geq 0$ being integers. In connection with the expansion of the solution u in the dihedral angle D (and in the infinite cone K) the following question arises : *Are the new right hand sides* F *of* (30.1) *and* $G_j^\pm$ *of* (30.2) *contained in* $V^{\ell_1, p_1}(D, \kappa_1(\cdot))$ *and in* $V^{\ell_1 + 2m - m_j^\pm - 1/p_1, \, p_1}(\Gamma^\pm, \kappa_1(\cdot))$, *respectively?* In a special case the following lemma gives an answer.

(ii) <u>LEMMA</u>. *If* $\ell = \ell_1$, $p = p_1$, $\beta(z) - 1 \leq \beta_1(z) \leq \beta(z)$ *for*

$z \in M \cap B_{R_0}$, $f \in V^{\ell, p}(D, \kappa_1(\cdot))$ *and* $g_j^\pm \in V^{\ell + 2m - m_j^\pm - 1/p, \, p}(\Gamma^\pm, \kappa_1(\cdot))$ *and*

$\dfrac{\partial u}{\partial z_i} \in V^{\ell + 2m - 1, p}(D, \kappa_1(\cdot))$ *for* $i = 1, \ldots, N-2$, *then* $F \in V^{\ell, p}(D, \kappa_1(\cdot))$ *and*

$G_j^\pm \in V^{\ell + 2m - m_j^\pm - 1/p, \, p}(\Gamma^\pm, \kappa_1(\cdot))$, $j = 1, 2, \ldots, m$.

P r o o f : We only have to consider the sums $F_1(y,z) = \sum\limits_{|\alpha_1| = 2m} [a_{(\alpha_1, 0)}(y,z)$

$- a_{(\alpha_1, 0)}(0,z)] \, D_y^{\alpha_1} u$ and $\sum\limits_{|\alpha_1| = m_j^\pm} [b_{j, (\alpha_1, 0)}^\pm(y,z) - b_{j, (\alpha_1, 0)}^\pm(0,z)] \, D_y^{\alpha_1} u$

$= G_{j,1}^\pm(y,z)$. For $|\delta| \leq \ell$, $\delta = (\delta_1, \ldots, \delta_N)$ we have

$\quad |D_x^\delta F_1(y,z)|$

$\quad = |\sum\limits_{|\alpha_1| = 2m} \sum\limits_{|\mu| \leq |\delta|} \binom{\delta}{\mu} D_x^\mu \big(a_{(\alpha_1, 0)}(y,z) - a_{(\alpha_1, 0)}(0,z)\big) D_x^{\delta - \mu}(D_y^{\alpha_1} u)|$

$\quad \leq c \big[\sum\limits_{\substack{|\alpha_1| = 2m \\ }} \sum\limits_{\substack{|\mu| \leq |\delta| \\ |\mu_1| = 0}} \binom{\delta}{\mu} |y| \, |D_x^{\delta - \mu} D_y^{\alpha_1} u| + \sum\limits_{|\alpha_1| = 2m} \sum\limits_{1 \leq |\mu| \leq |\delta|} \binom{\delta}{\mu} |D_x^{\delta - \mu} D_y^{\alpha_1} u| \big]$

where $D_x^\mu = D_y^{\mu_1} D_z^{\mu_2}$. Since a function v with bounded support satisfies

$$\|v; \, V^{\ell, p}(D, \, \kappa_1(\cdot) + 1\| \leq c \|v; \, V^{\ell, p}(D, \, \kappa(\cdot))\|$$

and $\qquad \|v; \, V^{\ell - 1, p}(D, \, \kappa_1(\cdot))\| \leq c \|v; \, V^{\ell, p}(D, \, \kappa(\cdot))\|$,

we get the assertion for $F(y,z)$. Analogously we can derive the result for $G_j^\pm(y,z)$, $j = 1, \ldots, m$.

We see that the crucial point is to clarify whether the solution u is differentiable in the tangential direction. V. G. MAZ'JA, J. ROSSMANN [2] have studied this problem in the general case, V. A. KONDRAT'EV [3] has investigated it for the Dirichlet problem for the differential operators of the second order. The result of V. G. MAZ'JA, J. ROSSMANN [2] reads as follows.

(iii) <u>THEOREM</u>. *Let* $u \in V^{\ell + 2m, p}(D, \kappa(\cdot))$ *be a solution of* (22.7) *with* $u \equiv 0$ *for* $|x| > R_0$, *where* $\kappa(\cdot) = \beta(\cdot) + 1 - \dfrac{2}{p} + \ell$ *is a smooth function, defined for* $z \in M$, $\ell \geq 0$ *is an integer,* $1 < p < \infty$. *Assume that the conditions* (25.5) *and* (25.6) *are satisfied for all* $z \in M \cap B_{R_0}$ *and that the*

right hand sides of (22.7) satisfy $f \in V^{\ell,p}(D, \kappa(\cdot)) \cap V^{\ell_1,p_1}(D, \kappa_1(\cdot))$ *and*
$g_j^\pm \in V^{2m+\ell-m_j^\pm-1/p,\, p}(\Gamma^\pm,\kappa(\cdot)) \cap V^{2m+\ell_1-m_j^\pm-1/p_1,\, p_1}(\Gamma^\pm,\kappa_1(\cdot))$, $j = 1,2,\ldots,m$.
Assume that $\kappa_1(\cdot) = \beta_1(\cdot) + 1 - 2/p_1 + \ell_1 \in C^\infty(M)$ *and*

$$(30.3) \qquad \mathrm{Im}\,\lambda_-(z) - 1 < \beta_1(z) + 1 - 2m < \mathrm{Im}\,\lambda_+(z) \quad \text{for every } z \in M \cap B_{R_0}$$

$(\lambda_-(z)$ *and* $\lambda_+(z)$ *are defined in 25.1).* *Then* $\dfrac{\partial u}{\partial z_i} \in V^{\ell_1+2m-1,\,p_1}(D,\kappa_1(\cdot))$ *for*
$i = 1,2,\ldots,N-2$ *and*

$$
(30.4) \qquad
\begin{aligned}
&\left\| \frac{\partial u}{\partial z_i}; V^{\ell_1+2m-1,\,p_1}(D,\kappa_1(\cdot)) \right\| \leq c\left[\left\| f; V^{\ell_1,p_1}(D, \kappa_1(\cdot)) \right\| \right.\\
&\quad + \sum_{j=1}^{m} \left\| g_j^\pm; V^{\ell_1+2m-m_j^\pm-1/p_1,\, p_1}(\Gamma^\pm,\kappa_1(\cdot)) \right\| + \left. \left\| u; V^{\ell+2m,p}(D,\kappa(\cdot)) \right\| \right] .
\end{aligned}
$$

(iv) <u>COROLLARY</u>. *Let the assumptions of the above Theorem (iii) be satis-fied. Then the right hand side* $F(y,z)$ *of (30.1) is contained in*
$V^{\ell_1,p_1}(D,\kappa_1(\cdot))$ *and the right hand sides* $G_j^\pm(y,z)$, $j = 1,\ldots,m$, *of (30.2)*
are contained in $V^{\ell_1+2m-m_j^\pm-1/p_1,\, p_1}(\Gamma^\pm,\kappa_1(\cdot))$.

P r o o f : The regularity result 25.3 (i) implies that
$u \in V^{\ell_1+2m,p_1}(D, \kappa_1(\cdot) + 1)$. Using the above Lemma (ii) we obtain the assertion.

30.2. <u>EXPANSIONS WITHOUT ADDITIONAL TANGENTIAL SMOOTHNESS CONDITIONS.</u>

(i) <u>Introducing remarks.</u> We consider again the boundary value problem
(22.7) and assume that the solution satisfies $u \in V^{\ell+2m,p}(D,\kappa(\cdot))$ while the
right hand sides satisfy $f \in V^{\ell_1,p_1}(D,\kappa_1(\cdot))$ and
$g_j^\pm \in V^{\ell_1+2m-m_j^\pm-1/p_1,\, p_1}(\Gamma^\pm,\kappa_1(\cdot))$. If $\kappa(\cdot) = \beta(\cdot) + 1 - \dfrac{2}{p} + \ell$ and $\kappa_1(\cdot)$
$= \beta_1(\cdot) + 1 - \dfrac{2}{p_1} + \ell_1$ and if (30.3) holds, then in general
$u \notin V^{\ell_1+2m,p_1}(D,\kappa_1(\cdot))$.

Our goal is to derive an asymptotic expansion of the solution u as a sum
of "singular" terms and a "regular" term from $V^{\ell_1+2m,p_1}(D,\kappa_1(\cdot))$. We want to
use the results for the boundary value problems in domains with conical points.
Therefore we start from the problem (22.7) in the form (30.1) and (30.2). As-
sume that $F \in V^{\ell_1,p_1}(D,\kappa_1(\cdot))$ and $G_j^\pm \in V^{\ell_1+2m-m_j^\pm-1/p_1,\, p_1}(\Gamma^\pm,\kappa_1(\cdot))$. Fixing
a point $z = z_0 \in M \cap B_{R_0}$ we obtain the following plane boundary value problems
for a.e. $z_0 \in M \cap B_{R_0}$:

$$(30.5) \qquad A_0(0,z_0,D_y,0)u = F(y,z_0) \quad \text{in } K_{z_0} = K \times \{z_0\} ,$$

$$(30.6) \qquad B_{j,0}^\pm(0,z_0,D_y,0)u = G_{j,0}^\pm(y,z_0) \quad \text{on } \gamma_{z_0}^\pm = \gamma^\pm \times \{z_0\} ,$$

where $F(\cdot,z_0) \in V^{\ell_1,p_1}(K_{z_0},\kappa_1(z_0))$, $G_j^\pm(\cdot,z_0) \in V^{\ell_1+2m-m_j^\pm-1/p_1,\,p_1}(\gamma_{z_0}^\pm,\kappa_1(z_0))$.

Theorem 7.4 implies that

$$(30.7) \qquad u(y,z_0) = \bar{u}(r,\omega,z_0) = \sum_{\gamma \in I} c_\gamma(z_0) u_\gamma(r,\omega,z_0) + w(r,\omega,z_0)$$

where $w(\cdot,z_0) \in V^{\ell_1+2m,p_1}(K_{z_0},\kappa_1(z_0))$ and $h_1(z_0) = \kappa_1(z_0) + \frac{2}{p_1} - \ell_1 - 2m$

$= \beta_1(z_0) + 1 - 2m < h(z_0) = \kappa(z_0) + \frac{2}{p} - \ell - 2m = \beta(z_0) + 1 - 2m$ provided no

eigenvalues of $\mathfrak{A}_0(z_0,\lambda)$ are situated on the line $\mathrm{Im}\,\lambda(z_0) = h_1(z_0)$.

The following questions immediately arise : *Are the coefficients* $c_\gamma = c_\gamma(z)$ *sufficiently smooth and does* $w = w(y,z) \in V^{2m+\ell_1,p_1}(D,\kappa_1(\cdot))$ *hold?*
The answers are given by V. G. MAZ'JA, B. A. PLAMENEVSKIĬ [1] provided the
right hand sides and the solution of (22.7) are smooth in the tangential direction, that means, with respect to $z \in M$. If we do not require this smoothness
we have to consider an expansion similar to (30.7) with coefficients $\hat{c}_\gamma = \hat{c}_\gamma(x)$
instead of $c_\gamma = c_\gamma(z)$. The last problem was studied by V. A. KONDRAT'EV [4]
and V. A. NIKISHKIN [1] for the Dirichlet problem for the second order equations and by V. G. MAZ'JA, J. ROSSMANN [1], [2] for elliptic boundary value
problems of higher orders. Let us present the result of V. G. MAZ'JA, J. ROSS-
MANN [2].

(ii) __THEOREM.__ *Let the assumptions of Theorem (iii) of 30.1 be satisfied,
let the eigenvalues* $\lambda_\mu(z)$ *of* $\mathfrak{A}(z,\lambda)$ *in the strip* $h_1(z) = \kappa_1(z) - \ell_1 - 2m$
$+ \frac{2}{p_1} < \mathrm{Im}\,\lambda_\mu(z) < \kappa(z) - \ell - 2m + \frac{2}{p} = h(z)$ *have a constant multiplicity for*
$z \in B_{R_0} \cap M$ *and let* $\lambda_\mu(z) \neq \lambda_\nu(z)$ *for* $\mu \neq \nu$. *If no eigenvalues of* $\mathfrak{A}(z,\lambda)$
are situated on the line $h_1(z) = \mathrm{Im}\,\lambda(z)$ *then*

$$(30.8) \qquad u(x) = \sum_{\gamma \in I} \hat{c}_\gamma(x) u_\gamma(x) + w(x) ,$$

where $u_\gamma(x) = \bar{u}_\gamma(r,\omega,z)$ *is defined by* (7.9) *for* $z \in M \cap B_{R_0}$,
$\hat{c}_\gamma \in \bigcap_{\nu \geq 1} W^{\nu,p_1}(D,d_M, \kappa_1(\cdot) - \ell_1 - 2m - \mathrm{Im}\,\lambda_\mu(\cdot) + \nu + \varepsilon)$, $\varepsilon > 0$ *small, and*
$w \in V^{\ell_1+2m,p_1}(D,\kappa_1(\cdot) + \varepsilon)$.

If the eigenvalues are simple, then $\varepsilon = 0$.

P r o o f : We can restrict ourselves to $p = p_1$, $\ell = \ell_1$, $\beta(z) - \beta_1(z) < 1$
as a result of 25.3 (i). Further, we prove this theorem only if $N = 3$ and if
the eigenvalues are simple. That means, we will prove the expansion (cf. (7.4))

$$(30.9) \qquad u(x) = \bar{u}(r,\omega,z) = \sum_{\mu=1}^{N_0} \hat{c}_\mu(x)\, r^{i\lambda_\mu(z)} \phi_\mu(\omega,z) + w ,$$

where $w \in V^{\ell+2m,p}(D,\kappa_1(\cdot))$ and $\phi_\mu(\omega,z) = u_\mu^{0,1}(\omega,z)$.

First step. Let us start with the expansion (30.7), namely,

$$(30.10) \qquad u(y,z_0) = \sum_{\mu=1}^{N_0} c_\mu(z_0) \, r^{i\lambda_\mu(z_0)} \phi_\mu(\omega,z_0) + w(y,z_0) \; .$$

The estimate (7.11) yields

$$(30.11) \qquad \|w(\cdot,z_0); V^{\ell+2m,P}(K_{z_0},\kappa_1(z_0))\|^P \leq c_1(z_0)\Big(\|F(\cdot,z_0); V^{\ell,P}(K_{z_0},\kappa_1(z_0))\|^P$$

$$+ \sum_{j=1}^{m} \|G_j^\pm(\cdot,z_0); V^{\ell+2m-m_j^\pm-1/p,p}(\gamma_{z_0}^\pm,\kappa_1(z_0))\|^P\Big)$$

$$\leq c_2(z_0)\Big[\|f(\cdot,z_0); V^{\ell,P}(K_{z_0},\kappa_1(z_0))\|^P$$

$$+ \Big\|\tfrac{\partial u}{\partial z}(\cdot,z_0); V^{\ell+2m-1,P}(K_{z_0},\kappa_1(z_0))\Big\|^P + \|u(\cdot,z_0); V^{\ell+2m,P}(K_{z_0},\kappa(z_0))\|^P$$

$$+ \sum_{j=1}^{m} \|g_j^\pm(\cdot,z_0); V^{\ell+2m-m_j^\pm-1/p,p}(\gamma_{z_0}^\pm,\kappa_1(z_0))\|^P\Big] \; .$$

If the support of u is sufficiently small (otherwise we take a suitable partition of unity) we can replace $c_1(z_0)$ and $c_2(z_0)$ by a constant which is independent of z_0 (this follows from the fact that the norm of the operator $\mathcal{U}_0(0,z,D_y,0) - \mathcal{U}_0(0,z_0,D_y,0)$ is small for $z, z_0 \in B_{R_0} \cap M$). Multiplying (30.10) by $\psi_\nu(\omega,z_0)r^{-i\lambda_\nu(z_0)}$, $\nu = 1,2,\ldots,N_0$, where $\displaystyle\int_0^{\omega_0} \phi_\mu(\omega,z_0)\psi_\nu(\omega,z_0)d\omega = \delta_{\nu\mu}$ ($\delta_{\nu\mu}$ the Kronecker symbol), and integrating with respect to ω we obtain

$$(30.12) \qquad v_\nu(r,z_0) = c_\nu(z_0) + w_\nu(r,z_0) \; , \quad \nu = 1,2,\ldots,N_0 \; ,$$

where

$$v_\nu(r,z_0) = \int_0^{\omega_0} u(r,\omega,z_0) \, \psi_\nu(\omega,z_0) \, r^{-i\lambda_\nu(z_0)} \, d\omega$$

$$w_\nu(r,z_0) = \int_0^{\omega_0} w(r,\omega,z_0) \, \psi_\nu(\omega,z_0) \, r^{-i\lambda_\nu(z_0)} \, d\omega \; .$$

Second step. We show that $v_\nu \in W^{1,P}(D, d_M, \kappa_1(\cdot) - \ell - 2m - \operatorname{Im}\lambda_\nu(\cdot) + 1)$ and that c_ν are their traces on M . We have

$$(30.13) \qquad \int_D r^{p(\kappa_1(z) - \operatorname{Im}\lambda_\nu(z) - \ell - 2m + 1)} |v_\nu(r,z)|^P \, dx$$

$$\leq c \int_D r^{p(\kappa_1(z) - \ell - 2m + 1)} |u(x)|^P \, dx \leq c\|u; V^{\ell+2m,P}(D,\kappa(\cdot))\|^P < \infty \; ,$$

and since $\dfrac{\partial v_\nu}{\partial r} = \dfrac{\partial w_\nu}{\partial r}$, (see (30.12)),

122

$$(30.14) \quad \int_D r^{p(\kappa_1(z) - \operatorname{Im} \lambda_\nu(z) - \ell - 2m + 1)} \left| \frac{\partial v_\nu(r,z)}{\partial r} \right|^p dx$$

$$\leq c \left[\int_D r^{p(\kappa_1(z) - \ell - 2m)} |w|^p \, dx + \int_D r^{p(\kappa_1(z) - \ell - 2m + 1)} \left| \frac{\partial w}{\partial r} \right|^p dx \right]$$

$$\leq c \| w; \ V^{\ell+2m,p}(D, \kappa_1(\cdot)) \|^p \leq c \big[\| f; \ V^{\ell+2m,p}(D, \kappa_1(\cdot)) \|^p$$

$$+ \sum_{j=1}^m \| g_j^\pm; \ V^{\ell+2m-m_j^\pm - 1/p, \ p}(\Gamma^\pm, \kappa_1(\cdot)) \|^p + \| u; \ V^{\ell+2m,p}(D, \kappa(\cdot)) \|^p$$

$$+ \left\| \frac{\partial u}{\partial z}; \ V^{\ell+2m-1,p}(D, \kappa_1(\cdot)) \right\|^p \big] < \infty \ .$$

We have used the estimate (30.11), integrating it with respect to z . Further, we have

$$(30.15) \quad \int_D r^{p(\kappa_1(z) - \operatorname{Im} \lambda_\nu(z) - \ell - 2m + 1)} \left| \frac{\partial v_\nu(r,z)}{\partial z} \right|^p dx$$

$$\leq c \left(\left\| \frac{\partial u}{\partial z}; \ V^{\ell+2m-1,p}(D, \kappa_1(\cdot)) \right\|^p + \| u; \ V^{\ell+2m,p}(D, \kappa(\cdot)) \|^p \right) < \infty \ .$$

The estimates (30.13), (30.14) and (30.15) imply that $v_\nu \in W^{1,p}(D, d_M, \kappa_1(\cdot) - \ell - 2m - \operatorname{Im} \lambda_\nu(\cdot) + 1)$. Lemma (iii) of 29.2 implies that the traces of $v_\nu \in W^{1,p}(D, d_M, \kappa_1(\cdot) - \ell - 2m + 1 - \operatorname{Im} \lambda_\nu(\cdot))$ exist. Since

$$\int_D r^{p(\kappa_1(z) - \operatorname{Im} \lambda_\nu(z) - \ell - 2m)} |v_\nu(r,z) - c_\nu(z)|^p \, dx$$

$$= \int_D r^{p(\kappa_1(z) - \operatorname{Im} \lambda_\nu(z) - \ell - 2m)} |w_\nu(r,z)|^p \, dx$$

$$\leq c \int_D r^{p(\kappa_1(z) - \ell - 2m)} |w|^p \, dx < \infty$$

(we have used (30.11) again) the traces of v_ν are

$$c_\nu \in B^{-\kappa_1(\cdot) + \ell + 2m + \operatorname{Im} \lambda_\nu(\cdot) - 2/p, \ p}(M) \ .$$

<u>Third step</u>. It follows from (vi) of 29.2 that there is an extension $\hat{c}_\nu \in \bigcap_{k \geq 1} W^{k,p}(D, \ d_M, \ \kappa_1(\cdot) - \ell - 2m - \operatorname{Im} \lambda_\nu(\cdot) + k)$ of c_ν with

$$(30.16) \quad \int_D r^{p(\kappa_1(z) - \ell - 2m - \operatorname{Im} \lambda_\nu(z))} |\hat{c}_\nu(x) - c_\nu(z)|^p \, dx < \infty \ .$$

We write the expansion (30.10) in the following manner :

$$u(y,z) = \sum_{\mu=1}^{N_0} \hat{c}_\mu(x) \, r^{i\lambda_\mu(z)} \, \phi_\mu(\omega,z) + w(y,z) -$$

$$- \sum_{\mu=1}^{N_0} \left(\hat{c}_\mu(x) - c_\mu(z) \right) r^{i\lambda_\mu(z)} \, \phi_\mu(\omega,z)$$

$$= \sum_{\mu=1}^{N_0} \hat{c}_\mu(x) \, r^{i\lambda_\mu(z)} \, \phi_\mu(\omega,z) + w_0(y,z) \; .$$

We indicate the proof of the inclusion $w_0 \in V^{\ell+2m,P}(D, \kappa_1(\cdot))$. From (30.11) and (30.16) we obtain, that $w_0 \in V^{0,P}(D, \kappa_1(\cdot) - \ell - 2m)$. A lengthy calculation shows that $Aw_0 \in V^{\ell,P}(D, \kappa_1(\cdot))$ and $B_j^\pm w_0 \in V^{\ell+2m-m_j^\pm-1/p, \, P}(\Gamma^\pm, \kappa_1(\cdot))$, $j = 1,2,\ldots,m$. Lemma (i) of 25.3 yields that $w_0 \in V^{\ell+2m,P}(D, \kappa_1(\cdot))$.

30.3. EXPANSIONS WITH ADDITIONAL TANGENTIAL SMOOTHNESS CONDITIONS.

(i) We will investigate the influence of the smoothness in the z-dimension of the right hand sides of (22.7) on the smoothness of the coefficients $\hat{c}_\gamma$ appearing in (30.8). Roughly speaking, the connection between them can be expressed as follows : If

$$(30.17) \qquad D_z^\alpha f \in V^{\ell_1,P_1}(D, \kappa_1(\cdot) + \varepsilon) \quad \text{and}$$

$$D_z^\alpha g_j \in V^{\ell_1+2m-m_j^\pm-1/P_1, \, P_1}(\Gamma^\pm, \kappa_1(\cdot) + \varepsilon) \quad \text{for} \quad |\alpha| \leq s \; ,$$

$$(30.18) \quad D_z^\alpha u \in V^{\ell_1+2m,P_1}(D, \kappa_1(\cdot) + 1 + \varepsilon) \quad \text{for} \quad |\alpha| \leq s \; ,$$

$$(30.19) \quad D_z^\alpha u \in V^{\ell_1+2m-1,P_1}(D, \kappa_1(\cdot) + \varepsilon) \quad \text{for} \quad |\alpha| = s + 1 \; ,$$

then the traces c_γ of $\hat{c}_\gamma$ are from $B^{-\kappa_1(\cdot)+\ell_1+2m+ \, \mathrm{Im} \, \lambda_\mu(\cdot) \, +s \, -2/P_1 \, -\varepsilon, P_1}(M)$ ($\varepsilon > 0$ is a small real number). If the assumptions of Theorem (ii) from 30.2 are satisfied, then the conditions (30.18) and (30.19) follow from (30.17).

(ii) <u>REMARK</u>. Let us start with a remark to the idea of the proof of the above mentioned result on the traces c_γ . As in the proof of Theorem (ii) from 30.2 we can restrict ourselves to the case $\ell = \ell_1$, $p = p_1$ and $\kappa_1(z) + 1 > \kappa(z) > \kappa_1(z)$. The following conclusion was important in the proof of Theorem (ii) from 30.2 : Since (30.13), (30.14) and (30.15) hold, the function v_ν , introduced in formula (30.12), is contained in $W^{1,P}(D,d_M, \kappa_1(\cdot) - \ell - 2m - \mathrm{Im} \, \lambda_\nu(\cdot) + 1)$. If (30.17), (30.18) and (30.19) are satisfied, we cannot derive an analogous result, that is, $v_\nu \in W^{s+1,P}(D,d_M, \kappa_1(\cdot) - \ell - 2m - \mathrm{Im} \, \lambda_\nu(\cdot) + 1)$ for $s > 0$. Therefore we have to proceed in another way, namely, to use a result which was proved by V. G. MAZ'JA, J. ROSSMANN [1] by induction with respect to s .

(iii) <u>LEMMA</u>. *Let* $u \in V^{\ell_1+2m,P_1}(D, \kappa_1(\cdot) + 1)$ *be a solution of problem*

(22.7) *with bounded support* $(\ell \geq 0)$. *Let the conditions* (30.17), (30.18) *and*
(30.19) *and the assumptions of Theorem* (ii) *from* 30.2 *for the eigenvalues of*
$\mathcal{U}_0(z,\lambda)$ *be satisfied. Then we have, for* $|\alpha| \leq s$,

$$(30.20) \qquad D_z^\alpha u = \sum_{\gamma \in I} \sum_{\alpha' \leq \alpha} \binom{\alpha}{\alpha'} D_z^{\alpha'} \hat{c}_\gamma(x) \, D_z^{\alpha-\alpha'} u_\gamma + D_z^\alpha w(x)$$

where $D_z^\alpha w \in V^{\ell_1+2m, p_1}(D, \kappa_1(\cdot) + \varepsilon)$, *and* $\hat{c}_\gamma \in W^{s,p_1}(D, d_M, \kappa_1(\cdot) - \ell_1 - 2m$

$- \operatorname{Im} \lambda_\mu(\cdot) + \varepsilon)$ *are the extensions of the functions*

$c_\gamma \in B^{-\kappa_1(\cdot) + \ell_1 + 2m + \operatorname{Im} \lambda_\mu(\cdot) + s - 2/p_1 - \varepsilon, \, p_1}(M)$.

    (iv) <u>COROLLARY</u>. *Assume that the condition* (30.17) *is satisfied for*
$s = \ell_1 + 2m$ *and let the assumptions of Theorem* (ii) *of* 30.2 *be fulfilled.*
Then the following expansion holds :

$$(30.21) \qquad u(x) = \bar{u}(r,\omega,z) = \sum_{\gamma \in I} c_\gamma(z) u_\gamma(r,\omega,z) + \bar{w}_1(r,\omega,z)$$

where $c_\gamma \in B^{-\kappa_1(\cdot) + \operatorname{Im} \lambda_\mu(\cdot) + 2\ell_1 + 4m - 2/p_1 - \varepsilon, \, p_1}(M)$ *and*

$w_1 \in V^{\ell_1+2m, p_1}(D, \kappa_1(\cdot) + \varepsilon)$. $(\varepsilon > 0$ is a small real number. If $\mathcal{U}_0(\lambda,z)$
has only simple eigenvalues then $\varepsilon = 0$.)

P r o o f : We use formula (30.20) setting $\alpha = (0,0,\ldots,0)$ and $\hat{c}_\gamma = \hat{c}_\gamma - c_\gamma +$
$+ c_\gamma$. One can calculate (see V. G. MAZ'JA, J. ROSSMANN [1]) that $w_1 = w +$

$$+ \sum_{\gamma \in I} (\hat{c}_\gamma - c_\gamma) u_\gamma \in V^{\ell_1+2m, p_1}(D, \kappa_1(\cdot) + \varepsilon) .$$

    (v) <u>REMARK</u>. Lemma (iii) from (30.3) yields the possibility to improve
the expansion (30.8), that is, to add more singular functions and to get a
smoother remainder w , provided the right hand sides of (22.7) are smooth
enough. We refer to V. G. MAZ'JA, J. ROSSMANN [1], [2] in this connection.

<u>30.4. EXAMPLE</u>. (i) We consider the mixed boundary value problem

$$(30.22) \qquad - \Delta u = f \ \text{ in } \ D , \quad u|_{\Gamma^-} = 0 , \quad \frac{\partial u}{\partial n}\Big|_{\Gamma^+} = 0$$

where $f \in L^2(D)$ and has bounded support. Let $u \in V^{2,2}(D,1)$ be a solution of
(30.22) with bounded support.

    (ii) Let us verify the assumptions of Theorem (ii) from 30.2.

- The functions $\kappa(z) = \beta(z) + 1 - \frac{2}{p} + \ell = 1 + 1 - 1 + 0 = 1$ and $\kappa_1(z) =$
  $= \beta_1(z) + 1 - 2/p_1 + \ell_1 = 0 + 1 - 1 + 0 = 0$ are smooth functions defined
  on M .
- The condition (25.5) is valid, since the eigenvalues of $\mathcal{U}_0(z,\lambda)$ have
  the form $\lambda_\mu(z) = \lambda_\mu = i(\mu \pm \tfrac{1}{2})\dfrac{\pi}{\omega_0}$, $\mu = 0, \pm 1, \pm 2, \ldots$ (cf.2.3) and therefore

no eigenvalue is situated on the line $\mathrm{Im}\ \lambda = \beta(z) + 1 - 2m = 0$.

- The condition (25.6) is satisfied, since the boundary value problem is coercive in a bounded domain (cf. § 28).

- $f \in V^{0,2}(D,1) \cap V^{0,2}(D,0)$.

- The condition (30.3) is valid : $\mathrm{Im}\ \lambda_-(z) - 1 = -\frac{1}{2}\frac{\pi}{\omega_0} - 1 < - 1 < \frac{1}{2}\frac{\pi}{\omega_0} = \mathrm{Im}\ \lambda_+(z)$.

- The eigenvalues are simple and $\lambda_\mu(z) \neq \lambda_\nu(z)$ for $\mu \neq \nu$.

- If $\omega_0(z) = \omega_0 \notin \{\frac{\pi}{2}, \frac{3}{2}\pi\}$, then no eigenvalue of $\mathcal{Q}_0(z,\lambda)$ is situated on the line $\mathrm{Im}\ \lambda = h_1(z) = - 1$.

(iii) Theorem (ii) from 30.2 and Corollary (iv) from 30.3 yield the expansion of the solution u : If $\omega_0 \neq \frac{\pi}{2}$ and $\omega_0 \neq \frac{3}{2}\pi$ and $f \in L^2(D)$ then

$$(30.23) \qquad u(x) = \sum_{-1 < \mathrm{Im}\ \lambda_\mu < 0} \hat{c}_\mu(x)\ r^{i\lambda_\mu}\ \phi_\mu(\omega) + w(x)$$

$$= \sum_{-1 < \mathrm{Im}\ \lambda_\mu < 0} \hat{c}_\mu(x)\ r^{i\lambda_\mu}\ \cos(i\lambda_\mu)\omega + w(x) ,$$

where $w \in V^{2,2}(D,0)$ and $\hat{c}_\mu \in W^{2,2}(D, d_M, -\mathrm{Im}\ \lambda_\mu)$. If $\omega_0 \neq \frac{\pi}{2}$ and $\omega_0 \neq \frac{3}{2}\pi$ and $D_z^\alpha f \in L^2(D)$ for $|\alpha| \leq 2$ then

$$(30.24) \qquad u(x) = \sum_{-1 < \mathrm{Im}\ \lambda_\mu < 0} c_\mu(z)\ r^{i\lambda_\mu}\ \cos(i\lambda_\mu)\omega + w_1(x)$$

where $c_\gamma \in B^{\mathrm{Im}\ \lambda_\mu(\cdot) + 3,\ 2}(M)$ and $w_1 \in V^{2,2}(D,0)$.

§ 31 . E x p a n s i o n s   i n   a   b o u n d e d   d o m a i n

31.1. REMARKS. We are now concerned with problem (22.6) with the corresponding operator $\mathcal{Q}(x,D_x)$ defined by (22.5). We resume the considerations of § 26 : We have formulated the conditions (26.4) and (26.5) which play an important role in connection with the solvability and regularity of our boundary value problem. If they are satisfied and $\mathrm{Im}\ \lambda_-(\zeta) < \beta_1(\zeta) + 1 - 2m < \mathrm{Im}\ \lambda_+(\zeta)$, that means no eigenvalues of $\mathcal{Q}(\zeta,\lambda)$ are situated in the strip

$$(31.1) \qquad h_1(\zeta) = \beta_1(\zeta) + 1 - 2m < \mathrm{Im}\ \lambda < \beta(\zeta) + 1 - 2m = h(\zeta) , \quad \zeta \in M ,$$

then the assertion of the regularity theorem 26.3 is valid. We now investigate the structure of the solution, if there are eigenvalues of $\mathcal{Q}_0(\zeta,\lambda)$ situated in the strip (31.1). The following result describes this situation (cf. V. G. MAZ'JA, J. ROSSMANN [1], [2]).

31.2. THEOREM. *Let* $u \in V^{\ell+2m,p}(\Omega,\kappa(\cdot))$ *be a solution of the boundary value problem (22.6), where* $\kappa(\cdot) = \beta(\cdot) + \ell + 1 - 2/p$ *is a smooth function, defined on the edge set* M *of* $\partial\Omega$. *Assume that* $f \in V^{\ell_1,p_1}(\Omega,\kappa_1(\cdot))$ *and* $g_j^{(q)} \in V^{\ell_1+2m-mq_j\ -1/p_1,\ p_1}(\Gamma_q,\kappa_1(\cdot))$, $q = 1,\ldots,T$, *where*

$\kappa_1(\cdot) = \beta_1(\cdot) + \ell_1 + 1 - 2/p_1 \in C^\infty(M)$ *and*

(31.2) $\mathrm{Im}\, \lambda_-(\zeta) - 1 < \beta_1(\zeta) + 1 - 2m < \mathrm{Im}\, \lambda_+(\zeta)$ *for every* $\zeta \in M$.

Further, assume that the conditions (26.4) and (26.5) are satisfied and that the eigenvalues $\lambda_\mu(\zeta)$ *of* $\mathfrak{A}_0(\zeta,\lambda)$ *from the strip*

(31.3) $h_1(\zeta) = \beta_1(\zeta) + 1 - 2m < \mathrm{Im}\, \lambda < \beta(\zeta) + 1 - 2m = h(\zeta)$

have a constant multiplicity, $\lambda_\mu(\zeta) \neq \lambda_\nu(\zeta)$ *for* $\zeta \in M$ *and* $\mu \neq \nu$ *and no eigenvalue of* $\mathfrak{A}_0(\zeta,\lambda)$ *lies on the lines* $\mathrm{Im}\, \lambda = h_1(\zeta)$ *and* $\mathrm{Im}\, \lambda = h(\zeta)$. *Then*

(31.4) $u(x) = \sum\limits_{\gamma \in I} \eta(x)\, \hat{c}_\gamma(x)\, u_\gamma(x) + w(x)$

where $w \in V^{\ell_1+2m,p_1}(\Omega, \kappa_1(\cdot) + \varepsilon)$, $\hat{c}_\gamma \in \bigcap\limits_{\nu \geq 1} W^{\nu,p_1}(\Omega, d_M, \kappa_1(\cdot) - \ell_1 - 2m$

$- \mathrm{Im}\, \lambda_\mu(\cdot) + \nu + \varepsilon)$, $\eta \in C_0^\infty(R^N)$, $\eta(x) \equiv 1$ *if* $\mathrm{dist}(x,M) < \frac{\delta}{2}$, $\eta(x) \equiv 0$

if $\mathrm{dist}(x,M) > \delta$, $\varepsilon > 0$ *and* $\delta > 0$ *being small real numbers.* ($u_\gamma(x)$ *is defined by (7.9) for every* $x \in \mathrm{supp}\, \eta$.)

P r o o f :   (i) We consider a sufficiently fine covering $\Omega \subset \bigcup\limits_{i=1}^{T_2} U_i$ and a corresponding partition of unity $\sum\limits_{i=1}^{T_2} \eta_i = 1$ as in 26.1. Assume again that $U_i \cap M \neq \emptyset$ for $i = 1,\dots,T_1$ and, for simplicity, that Ω has only one edge. Let us take an element U_{i_0} of the covering with $U_{i_0} \cap M \neq \emptyset$ and a point $\zeta_0 \in M \cap U_{i_0}$. Similarly to 26.1 (ii) we consider the dihedral angle D_{ζ_0} with the tangential planes at ζ_0 as faces. We map $U_{i_0} \cap \bar{\Omega}$ diffeomorphically into D_{ζ_0} , taking into account the local coordinates introduced in 26.1 (ii). We write the operator $\mathfrak{A}(x,D_x)$ in the new coordinates $\tilde{x} = (\tilde{x}_1,\dots,\tilde{x}_N) =$ $= (\tilde{y}_1,\tilde{y}_2,\zeta_1,\dots,\zeta_{N-2})$. Notice that the different edge angles of points from $M \cap U_{i_0}$ are preserved in the new variable coefficients, and that the conditions (26.4) and (26.5) are in accordance with the conditions (25.5) and (25.6) for the transformed operator.

     (ii) In this way we obtain a new boundary value problem for the function $\eta_{i_0}(x)u(x) = u_0(x) = \tilde{u}_0(\tilde{x})$ in the dihedral angle D_{ζ_0} , namely,

(31.5)
$$\tilde{A}(\tilde{x},D_{\tilde{x}})\tilde{u}_0 = \tilde{f} \quad \text{in} \ D_{\zeta_0} ,$$
$$\tilde{B}_j^\pm(\tilde{x},D_{\tilde{x}})\tilde{u}_0 = \tilde{g}_j^\pm \quad \text{on} \ \Gamma_{\zeta_0}^\pm , \quad j = 1,2,\dots,m ,$$

where $\tilde{f} \in V^{\ell_1,p_1}(D_{\zeta_0},\tilde{\kappa}_1(\cdot))$, $\tilde{g}_j^\pm \in V^{\ell_1+2m-m_j^\pm-1/p_1,\, p_1}(\Gamma_{\zeta_0}^\pm,\tilde{\kappa}_1(\cdot))$. Theorem (ii) from 30.2 yields that

(31.6) $u_0(x) = \tilde{u}_0(\tilde{x}) = \sum\limits_{\gamma \in I} \tilde{\hat{c}}_{\gamma,0}(\tilde{x})\, \tilde{u}_{\gamma,0}(\tilde{x}) + \tilde{w}_0(\tilde{x})$

$$=: \sum\limits_{\gamma \in I} \hat{c}_{\gamma,0}(x)\, u_{\gamma,0}(x) + w_0(x) ,$$

where the right hand sides are determined by the inverse coordinate transformations. We obtain that $\hat{c}_{\gamma,0} \in \bigcap\limits_{\nu \geq 1} W^{\nu,p_1}(\Omega, d_M, \kappa_1(\cdot) - \ell_1 - 2m - \operatorname{Im} \lambda_\mu(\cdot) + \nu + \varepsilon)$ and $w_0 \in V^{\ell_1+2m,p_1}(\Omega, \kappa_1(\cdot) + \varepsilon)$.

(iii) We choose the function η in such a way that
$$\eta \eta_i = \eta_i \quad \text{for} \quad i = 1,\ldots,T_1 \ .$$
Then we obtain, in virtue of (31.6), that
$$u(x) = \eta(x) \sum_{i=1}^{T_1} \eta_i(x)u(x) + \sum_{i=T_1+1}^{T_2} \eta_i(x)u(x)$$

$$= \eta(x) \sum_{i=1}^{T_1} \sum_{\gamma \in I} \hat{c}_{\gamma,i}(x)u_{\gamma,i}(x) + \sum_{i=1}^{T_1} w_i(x) + \sum_{i=T_1+1}^{T_2} \eta_i(x)u(x)$$

$$= \eta(x) \sum_{\gamma \in I} \hat{c}_\gamma(x)u_\gamma(x) + w(x)$$

where $\hat{c}_\gamma(x)u_\gamma(x) = \sum\limits_{i=1}^{T_1} \hat{c}_{\gamma,i}(x)u_{\gamma,i}(x)$. Notice that $I = I(x)$, but I is independent of i .

31.3. <u>REMARKS</u>. (i) The complete asymptotic expansion of the solution u can be found in V. G. MAZ'JA, J. ROSSMANN [1], [2].

(ii) If we assume in addition that the right hand sides of (22.6) are sufficiently smooth in the tangential direction then we obtain an expansion of the solution in which the coefficients depend only on the edge points.

§ 32 . E x a m p l e

32.1. <u>THE PROBLEM</u>. Let Ω be a bounded domain in R^3 with only one edge M and $\partial\Omega = M \cup \Gamma_1 \cup \Gamma_2$ (see Fig. 21). We consider the mixed boundary value problem

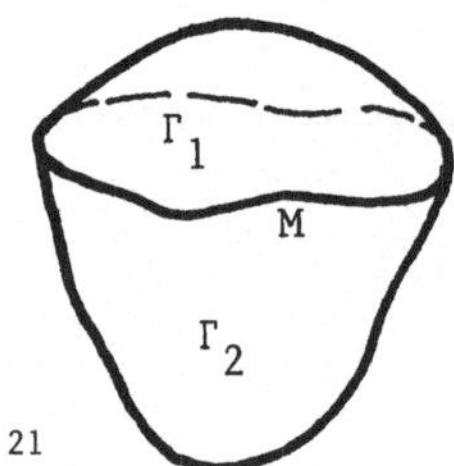

Fig. 21

$$-\Delta u = f \quad \text{in} \quad \Omega \ ,$$
$$(32.1)$$
$$u\big|_{\Gamma_1} = 0 \ , \quad \frac{\partial u}{\partial n}\Big|_{\Gamma_2} = 0 \ ,$$

where $f \in L^{p_1}(\Omega)$, $p_1 \geq 2$. We are interested in the structure of the weak solution u . Theorem 31.2 describes solutions from the weighted spaces

$V^{\ell+2m,p}(\Omega,\kappa(\cdot))$. The problem now is to clarify whether we can use this theorem also for a solution $u \in W^{1,2}(\Omega)$ and obtain an asymptotic expansion near the edge as in (31.4). The conditions (26.4) and (26.5) play an important role in the assumptions of Theorem 31.2. Let us start with these conditions.

32.2. **THE CONDITIONS (26.4) AND (26.5)**. (i) The Lax–Milgram Theorem yields
that for every $f \in \left[W^{1,2}_{\Gamma_1}(\Omega, d_{\Gamma_1}, 0) \right]^*$ there exists a uniquely determined weak
solution $u \in W^{1,2}_{\Gamma_1}(\Omega, d_{\Gamma_1}, 0) = V$. (For the definition of this space see (0.15).)

(ii) We have

$$V^{0,2}(\Omega, 1) \subset \left[W^{1,2}_{\Gamma_1}(\Omega, d_{\Gamma_1}, 0) \right]^* ,$$

because 0.12 yields

$$W^{1,2}_{\Gamma_1}(\Omega, d_{\Gamma_1}, 0) = H^{1,2}_{\Gamma_1}(\Omega, d_{\Gamma_1}, 0) \subset H^{0,2}_{\Gamma_1}(\Omega, d_{\Gamma_1}, -2)$$

and, on the other hand,

$$V^{0,2}(\Omega, 1) \subset H^{0,2}_{\Gamma_1}(\Omega, d_{\Gamma_1}, 2) \subset \left[H^{0,2}_{\Gamma_1}(\Omega, d_{\Gamma_1}, -2) \right]^* \subset \left[W^{1,2}_{\Gamma_1}(\Omega, d_{\Gamma_1}, 0) \right]^* .$$

(iii) Since $H^{1,2}_{\Gamma_1}(\Omega, d_{\Gamma_1}, 0) \subset V^{1,2}(\Omega, 0)$, it follows from (ii) that for
every $f \in V^{0,2}(\Omega, 1)$ there is a weak solution $u \in H^{1,2}_{\Gamma_1}(\Omega, d_{\Gamma_1}, 0) \subset V^{1,2}(\Omega, 0)$
of (32.1).

(iv) Furthermore, the solution u is contained in $\tilde{V}^{1,2}(\Omega, 0) \subset V^{1,2}(\Omega, 0)$
(for the definition see 0.9), because the norm of $\tilde{V}^{1,2}(\Omega, 0)$ is equivalent to
the norm

$$\| u; \, V^{1,2}(\Omega, 0) \| + \| \Delta u; \, V^{-1,2}(\Omega, 0) \| .$$

The last assertion was proved by J. ROSSMANN [1]. (The space $V^{-1,2}(\Omega, 0)$ was
defined in 0.8.)

(v) The following regularity result, not involving the assumptions (26.4)
and (26.5), is valid (see J. ROSSMANN [1]) : *Let* $u \in \tilde{V}^{1,2}(\Omega, 0)$ *be a solution
of the problem* (32.1) *and let* $f \in V^{0,2}(\Omega, 1)$. *Then* $u \in \tilde{V}^{2,2}(\Omega, 1) = V^{2,2}(\Omega, 1)$.

(vi) We now use the connection between the boundary value problems with
vanishing and nonvanishing boundary data. Namely, the boundary value problem
(32.1) is uniquely solvable in $V^{2,2}(\Omega, 1)$ for every $f \in V^{0,2}(\Omega, 1)$ if and
only if the problem

$$- \Delta u = f \quad \text{in} \quad \Omega \, ,$$

$$u|_{\Gamma_1} = g_1 \, , \quad \left. \frac{\partial u}{\partial n} \right|_{\Gamma_2} = g_2$$

is uniquely solvable for every $f \in V^{0,2}(\Omega, 1)$, $g_1 \in V^{3/2, \, 2}(\Gamma_1, 1)$,
$g_2 \in V^{1/2, \, 2}(\Gamma_2, 1)$. (See V. G. MAZ'JA, B. A. PLAMENEVSKIĬ [6], J. ROSSMANN [1].)

(vii) Now we can use the solvability theorem 26.2 and obtain finally that
the conditions (26.4) and (26.5) are satisfied for $\beta(\zeta) \equiv 1$ for all $\zeta \in M$.

32.3. **THE ASYMPTOTIC EXPANSION OF THE WEAK SOLUTION**. (i) We consider the
original problem (32.1), where $f \in L^{p_1}(\Omega)$, $p_1 \geq 2$. Since $L^{p_1}(\Omega) \subset V^{0,2}(\Omega, 1)$
it follows from 32.2 that there exists a uniquely determined solution

$u \in V^{2,2}(\Omega,1) \subset W^{1,2}(\Omega)$. Let us investigate the assumptions of Theorem 31.2 for $\ell = 0$, $p = 2$, $\kappa(\zeta) = \beta(\zeta) \equiv 1$ for every $\zeta \in M$, $\ell_1 = 0$ and $\kappa_1(\zeta) = \beta_1(\zeta) + 1 - 2/p_1 = 0$ for every $\zeta \in M$. We denote by $\omega_0(\zeta)$ the edge angle at $\zeta \in M$. We have $\operatorname{Im} \lambda_-(\zeta) = -\frac{1}{2}\frac{\pi}{\omega_0(\zeta)}$ (see 2.3) and assume that no eigenvalues of $\mathcal{U}_0(\zeta,\lambda)$ are situated on the line $\operatorname{Im} \lambda(\zeta) = \frac{2}{p_1} - 2$. Theorem 31.2 yields :

(ii) If $\frac{2}{p_1} - 2 < - \frac{\pi}{2\omega_0(\zeta)}$, then

$$(32.2) \qquad u(x) = \sum_{\substack{0 < (\mu+1/2)\,\pi/\omega_0(\zeta) < 2-2/p_1 \\ \mu = 0,1,2,\ldots}} \hat{c}_\mu(x)\, r^{(\mu+1/2)\pi/\omega_0(\zeta)}$$

$$\cdot \cos\left(\mu + \tfrac{1}{2}\right)\frac{\pi}{\omega_0(\zeta)} + w(x)$$

where $w \in V^{2,p_1}(\Omega,0)$, $\hat{c}_\mu \in W^{1,p_1}\!\left(\Omega,\ d_M,\ -1 + (\mu + \tfrac{1}{2})\frac{\pi}{\omega_0(\cdot)}\right)$. If $\frac{2}{p_1} - 2 > - \frac{\pi}{2\omega_0(\zeta)}$ then $u \in V^{2,p_1}(\Omega,0)$.

(iii) <u>REMARK</u>. If $f \in W^{2,p_1}(\Omega)$ then instead of the coefficients $\hat{c}_\mu$ we have in (32.2) the coefficients $c_\mu \in B^{-(\mu+1/2)\pi/\omega_0(\cdot)\,+\,4\,-2/p_1,\ p_1}(M)$.

Section 10 : *C a l c u l a t i o n   o f   t h e   c o e f f i c i e n t s*

Our goal is to calculate the coefficients $\hat{c}_\gamma(x)$ in (30.8) or $c_\gamma(z)$ in (30.21) and then the coefficients $\hat{c}_\gamma(x)$ in (31.4). The method is similar to that which we have used for boundary value problems in domains with conical points in Section 4.

§ 33 . T h e   c o e f f i c i e n t   f o r m u l a   i n
        a   d i h e d r a l   a n g l e

<u>33.1. A BOUNDARY VALUE PROBLEM WITH A PARAMETER</u>. (i) Let us start with the special boundary value problem (22.9), where the corresponding operator

$$\mathcal{U}_0(D_x) \quad \text{maps} \quad V^{\ell+2m,p}(D,\kappa) \quad \text{into} \quad V^{\ell,p}(D,\kappa) \times \sum_{\pm}\prod_{j=1}^{m} V^{\ell+2m-m_j^{\pm}-1/p,\,p}_{(\Gamma^{\pm},\kappa)} ,$$

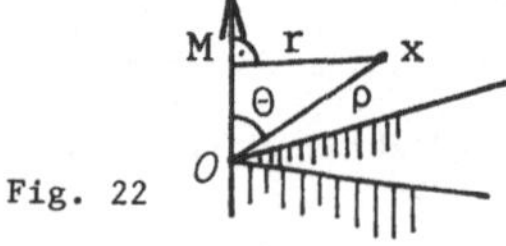

$\kappa = \beta + 1 - \frac{2}{p} + \ell$ is a real number. We consider the spherical coordinates $(\rho,\bar{\omega})$ in R^N , $\rho = |x|$, $\bar{\omega} = (\omega_1,\ldots,\omega_{N-1})$, in contrast to the spherical coordinates (r,ω) in R^2 used in § 22. We define Θ by the relation

Fig. 22

$$(33.1) \qquad \rho \sin \Theta = r ,$$

which means that $\sin \Theta = \sin \omega_1 \ldots \sin \omega_{N-2}$. (See Fig. 22 for $N = 3$.) We write (22.9) in the new coordinates as

$$(33.2) \qquad \begin{aligned} A_0(D_x)u &= \rho^{-2m} \mathcal{L}(\overline{\omega}, D_{\overline{\omega}}, \rho D_\rho)\overline{u} = \overline{f}(\rho, \overline{\omega}) \ , \\ B_{j,0}^{\pm}(D_x)u &= \rho^{-m_j^{\pm}} \mathcal{M}_j^{\pm}(\overline{\omega}, D_{\overline{\omega}}, \rho D_\rho)\overline{u} = \overline{g}_j^{\pm}(\rho, \overline{\omega}) \ , \end{aligned}$$

where $D_\rho = \dfrac{1}{i}\dfrac{\partial}{\partial\rho}$, $j = 1,2,\ldots,m$. Transforming (33.2) analogously to 6.1 we obtain a boundary value problem with a complex parameter λ :

$$\mathcal{L}(\overline{\omega}, D_{\overline{\omega}}, \lambda)\ \tilde{u}(\lambda, \overline{\omega}) = \sum_{|\alpha|+s\leq 2m} \tilde{a}_{\alpha s}(\overline{\omega})\ \lambda^s\ D_{\overline{\omega}}^\alpha\ \tilde{u} = \tilde{F}(\lambda, \overline{\omega}) \quad \text{for}\ \ \overline{\omega} \in Q\ ,$$

$$(33.3) \qquad \mathcal{M}_j^{\pm}(\overline{\omega}, D_{\overline{\omega}}, \lambda)\ \tilde{u}(\lambda, \overline{\omega}) = \sum_{|\alpha|+s\leq m_j^{\pm}} \tilde{b}_{j,\alpha s}^{\pm}(\overline{\omega})\ \lambda^s\ D_{\overline{\omega}}^\alpha\ \tilde{u} = \tilde{G}_j^{\pm}(\lambda, \overline{\omega}) \quad \text{for}\ \ \overline{\omega} \in \partial Q^{\pm},$$

$$j = 1,2,\ldots,m\ ,$$

where $Q = D \cap S^{N-1}$, $\partial Q^{\pm} = \Gamma^{\pm} \cap S^{N-1}$, S^{N-1} is the surface of the N-dimensional unit ball. Q is a spherical bounded domain with two corner points if

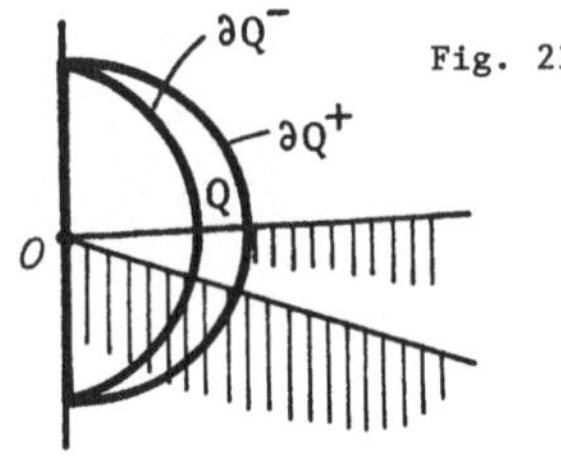

Fig. 23

$N = 3$ and with an (N-3)-dimensional edge M_Q if $N > 3$. (See Fig. 23 for $N = 3$.)

(ii) For every λ the problem (33.3) is a boundary value problem with variable coefficients in a bounded domain Q with an edge (or two corner points if $N = 3$). Let us denote by $\mathcal{U}_Q(\overline{\omega}, \lambda)$ the operator of problem (33.3) for a fixed λ :

$$(33.4) \qquad \mathcal{U}_Q(\overline{\omega}, \lambda) : V^{\ell+2m,p}(Q, \kappa(\cdot)) \to V^{\ell,p}(Q, \kappa(\cdot)) \times$$

$$\times \sum_{\pm} \prod_{j=1}^{m} V^{\ell+2m-m_j^{\pm}-1/p,\ p}(\partial Q^{\pm}, \kappa(\cdot))\ .$$

Here $\kappa(\cdot) \equiv \kappa = \beta + 1 - \dfrac{2}{p} + \ell$ and $V^{k,p}(Q, \kappa)$ is equipped with the norm

$$(33.5) \qquad \|u;\ V^{k,p}(Q,\kappa)\| = \Big(\int_Q \sum_{|\alpha|\leq k} (\sin\theta)^{p(\kappa-k+|\alpha|)}\ |D_{\overline{\omega}}^\alpha u|^p\ d\overline{\omega}\Big)^{1/p}\ ,$$

$$d\overline{\omega} = \sin^{N-2}\omega_1 \cdots \sin\omega_{N-2}\ d\omega_1 \cdots d\omega_{N-1}\ .$$

The trace spaces are defined analogously to (0.20). We can use the ideas of § 26 in order to investigate the properties of the operator (33.4) (reduction to special boundary value problems in a (N-1)-dimensional dihedral angle). The following result was proved by V. G. MAZ'JA, J. ROSSMANN [1] :

(iii) **THEOREM.** *Assume the conditions* (23.6) *and* (23.7) *are satisfied and* λ *is no eigenvalue of* $\mathcal{U}_Q(\overline{\omega}, \lambda)$. *Then the operator* (33.4) *is an isomorphism.*

(iv) The calculation of some eigenvalues of $\mathcal{U}_Q(\overline{\omega}, \lambda)$ and the corresponding system of Jordan chains (see 7.1 for the definition) is easy provided we know the eigenvalues and the Jordan chains of $\mathcal{U}_0(\lambda)$ ($\mathcal{U}_0(\lambda)$ was defined by (23.3)). The following lemma describes the situation.

(v) <u>LEMMA</u>. *Let* λ_0 *be an eigenvalue of* $\mathcal{U}_0(\lambda)$ *with* $\mathrm{Im}\ \lambda_0 < \beta + 1 - 2m$, *and* $\{u^{k,\sigma}\}_{\substack{k=0,\dots,\kappa_\sigma-1 \\ \sigma=1,\dots,I}}$ *a system of Jordan chains of* $\mathcal{U}_0(\lambda)$ *with respect to* λ_0 *(cf. (7.10)). Then* λ_0 *is also an eigenvalue of* $\mathcal{U}_Q(\overline{\omega},\lambda)$,

$$\{y^{k,\sigma}\}_{\substack{k=0,\dots,\kappa_\sigma-1 \\ \sigma=1,\dots,I}}$$

(33.6)

$$= \{ \sum_{s=0}^{k} \frac{1}{s!}(\sin\theta)^{i\lambda_0}\bigl(i\ \log(\sin\theta)\bigr)^s\ u^{k-s,\sigma}(\omega)\}_{\substack{k=0,\dots,\kappa_\sigma-1 \\ \sigma=1,\dots,I}}$$

is a system of Jordan chains of $\mathcal{U}_Q(\overline{\omega},\lambda)$ *with respect to* λ_0, *and* $y^{k,\sigma} \in V^{2m+\ell,p}(Q,\kappa)$ *for* $k = 0,\dots,\kappa_\sigma-1$, $\sigma = 1,\dots,I$.

P r o o f : We have to show that

(33.7) $$\sum_{\mu=0}^{k} \frac{1}{\mu!} \frac{\partial^\mu \mathcal{U}_Q(\overline{\omega},\lambda_0)}{\partial\lambda^\mu} y^{k-\mu,\sigma}(\overline{\omega}) = 0 \quad \text{for}\quad k = 0,\dots,\kappa_\sigma-1\ ,\ \sigma = 1,\dots,I\ .$$

We know (see 7.2) that the functions

$$u_{\sigma,k}(r,\omega) = r^{i\lambda_0} \sum_{s=0}^{k} \frac{1}{s!}\ (i\ \log r)^s\ u^{k-s,\sigma}(\omega)$$

are solutions of the boundary value problem (23.1)

$$A_0(D_y,0)\ u = 0 \quad \text{in}\quad K\ ,$$
$$B_{j,0}^{\pm}(D_y,0)\ u = 0 \quad \text{on}\quad \gamma^{\pm}\ ,\quad j = 1,\dots,m\ .$$

(Recall that $x = (y,z)$, $y \in K$, $z \in M = R^{N-2}$.) Since $u_{\sigma,k}$ is independent of z we have

(33.8) $$A_0(D_y,D_z)\ u_{\sigma,k} = 0 \quad \text{in}\quad D\ ,$$
$$B_{j,0}^{\pm}(D_y,D_z)\ u_{\sigma,k} = 0 \quad \text{on}\quad \Gamma^{\pm}\ ,\quad j = 1,\dots,m\ .$$

Setting $r = \rho \sin\theta$ we obtain

(33.9) $$u_{\sigma,k}(r,\omega) = U_{\sigma,k}(\rho,\overline{\omega}) = \rho^{i\lambda_0} \sum_{s=0}^{k} \frac{1}{s!}(i\ \log\rho)^s\ y^{k-s,\sigma}(\overline{\omega})\ .$$

The functions $U_{\sigma,k}(\rho,\overline{\omega})$ are solutions of

(33.10) $$A_0(D_x)\ u = \rho^{-2m}\ \mathcal{L}(\overline{\omega},D_{\overline{\omega}},\rho D_\rho)\ u = 0 \quad \text{in}\quad D\ ,$$
$$B_{j,0}^{\pm}(D_x)\ u = \rho^{-m_j^{\pm}}\ \mathcal{M}_j^{\pm}(\overline{\omega},D_{\overline{\omega}},\rho D_\rho)\ u = 0 \quad \text{on}\quad \Gamma^{\pm}\ ,\ j = 1,\dots,m\ .$$

Taking into account the identity

$$(\rho D_\rho)^q U_{\sigma,k} = \rho^{i\lambda_0} \sum_{s=0}^{k} \frac{1}{s!}(i\ \log\rho)^s \sum_{\mu=0}^{k-s} \binom{q}{\mu} \lambda^{q-\mu}\ y^{k-s-\mu,\sigma}(\overline{\omega})$$

we obtain from (33.10) that

$$\mathcal{L}(\overline{\omega}, D_{\overline{\omega}}, \rho D_\rho)\, U_{\sigma,k}$$

$$= \rho^{i\lambda_0} \sum_{s=0}^{k} \frac{1}{s!}(i \log \rho)^s \sum_{\mu=0}^{k-s} \frac{1}{\mu!} \frac{\partial^\mu \mathcal{L}(\lambda_0)}{\partial \lambda^\mu}\, y^{k-s-\mu,\sigma}(\overline{\omega}) = 0 \ ,$$

(33.11)

$$\mathcal{M}_j^\pm(\overline{\omega}, D_{\overline{\omega}}, \rho D_\rho)\, U_{\sigma,k}$$

$$= \rho^{i\lambda_0} \sum_{s=0}^{k} \frac{1}{s!}(i \log \rho)^s \sum_{\mu=0}^{k-s} \frac{1}{\mu!} \frac{\partial^\mu \mathcal{M}_j^\pm(\lambda_0)}{\partial \lambda^\mu}\, y^{k-s-\mu,\sigma}(\overline{\omega}) = 0 \ ,$$

$$j = 1,\ldots,m \ .$$

(33.11) yields the relation (33.7). Since $u^{k,\sigma}$ are smooth functions and $\mathrm{Im}\,\lambda_0 < \beta + 1 - 2m$ the functions $y^{k,\sigma}$, $k = 0,\ldots,\kappa_\sigma - 1$, $\sigma = 1,\ldots,I$ are contained in $V^{\ell+2m,p}(Q,\kappa)$.

<u>33.2. THE ADJOINT PROBLEM.</u> (i) Let us recall the coefficient formula (12.14) for the special boundary value problem in an infinite cone. The solutions v_γ of the adjoint boundary value problem have played an essential role in this formula. The functions v_γ were determined by the eigenvalues and the corresponding Jordan chains of the adjoint operator $\mathcal{O}\mathcal{U}_0^*(\lambda^*)$.

The coefficient formula for the special boundary value problem in a dihedral angle will be similar to (12.14). We only have to replace v_γ by the solutions of the adjoint boundary value problem in the dihedral angle, which will be determined by the eigenvalues and the corresponding Jordan chains of the adjoint operator $\mathcal{O}\mathcal{U}_Q^*(\lambda^*)$ (see below).

(ii) Let $\{B_{j,0}^\pm\}_{j=1,\ldots,m}$ be normal systems (see J. WLOKA [1] or J. L. LIONS, E. MAGENES [1]) of boundary operators such that Green's formula holds :

(33.12)
$$\int_D A_0 u\, \overline{v}\, dx + \sum_\pm \sum_{j=1}^{m} \int_{\Gamma^\pm} B_{j,0}^\pm u\, \overline{T_{j,0}^\pm v}\, d\sigma$$

$$= \int_D u\, \overline{A_0^* v}\, dx + \sum_\pm \sum_{j=1}^{m} \int_{\Gamma^\pm} S_{j,0}^\pm u\, \overline{B_{j,0}^{*\pm} v}\, d\sigma$$

for all $u \in V^{2m,p}(D,\, \kappa - \ell)$ and $v \in V^{2m,q}(D,\, -\kappa + \ell + 2m)$, $\frac{1}{p} + \frac{1}{q} = 1$. Here A_0^* , $T_{j,0}^\pm$, $S_{j,0}^\pm$, $B_{j,0}^{*\pm}$ are operators with constant coefficients and of orders $2m$, $2m - m_j^\pm - 1$, $2m - 1 - \mu_j^\pm$, $\mu_j^\pm$, respectively. The adjoint problem to (22.9) now is

(33.13)
$$A_0^* v = f \quad \text{in } D \ ,$$
$$B_{j,0}^{*\pm} v = g_j^\pm \quad \text{on } \Gamma^\pm \ , \quad j = 1,2,\ldots,m \ .$$

For the operator $\mathcal{O}\mathcal{U}_0^* = \{A_0^*, B_{j,0}^{*\pm}\}_{j=1,\ldots,m}$ we have

(33.14) $\quad \mathcal{O}\mathcal{U}_0^* : V^{2m,q}(D,\, -\kappa + \ell + 2m) \to$

$$\to V^{0,q}(D,\, -\kappa + \ell + 2m) \times \sum_\pm \prod_{j=1}^{m} V^{2m-\mu_j^\pm-1/q,\,q}(\Gamma^\pm,\, -\kappa + \ell + 2m) \ .$$

(iii)   We write $\mathcal{O\!\!L}\,_0^*$ in the spherical coordinates $(\rho,\bar{\omega})$ in the form

$$A_0^*(D_x) = \rho^{-2m}\,\mathcal{L}^*(\bar{\omega},D_{\bar{\omega}},\rho D_\rho)\ ,$$

$$B_{j,0}^{*\pm}(D_x) = \rho^{-\mu_j^{\pm}}\,\mathcal{M}_j^{*\pm}(\bar{\omega},D_{\bar{\omega}},\rho D_\rho)\ ,\quad j = 1,2,\ldots,m\ ,$$

and consider the boundary value problem (cf. (33.3))

$$(33.15)\qquad
\begin{aligned}
\mathcal{L}^*(\bar{\omega},D_{\bar{\omega}},\lambda)\ \tilde{v}(\bar{\omega},\lambda) &= \tilde{F}(\bar{\omega},\lambda)\quad \text{for }\ \bar{\omega}\in Q\ ,\\[4pt]
\mathcal{M}_j^{*\pm}(\bar{\omega},D_{\bar{\omega}},\lambda)\ \tilde{v}(\bar{\omega},\lambda) &= \tilde{G}_j^{\pm}(\bar{\omega},\lambda)\quad \text{for }\ \bar{\omega}\in \partial Q^{\pm}\ ,\quad j = 1,\ldots,m\ .
\end{aligned}$$

For the operator $\mathcal{O\!\!L}\,_Q^*(\lambda^*) = \left\{\mathcal{L}^*(\lambda^*),\mathcal{M}_j^{*\pm}(\lambda^*)\right\}_{j=1,\ldots,m}$ we have

$$(33.16)\qquad
\begin{aligned}
\mathcal{O\!\!L}\,_Q^*(\lambda^*) :\ &V^{2m,q}(Q,\ -\kappa+\ell+2m) \to\\[4pt]
&\to V^{0,q}(Q,\ -\kappa+\ell+2m)\times \sum_{\pm}\prod_{j=1}^{m} V^{2m-m_j^{\pm}-1/q,\ q}(\partial Q^{\pm},\ -\kappa+\ell+2m)\ .
\end{aligned}$$

(iv)   We can calculate some eigenvalues of $\mathcal{O\!\!L}\,_Q^*(\lambda^*)$ using the ideas of the proof of Lemma 12.2 for conical points. If we replace r by ρ , ω by $\bar{\omega}$ and G by Q in Lemma 12.2 we obtain

(v)   <u>LEMMA</u>. *If λ_0 is an eigenvalue of $\mathcal{O\!\!L}\,_Q(\lambda)$ then $\lambda_0^* = \bar{\lambda}_0 + i(N - 2m)$ is an eigenvalue of $\mathcal{O\!\!L}\,_Q^*(\lambda^*)$.*

(vi)   Let λ_μ be an eigenvalue of $\mathcal{O\!\!L}\,_0(\lambda)$. Lemma (v) yields that $\lambda_\mu^* = \bar{\lambda}_\mu + i(N - 2m)$ is an eigenvalue of $\mathcal{O\!\!L}\,_Q^*(\lambda^*)$. We consider a Jordan chain of $\mathcal{O\!\!L}\,_Q^*(\lambda^*)$ with respect to λ_μ^* ,

$$(33.17)\qquad
\begin{bmatrix}
w_\mu^{0,\sigma}\\
\vdots\\
w_\mu^{\kappa_\mu-1,\sigma}
\end{bmatrix}\ .$$

For $\gamma = (\mu,\sigma,k)$ we denote

$$(33.18)\qquad V_\gamma(0) = V_\gamma(0,\rho,\bar{\omega}) = \rho^{i\lambda_\mu^*}\sum_{s=0}^{k}\frac{1}{s!}(i\log\rho)^s\, w_\mu^{k-s,\sigma}(\bar{\omega})\ ,$$

and for $\gamma' = (\mu,\sigma,k') = (\mu,\sigma,\kappa_{\mu_\sigma}-1-k)$ we write

$$(33.19)\qquad V_{\gamma'}(z) = |x - z|^{i\lambda_\mu^*}\sum_{s=0}^{k'}\frac{1}{s!}(i\log|x - z|)^s\, w_\mu^{k'-s,\sigma}(\bar{\omega})\quad \text{for }\ z\in M\ .$$

The functions $V_\gamma(0)$ and $V_{\gamma'}(z)$ are solutions of

$$A_0^*v = 0\quad \text{in }\ D\ ,$$

$$B_{j,0}^{*\pm}v = 0\quad \text{on }\ \Gamma^{\pm}\ .$$

(vii)   We choose a canonical system of Jordan chains of $\mathcal{O\!\!L}\,_Q(\lambda)$ with respect to the eigenvalue λ_μ (cf. (7.10)), $\{y_\mu^{\sigma,k}\}$, and a canonical system of Jordan chains of $\mathcal{O\!\!L}\,_Q^*(\lambda^*)$ with respect to the eigenvalue $\lambda_\mu^* = \bar{\lambda}_\mu + i(N-2m)$,

$\{w_\mu^{\sigma,k}\}$, in such a way that the following *biorthonormality condition* is satisfied :

$$\sum_{\nu=0}^{k} \sum_{t=\nu+1}^{\tau+\nu+1} \frac{1}{t!} \left(\frac{\partial^t \mathcal{L}(\lambda_\mu)}{\partial \lambda^t} y_\mu^{k-\nu,\sigma} \, , \, w_\mu^{\tau-t+\nu+1,\sigma'} \right)_Q + \sum_{\pm} \sum_{j=1}^{m} \sum_{\nu=0}^{k} \sum_{t=\nu+1}^{\tau+\nu+1} \frac{1}{t!}$$

$$(33.20) \qquad \cdot \left(\frac{\partial^t \mathcal{M}_j^\pm(\lambda_\mu)}{\partial \lambda^t} y_\mu^{k-\nu,\sigma} \, , \, \sum_{s=0}^{\tau-t+\nu+1} \frac{1}{s!} \frac{\partial^s Q_j^\pm}{\partial \lambda^s} \big(\overline{\lambda}_\mu + i(N-2m)\big) w_\mu^{\tau-t+\nu+1,\sigma'} \right)_{\partial Q^\pm}$$

$$= \delta_{\sigma\sigma'} \, \delta_{\kappa_{\mu_\sigma}-1-k,\tau} \, ,$$

$$k = 0,\ldots,\kappa_{\mu_\sigma}-1 \, , \quad \sigma = 1,\ldots,I_\mu \, , \quad \tau = 0,\ldots,\kappa_{\mu_\sigma}-1 \, , \quad \sigma' = 1,\ldots,I_\mu \, , \quad (f,g)_Q$$

$$= \int_Q f \, \overline{g} \, d\overline{\omega} \, , \quad (f,g)_{\partial Q^\pm} = \int_{\partial Q^\pm} f \, \overline{g} \, d\sigma_{\overline{\omega}} \, , \quad T_{j,0}^\pm(D_x) = \rho^{2m-m_j^\pm-1} Q_j(\overline{\omega},D_{\overline{\omega}},\rho D_\rho) \, .$$

Now we are able to present the coefficient formula for the special boundary value problem in a dihedral angle.

33.3. **THEOREM** (coefficient formula). *Let the assumptions of Theorem (ii) from 30.2 be satisfied for* $\kappa(\cdot) \equiv \kappa$ *and* $\kappa_1(\cdot) \equiv \kappa_1$ *, and let only one eigenvalue of* $\mathcal{U}_0(\lambda)$ *be situated in the strip*

$$h_1 = \beta_1 + 1 - 2m < \text{Im } \lambda < \beta + 1 - 2m = h \, .$$

Then the traces of the coefficients $\hat{c}_\gamma$ *in the expansion* (30.8)

$$u(x) = \sum_{\gamma \in I} \hat{c}_\gamma(x) u_\gamma(x) + w(x)$$

are given by the formula

$$(33.21) \qquad c_\gamma(z) = \big(f, \, iv_{\gamma'}(z)\big)_D + \sum_{\pm} \sum_{j=1}^{m} \big(g_j^\pm, \, iT_j^\pm \, v_{\gamma'}(z)\big)_{\Gamma^\pm}$$

provided (33.20) *is valid.* ($\hat{c}_\gamma$ *can be constructed by a suitable extension of* c_γ *.*)
For the p r o o f we refer to V. G. MAZ'JA, J. ROSSMANN [1].

33.4. **EXAMPLE:** The Dirichlet problem. (i) We consider the Dirichlet problem

$$(33.22) \qquad \begin{aligned} -\Delta u &= f \quad \text{in } D \, , \\ u &= 0 \quad \text{on } \Gamma^\pm \, , \end{aligned}$$

where $f \in L^{p_1}(D) = V^{0,p_1}(D,0)$, $D \subset R^3$ is a dihedral angle with the edge angle ω_0 . Assume that

$$(33.23) \qquad -\frac{2\pi}{\omega_0} < \frac{2}{p_1} - 2 < -\frac{\pi}{\omega_0} \quad \text{and} \quad -\frac{\pi}{\omega_0} - 1 < \frac{2}{p_1} - 2$$

and that $u \in V^{2,2}(D,1)$ is a solution of (33.22) with bounded support.

(ii) Since the eigenvalues of $\mathcal{U}_0(\lambda)$ are $\lambda_\mu = -i\frac{\mu\pi}{\omega_0}$, $\mu = \pm 1, \pm 2, \ldots$
(33.23) guarantees that all assumptions of Theorem 33.3 are satisfied for
$\beta \equiv 1$, $\beta_1 \equiv \frac{2}{p_1} - 1$, $\ell_1 = \ell = 0$. The conditions (23.6) and (23.7) are

easily verified by using the ideas of 32.2. Therefore the expansion

$$(33.24) \qquad u(x) = \hat{c}_1(x) \, r^{\pi/\omega_0} \sin \frac{\pi}{\omega_0} \omega + w(x)$$

holds, where $r = \sqrt{y_1^2 + y_2^2}$ and $w \in V^{2,p_1}(D,0)$.

(iii) For the calculation of the trace of $\hat{c}_1$ we need the function $V_{\gamma'}(z) = |x - z|^{-\pi/\omega_0 - 1} w_1^{0,1}(\overline{\omega})$, where $w_1^{0,1}(\overline{\omega}) = w_1(\omega,\theta) = w_1(\overline{\omega})$ is an eigenfunction of $\mathcal{U}_Q^*(\lambda^*)$ with respect to the eigenvalue $i(\pi/\omega_0 + 1)$, satisfying (33.20). That means

$$A_Q^*(\overline{\lambda}_1 + i(N - 2))w_1(\overline{\omega}) = \left[\delta_{\overline{\omega}} + \lambda(\lambda - i(N - 2))\right]\Big|_{\lambda = \overline{\lambda}_1 + i(N-2)} w_1(\overline{\omega}) = 0$$
$$(33.25) \qquad\qquad\qquad \text{in } Q ,$$
$$B_j^{*\pm} w_1(\overline{\omega}) = w_1(\overline{\omega}) = 0 \qquad \text{on } \partial Q^\pm ,$$

where $N = 3$ and $\delta_{\overline{\omega}}$ denotes the Laplace-Beltrami operator. Since $\mathcal{U}_Q(\lambda)$ is selfadjoint and $(\sin \theta)^{\pi/\omega_0} \sin \frac{\pi\omega}{\omega_0}$ is an eigenfunction of $\mathcal{U}_Q(\lambda)$ (cf. (33.6)), it follows that $w_1(\omega,\theta) = c(\sin \theta)^{\pi/\omega_0} \sin \frac{\pi\omega}{\omega_0}$, where the constant c is determined by (33.20)

(iv) The biorthonormality condition (33.20) has the form $(N = 3)$

$$\left(\frac{\partial \mathcal{L}}{\partial \lambda}(\lambda_1)y_1, \, w_1\right)_Q = c \int_Q \left(2\lambda_1 - i(N - 2)\right)(\sin \theta)^{2\pi/\omega_0}(\sin \frac{\pi}{\omega_0} \omega)^2 \, d\overline{\omega}$$

$$= (-i) \, c\left[\frac{2\pi}{\omega_0} + N - 2\right] \int_Q (\sin \theta)^{2\pi/\omega_0 + 1} (\sin \frac{\pi}{\omega_0} \omega)^2 \, d\theta \, d\omega$$

$$= (-i) \, c\left[\frac{2\pi}{\omega_0} + N - 2\right] \frac{1}{2} \omega_0 \, 2 \int_0^{\pi/2} (\sin \theta)^{2\pi/\omega_0 + 1} \, d\theta = 1 .$$

Therefore we obtain

$$c = i \, \frac{\Gamma(1/2 + \pi/\omega_0)}{(\sqrt{\pi})^3 \, \Gamma(\pi/\omega_0)}$$

and, finally,

$$(33.26) \qquad c_1(z) = \frac{\Gamma(1/2 + \pi/\omega_0)}{(\sqrt{\pi})^3 \, \Gamma(\pi/\omega_0)} \int_D f(x) \, |x - z|^{-\pi/\omega_0 - 1} (\sin \theta)^{\pi/\omega_0} \sin \frac{\pi}{\omega_0} \omega \, dx$$

(Γ denotes the gamma function). For $N > 3$ we have

$$(33.27) \qquad c_1(z) = \frac{\Gamma\left((N-2)/2 + \pi/\omega_0\right)}{(\sqrt{\pi})^N \, \Gamma(\pi/\omega_0)} \int_D f(x) \, |x - z|^{-\pi/\omega_0 - N + 2}$$
$$\cdot (\sin \theta)^{\pi/\omega_0} \sin \frac{\pi\omega}{\omega_0} \, dx .$$

<u>33.5. EXAMPLE.</u> : <u>The mixed problem.</u> (i) We consider the mixed boundary value

problem

$$- \Delta u = f \quad \text{in} \quad D \ ,$$

(33.28)

$$u = 0 \quad \text{on} \quad \Gamma^+ \ , \quad \frac{\partial u}{\partial n} = 0 \quad \text{on} \quad \Gamma^- \ ,$$

where $f \in L^{p_1}(D) = V^{0,p_1}(D,0)$, $D \subset R^N$ is a dihedral angle with the edge angle ω_0 . Assume that

(33.29) $\quad -\frac{3}{2} \frac{\pi}{\omega_0} < \frac{2}{p_1} - 2 < -\frac{\pi}{2\omega_0}$ and $-\frac{\pi}{2\omega_0} - 1 < \frac{2}{p_1} - 2$

and that $u \in V^{2,2}(D,1)$ is a solution of (33.28) with bounded support.

(ii) Let us verify the assumptions of Theorem 33.3. To this end we consider a bounded domain Ω which coincides with the dihedral angle D in supp u . Assume that $\partial\Omega$ consists of the faces Γ_1 , Γ_2 and the edge M_Ω . It follows from 32.2 that the boundary value problem

$$- \Delta U = F \quad \text{in} \quad \Omega \ ,$$

$$U = 0 \quad \text{on} \quad \Gamma_1 \ , \quad \frac{\partial U}{\partial n} = 0 \quad \text{on} \quad \Gamma_2$$

has a uniquely determined solution $U \in V^{2,2}(\Omega,1)$ for every $F \in L^2(\Omega,1)$ $= V^{0,2}(\Omega,1)$. Therefore the conditions (26.4) and (26.5) are satisfied for all $\zeta \in M_\Omega$, $\beta \equiv 1$, and thus also for $\zeta \in M_D \cap$ supp u . Since the eigenvalues of $\mathcal{U}_0(\lambda)$ are $\lambda_\mu = - i(\mu + \frac{1}{2}) \frac{\pi}{\omega_0}$, $\mu = 0,\pm1,\pm2,\dots$, the condition (33.29) guarantees that all assumptions of 33.3 are valid for $\beta_1 \equiv \frac{2}{p_1} - 1$, $\ell_1 = \ell$ $= 0$. Therefore

(33.30 $\quad u(x) = \hat{c}_1(x) \, r^{\pi/2\omega_0} \cos \frac{\pi}{\omega_0} \omega + w(x)$

where $w \in V^{2,p_1}(D,0)$.

(iii) As in (iii) from 33.4 we obtain that the eigenfunction $w_1^{0,1}(\bar{\omega})$ $= w_1(\bar{\omega})$ of $\mathcal{U}_Q^*(\lambda^*)$ with respect to the eigenvalue $\lambda_1^* = \overline{\lambda}_0 + i(N - 2m)$ $= i(\frac{\pi}{2\omega_0} + N - 2)$ has the form

$$w_1(\bar{\omega}) = c(\sin \theta)^{\pi/2\omega_0} \cos \frac{\pi}{2\omega_0} \omega \ .$$

The constant c is to be determined from 33.20. Analogously to (iv) from 33.4 we obtain

(33.31) $\quad c = i \, \dfrac{\Gamma\big((N-2)/2 + \pi/2\omega_0\big)}{(\sqrt{\pi})^N \, \Gamma(\pi/2\omega_0)} \ .$

Therefore the trace of the coefficient $\hat{c}_1$ of the expansion (33.30) is given by

(33.32) $\quad c_1(z) = \dfrac{\Gamma\big((N-2)/2 + \pi/2\omega_0\big)}{(\sqrt{\pi})^N \, \Gamma(\pi/2\omega_0)} \displaystyle\int_D f(x) \, |x - z|^{-\pi/2\omega_0 \, -N+2}$

$$\cdot (\sin \theta)^{\pi/2\omega_0} \cos \frac{\pi}{2\omega_0} \omega \, dx \ .$$

33.6. <u>EXAMPLE : The Dirichlet problem for the biharmonic operator.</u>

(i) We consider the boundary value problem

$$\Delta^2 u = f \quad \text{in} \quad D ,$$

(33.33)

$$u = 0 \quad \text{on} \quad \Gamma^{\pm} , \quad \frac{\partial u}{\partial n} = 0 \quad \text{on} \quad \Gamma^{\pm} ,$$

where $f \in L^{p_1}(D)$, $D \subset R^N$ is a dihedral angle with the edge angle ω_0 . Assume that the eigenvalue $\lambda_{\sim}(\omega_0)$ for $\beta = 2$ is simple (cf. 3.2 and (24.1)). Let $\lambda_{=} = \lambda_{=}(\omega_0)$ be that eigenvalue of $\mathcal{O}_0(\lambda)$ for which the strip $\text{Im} \lambda_{=} < \text{Im} \lambda < \text{Im} \lambda_{-}$ contains no eigenvalues (see Fig. 6). Assume

(33.34) $\quad \text{Im} \lambda_{=}(\omega_0) < \dfrac{2}{p_1} - 4 < \text{Im} \lambda_{-}(\omega_0) \quad \text{and} \quad \text{Im} \lambda_{-}(\omega_0) - 1 < \dfrac{2}{p_1} - 4$

or

(33.35) $\quad \text{Im} \lambda_{=}(\omega_0) < \dfrac{2}{p_1} - 3 < \text{Im} \lambda_{-}(\omega_0) \quad \text{and} \quad \text{Im} \lambda_{-}(\omega_0) - 1 < \dfrac{2}{p_1} - 3 .$

(ii) Let $u \in V^{4,2}(D,2)$ be a solution of (33.33) with bounded support. The assumptions of Theorem 33.3 are satisfied for $\ell_1 = \ell = 0$, $\beta = \beta(\cdot) \equiv 2$ and $\beta_1 = \dfrac{2}{p_1} - 1$ if (33.34) is satisfied or $\beta_1 = \dfrac{2}{p_1}$ if (33.35) is satisfied. Therefore we have

(33.36) $\quad u(x) = \hat{c}_1(x) \, r^{i\lambda_-(\omega_0)} \, \phi_1(\omega) + w(x) ,$

where $w \in V^{4,p_1}(\Omega,0)$ if (33.34) is valid and $w \in V^{4,p_1}(D,1) \cap V^{3,p_1}(D,0)$ if (33.35) is valid; ϕ_1 is given by (3.4) and has the form (see P. GRISVARD [1])

(33.37)
$$\phi_1(\omega) = \left[\sin \alpha\omega_0 - \frac{\alpha}{\alpha - 2} \sin(\alpha-2)\omega_0\right]\left[\cos \alpha\omega - \cos(\alpha-2)\omega\right]$$
$$- \left[\cos \alpha\omega_0 - \cos(\alpha-2)\omega_0\right]\left[\sin \alpha\omega - \frac{\alpha}{\alpha - 2} \sin(\alpha-2)\omega\right] , \quad \alpha = i\lambda_-(\omega_0) .$$

(iii) Let us calculate $V_{\gamma'}(z) = |x - z|^{i\overline{\lambda_-(\omega_0)}-N+4} \, w_1^{0,1}(\overline{\omega})$ in the formula (33.21). Here $w_1^{0,1}(\overline{\omega}) = w_1(\overline{\omega})$ is an eigenfunction of $\mathcal{O}_{Q*}(\lambda^*)$ with respect to the eigenvalue $\lambda_-(\omega_0)$ and satisfies (33.20) Since

$$\mathcal{L}(\overline{\omega},D_{\overline{\omega}},\lambda) = \mathcal{L}^*(\overline{\omega},D_{\overline{\omega}},\lambda) = \delta_{\overline{\omega}}^2 - 2\delta_{\overline{\omega}}\left[(N - 4)(- i\lambda) + \lambda^2 + N - 4\right]$$
$$+ (i\lambda)(i\lambda + N - 4)(i\lambda - 2)(i\lambda + N - 2) ,$$

where $\delta_{\overline{\omega}}$ is the Laplace-Beltrami operator, we obtain for $\lambda^* = \overline{\lambda} + i(N - 4)$

$$\mathcal{L}^*(\overline{\omega},D_{\overline{\omega}},\lambda^*) = \delta_{\overline{\omega}}^2 - 2\delta_{\omega}\left[\overline{\lambda}^2 + i\overline{\lambda}(N - 4)\right] + N - 4$$
$$+ \overline{(i\lambda)(i\lambda + N - 4)(i\lambda - 2)(i\lambda + N - 2)} .$$

The boundary operators do not depend on λ . We consider the eigenfunction

$$y_1 = y_1(\overline{\omega}) = (\sin \theta)^{i\lambda_-(\omega_0)} \phi_1(\omega) \quad \text{of} \quad \mathcal{O}_Q(\lambda) \text{ with respect to the eigenvalue}$$
$\lambda_1 = \lambda_-(\omega_0)$ (see (33.6)). Since

$$\overline{\mathcal{L}(\overline{\omega},D_{\overline{\omega}},\lambda_1)y_1(\overline{\omega})} = \mathcal{L}\left(\overline{\omega}, D_{\overline{\omega}}, \overline{\lambda}_1 + i(N - 4)\right) \overline{y_1(\overline{\omega})} = 0$$

we have

$$w_1(\omega) = c \; \overline{y_1(\overline{\omega})}$$

where c is to be determined by formula (33.20). Therefore the trace of the coefficient $\hat{c}_1$ in (33.36) is given by

$$c_1(z) = \overline{c} \int_D f(x) \; i|x - z|^{\overline{i\lambda_-(\omega_0)}-N+4} \; y_1(\overline{\omega}) \; dx \; .$$

§ 34. The coefficient formula in a bounded domain

34.1. SOME REMARKS. (i) We deal with the calculation of the traces of the coefficients $\hat{c}_\gamma$ of the expansion (31.4) which belong to the eigenvalue with the biggest imaginary part.

(ii) Let us recall the structure of $\hat{c}_\gamma$. We have used a sufficiently fine covering $\{U_i\}_{i=1,\ldots,T_2}$ of Ω and a corresponding partition of unity $\sum_{i=1}^{T_2} \eta_i(x) = 1$ for all $x \in \Omega$ in the proof of Theorem 31.2. Then we have considered such a set U_{i_0} that $U_{i_0} \cap M \neq \emptyset$. Since $U_{i_0} \cap \Omega$ is diffeomorphic to a dihedral angle we have used the results of Theorem (ii) from 30.2 for $\eta_{i_0} u = u_0$. We have obtained

$$\hat{c}_\gamma(x) \; u_\gamma(x) = \sum_{i=1}^{T_1} \hat{c}_{\gamma,i}(x) \; u_{\gamma,i}(x) \; ,$$

where T_1 was the number of the sets U_i with $U_i \cap M \neq \emptyset$.

(iii) Let us proceed along these lines. We consider U_{i_0} as above and assume for simplicity

$$(34.1) \qquad U_{i_0} \cap \Omega = D_0 \cap B_{R_0}(0) \; ,$$

where D_0 is a dihedral angle with the edge angle ω_0 . Let $\mathcal{U}(x,D_x)$ $= \{A(x,D_x), \; B_j^{(q)}(x,D_x)\}_{\substack{j=1,\ldots,m \\ q=1,\ldots,T}}$ be the operator of the problem (22.6) and $u \in V^{\ell+2m,P}(\Omega,\kappa(\cdot))$ a solution of (22.6). Then $u_0 = \eta_{i_0} u$ is a solution of the following boundary value problem in the dihedral angle D_0 :

$$A(x,D_x) \; u_0(x) = \sum_{|\alpha|\leq 2m} a_\alpha(x) \; D_x^\alpha \; u_0(x) = f_0(x) \quad \text{in} \quad D_0 \; ,$$

$$(34.2) \qquad B_j^\pm(x,D_x) \; u_0(x) = \sum_{|\alpha|\leq m_j^\pm} b_{j,\alpha}^\pm(x) \; D_x^\alpha \; u_0(x) = g_{j,0}^\pm(x) \quad \text{on} \quad \Gamma_0^\pm \; ,$$
$$j = 1,2,\ldots,m \; ,$$

where $\pm$ denotes two corresponding indices q . Concerning the coefficients of (34.2) let us assume

$$a_\alpha(x) = a_\alpha(y,z) = a_\alpha(y) \quad \text{for} \quad x \in D_0 \cap B_{R_0}(0) \ , \quad |\alpha| = 2m \ ,$$

(34.3)
$$b_{j,\alpha}^\pm(x) = b_{j,\alpha}^\pm(y,z) = b_{j,\alpha}^\pm(y) \quad \text{for} \quad x \in \Gamma_0^\pm \cap B_{R_0}(0) \ , \quad |\alpha| = m_j^\pm \ ,$$
$$j = 1,\ldots,m \ ,$$

$(x = (y_1,y_2,z_1,\ldots,z_{N-2}) \ , \ (z_1,\ldots,z_{N-2}) \in M_{D_0})$. Let us write (34.2) in the form

$$A_0(0,D_x)\, u_0(x) = \sum_{|\alpha|=2m} a_\alpha(0) D_x^\alpha u_0(x) = f_0(x)$$

(34.4)
$$+ \sum_{|\alpha|=2m}\bigl(a_\alpha(0) - a_\alpha(y)\bigr)D_x^\alpha u_0 - \sum_{|\alpha|\le 2m-1} a_\alpha(x)D_x^\alpha u_0$$
$$= F_1(x) \quad \text{for} \quad x \in D_0 \cap B_{R_0}(0) \ ,$$

$$B_{j,0}^\pm(0,D_x)\, u_0(x) = \sum_{|\alpha|=m_j^\pm} b_{j,\alpha}^\pm(0) D_x^\alpha u_0(x) = g_{j,0}^\pm(x)$$

(34.5)
$$+ \sum_{|\alpha|=m_j^\pm}\bigl(b_{j,\alpha}^\pm(0) - b_{j,\alpha}^\pm(y)\bigr)D_x^\alpha u_0 - \sum_{|\alpha|\le m_j^\pm-1} b_{j,\alpha}^\pm(x)D_x^\alpha u_0$$
$$= G_{j,1}^\pm(x) \quad \text{for} \quad x \in \Gamma_0^\pm \cap B_{R_0}(0) \ .$$

34.2. THEOREM. *Let the assumptions of Theorem 31.2 be satisfied and assume that only one eigenvalue of $\mathcal{U}_0(\zeta,\lambda)$ is situated in the strip $h_1(\zeta) < \operatorname{Im}\lambda < h(\zeta)$ and that the conditions (34.1) and (34.3) are valid. Then the traces of the coefficients $\hat{c}_{\gamma,0}$ in the expansion*

$$u_0(x) = n_{i_0} u(x) = \sum_{\gamma \in I} \hat{c}_{\gamma,0}(x) u_{\gamma,0}(x) + w_0(x)$$

are given by

(34.6)
$$c_{\gamma,0}(z) = \bigl(F_1,\ iV_{\gamma',0}(z)\bigr)_{D_0} + \sum_{\pm}\sum_{j=1}^{m}\bigl(G_{j,1}^\pm,\ iT_j^\pm\, V_{\gamma',0}(z)\bigr)_{\Gamma_0^\pm} \ .$$

The functions $V_{\gamma',0}(z)$ are defined analogously to (33.19).

P r o o f : It follows from the assumptions that $F_1 \in V^{\ell_1,p_1}(D_0,\kappa_1(\cdot))$ and $G_{j,1}^\pm \in V^{\ell_1+2m-m_j^\pm-1/p_1,\ p_1}(\Gamma_0^\pm,\kappa_1(\cdot))$ (cf. 30.1 (iv)). Therefore we can use Theorem 33.3 for the boundary value problem (34.4), (34.5), thus obtaining (34.6).

34.3. REMARK. If we omit the condition (34.3) then a more complicated coefficient formula is valid (see V. G. MAZ'JA, J. ROSSMANN [1]).

PART TWO

E L L I P T I C   B O U N D A R Y   V A L U E   P R O B L E M S
W I T H   " N O N   R E G U L A R "   R I G H T   H A N D
S I D E S   A N D   C O E F F I C I E N T S

Chapter IV

ELLIPTIC PROBLEMS WITH "BAD" RIGHT HAND SIDES

In this chapter, we will proceed very closely to the scheme used in [I],
Chapter III. Therein we investigated the *Dirichlet problem* and used very substantially the properties of the spaces $W_0 = W_0^{k,2}(\Omega;s(d_M))$ (see Preliminaries,
formula (0.5) with the special choice $S = \{s(d_M(x)) \text{ for all } |\alpha| \leq k\}$; also
see Subsection 36.1 below). If we investigate other boundary value problems,
we use spaces V which satisfy

$$W_0 \subset V \subset W$$

with $W = W^{k,2}(\Omega;s(d_M))$. The space V used depends on the *type* of the boundary conditions considered; in this chapter, we will deal mainly with the
Neumann problem, and in this case we take for V the right limit space in the
above inclusions, namely the space W .

Nevertheless, before dealing with the Neumann problem, let us mention an
interesting result concerned with the Dirichlet problem.

Section 11 .   *T h e   D i r i c h l e t   p r o b l e m   i n   s p a c e s*
                   w i t h   p o w e r   t y p e   w e i g h t s

§ 35 .   B o u n d s   f o r   t h e   a d m i s s i b l e   p o w e r s

35.1.   INTRODUCTION.   Let us consider the differential operator of order $2k$
given by the formula

$$(35.1) \qquad (\mathscr{L} u)(x) = \sum_{|\alpha|,|\beta|\leq k} (-1)^{|\alpha|} D^{\alpha}\big(a_{\alpha\beta}(x) \, D^{\beta} \, u(x)\big) \, ,$$

together with the associated bilinear form

$$(35.2) \qquad a(u,v) = \sum_{|\alpha|,|\beta|\leq k} \int_{\Omega} a_{\alpha\beta}(x) \, D^{\beta}u(x) \, D^{\alpha}v(x) \, dx \, .$$

The coefficients $a_{\alpha\beta} = a_{\alpha\beta}(x)$ are defined for $x \in \Omega$, where Ω is a domain in $\mathbb{R}^N$ with a boundary $\partial\Omega$; α , β are N-dimensional multiindices, D^α is the differential operator of order $|\alpha|$ defined in Preliminaries - see p. 8.

In [I], we have investigated the solvability of the Dirichlet problem for the operator $\mathcal{L}$ in the weighted Sobolev space

$$(35.3) \qquad W^{k,2}(\Omega; d_M, \varepsilon)$$

defined as the set of functions $u = u(x)$, $x \in \Omega$, such that

$$(35.4) \qquad \| u; \ W^{k,2}(\Omega; d_m, \varepsilon) \|^2 = \sum_{|\alpha| \le k} \int_\Omega |D^\alpha u(x)|^2 \, d_M^\varepsilon(x) \ dx < \infty$$

with $d_M(x) = \operatorname{dist}(x,M)$, $M \subset \partial\Omega$, $\varepsilon \in \mathbb{R}$. It was shown that under the assumptions

$$(35.5) \qquad a_{\alpha\beta} \in L^\infty(\Omega) \qquad {}^{*)}$$

(= the *boundedness of the coefficients* of the differential operator $\mathcal{L}$) and

$$(35.6) \qquad a(u,u) \ge c \| u; W^{k,2}(\Omega) \|^2 \quad \text{for every} \ \ u \in C_0^\infty(\Omega)$$

(= the *ellipticity of the differential operator* $\mathcal{L}$; $W^{k,2}(\Omega)$ is the classical Sobolev space, $W^{k,2}(\Omega) = W^{k,2}(\Omega; d_M, 0)$, and $C_0^\infty(\Omega)$ is the set of infinitely differentiable functions on Ω with support contained in Ω) *there is an open interval* $J \subset \mathbb{R}$ *containing the origin such that for* $\varepsilon \in J$ *the Dirichlet problem for the operator* $\mathcal{L}$ *is uniquely (weakly) solvable in the weighted Sobolev space* $W^{k,2}(\Omega; d_M, \varepsilon)$.

[See [I], Lemma 14.3 and Theorem 14.4. More precisely, the last assertion means : *There is an interval* J *such that for*

$$(35.7) \qquad \varepsilon \in J$$

and for given $u_0 \in W^{k,2}(\Omega; d_M, \varepsilon)$ and $F \in \left(W_0^{k,2}(\Omega; d_M, -\varepsilon) \right)^*$ there is one and only one $u \in W^{k,2}(\Omega; d_M, \varepsilon)$ (= the w e a k solution) such that

$$u - u_0 \in W_0^{k,2}(\Omega; d_M, \varepsilon)$$

and

$$a(u,v) = <F,v> \quad \text{for every} \ \ v \in C_0^\infty(\Omega) \ .]$$

<u>35.2. THE SIZE OF THE INTERVAL J</u> . Let us concentrate our attention on the condition (35.7), i.e., on the set J of admissible values of the power ε

${}^{*)}$ This condition can be weakened to the form

$$(35.5^*) \qquad a_{\alpha\beta} d_M^{2k-|\alpha|-|\beta|} \in L^\infty(\Omega) \ , \quad |\alpha| \ , \ |\beta| \le k \ ;$$

see Subsection 37.5, formula (37.15) below.

in the weight function d_M^ε . It follows from the estimates and examples in [I] that the size of the interval J depends

    – on the coefficients $a_{\alpha\beta}$ of the operator $\mathscr{L}$ (mainly on their L^∞-norm $\|a_{\alpha\beta};L^\infty(\Omega)\|$ – see (35.5) – and on the constant c in the ellipticity condition (35.6)), and

    – on the properties of the domain Ω and of the set $M \subset \partial\Omega$.

Of course, we derived only *upper estimates* for J , i.e., for the admissible values of $|\varepsilon|$, but the following example which is due to J. VOLDŘICH [2] shows that, roughly speaking, *for every $\varepsilon \neq 0$, an elliptic differential operator can be constructed such that the corresponding Dirichlet problem is n o t   s o l v a b l e   in the weighted Sobolev space $W^{k,2}(\Omega;d_M,\varepsilon)$ with the weight function d_M^ε .*

35.3. <u>EXAMPLE.</u> We set $N = 1$ and choose the onedimensional interval $(0,1)$ for Ω . Let us consider the Dirichlet problem

$$(35.8) \qquad \begin{aligned} - u''(x) + \delta(\delta - 1)x^{-2}\, u(x) &= f(x) \quad \text{for } x \in (0,1) , \\ u(0) = u(1) &= 0 \end{aligned}$$

with a suitable constant δ ; the corresponding bilinear form $a(u,v)$ is then given by the formula

$$(35.9) \qquad a(u,v) = \int_0^1 u'(x)\, v'(x)\, dx + \delta(\delta - 1) \int_0^1 x^{-2}\, u(x)\, v(x)\, dx .$$

Further, we set $M = \{0\}$, so that $d_M(x) = x$, and consider the (weak) solvability of the Dirichlet problem in the weighted Sobolev space $W^{1,2}\big((0,1);x,\varepsilon\big)$ normed by

$$\big\|u;W^{1,2}\big((0,1);x,\varepsilon\big)\big\|^2 = \int_0^1 |u'(x)|^2\, x^\varepsilon\, dx + \int_0^1 |u(x)|^2\, x^\varepsilon\, dx .$$

    (i) The *boundedness condition* (35.5) is not fulfilled in general, since the coefficient $a_{00}(x) = \delta(\delta - 1)x^{-2}$ is   n o t   b o u n d e d , but in this case we can use the modified "boundedness" condition (35.5*) (see the footnote on p. 142) since

$$a_{00}(x)\big[d_M(x)\big]^2 = \delta(\delta - 1)x^{-2}\, x^2 = \delta(\delta - 1) \in L^\infty(\Omega)$$

(we have $k = 1$, $|\alpha| = |\beta| = 0$).

    (ii) *The ellipticity condition* (35.6) *is fulfilled if we assume that*

$$(35.10) \qquad |\delta(\delta - 1)| < \tfrac{1}{4} .$$

Indeed, Hardy's inequality (0.32) (with $p = 2$) yields that for every function

$u \in C_0^\infty((0,1))$ we have

$$(35.11) \qquad \int_0^1 |u(x)|^2 \, x^{\varepsilon-2} \, dx \leq \frac{4}{|1 - \varepsilon|^2} \int_0^1 |u'(x)|^2 \, x^\varepsilon \, dx$$

provided $\varepsilon \neq 1$. Using this inequality for $\varepsilon = 0$ we easily obtain

$$a(u,u) \geq \left(1 - 4|\delta(\delta - 1)|\right) \int_0^1 |u'(x)|^2 \, dx$$

which is (35.6) with $c = 1 - 4|\delta(\delta - 1)|$. [We have used the fact that the

expressions $\displaystyle\int_0^1 |u'(x)|^2 \, dx$ and $\displaystyle\int_0^1 |u(x)|^2 \, dx + \int_0^1 |u'(x)|^2 \, dx$ are equivalent

norms in $W^{1,2}((0,1))$ for $u \in C_0^\infty(\Omega)$, i.e., equivalent norms in $W_0^{1,2}((0,1))$.]

(iii) In view of the foregoing points (i) and (ii), the existence of a
certain interval J is guaranteed such that the Dirichlet problem (35.8) is
weakly solvable in $W^{1,2}(\Omega;x,\varepsilon)$, provided $\varepsilon \in J$. Now, let $\varepsilon \neq 0$ and let
$|\varepsilon|$ be sufficiently small. Put

$$\delta = \frac{1 - \varepsilon}{2}$$

and suppose that the right hand side f of the Dirichlet problem (35.8) is of
the form

$$(35.12) \qquad f(x) = \begin{cases} x^{\delta-2}(-\ln x)^{\omega-1}\left[2\omega\delta - \omega - \omega(\omega - 1)(-\ln x)^{-1}\right] & \text{for } x \in (0,\tfrac{1}{2}), \\ -w''(x) + \delta(\delta - 1)x^{-2} \, w(x) & \text{for } x \in [\tfrac{1}{2},1) \end{cases}$$

where $\omega \neq 0$, $|\omega| < \frac{1}{2}$ and w is a function twice continuously differentiable
in the interval $(\frac{1}{3}, \frac{4}{3})$, $w(x) = x^\delta(-\ln x)^\omega$ for $x \in (\frac{1}{3}, \frac{1}{2})$ and $w(1) = 0$.
Then *the functional* F *defined by*

$$<F,v> = \int_0^1 f(x) \, v(x) \, dx$$

belongs to $\left[W_0^{1,2}((0,1);x,-\varepsilon)\right]^*$. Indeed, inequality (35.11) implies

$$\left| \int_0^{1/2} x^{\delta-2}(-\ln x)^{\omega-1} \, v(x) \, dx \right|$$

$$\leq \frac{4}{|1 - \varepsilon|^2} \left(\int_0^1 |v'(x)|^2 \, x^{-\varepsilon} \, dx \right)^{1/2} \left(\int_0^{1/2} x^{-1}(-\ln x)^{2\omega-2} \, dx \right)^{1/2}$$

and consequently, we have

$$\left| \int_0^1 f(x) \, v(x) \, dx \right| \leq c_1 \left(\int_0^1 |v'(x)|^2 \, x^{-\varepsilon} \, dx \right)^{1/2} \leq c_2 \|v;W_0^{1,2}((0,1);x,-\varepsilon)\|$$

for every $v \in C_0^\infty(\Omega)$ (and so for every $v \in W_0^{1,2}((0,1);x,-\varepsilon)$).

(iv) One can easily verify that *the function*

$$(35.13) \quad u(x) = \begin{cases} 0 & \text{for } x = 0 , \\ x^\delta(-\ln x)^\omega & \text{for } x \in (0,1/2] , \\ w(x) & \text{for } x \in (1/2,1] \end{cases}$$

is a weak solution of the Dirichlet problem (35.8) [i.e., u fulfils the identity $a(u,v) = <F,v>$ for every $v \in C_0^\infty((0,1))$]. On the other hand, in view of the relations $2\delta - 2 + \varepsilon = -1$, $2\omega > -1$ we have

$$\int\limits_0^1 |u'(x)|^2 \, x^\varepsilon \, dx \geq c_3 \int\limits_0^{1/2} x^{2\delta-2+\varepsilon}(-\ln x)^{2\omega} \, dx = + \infty$$

and consequently,

$$(35.14) \quad u \notin W^{1,2}((0,1);x,\varepsilon) .$$

In view of the uniqueness, *there cannot be another weak solution* of the Dirichlet problem (35.8) belonging to the weighted space $W^{1,2}((0,1);x,\varepsilon)$.

Thus, we have shown that for $\varepsilon \neq 0$, $|\varepsilon|$ sufficiently small, the *condition* $\varepsilon \in J$ *is n o t fulfilled* if we choose $\delta = (1 - \varepsilon)/2$ in (35.8).

Section 12 . *T h e   N e u m a n n   p r o b l e m*

§ 36 . F o r m u l a t i o n   o f   t h e   p r o b l e m

36.1. <u>INTRODUCTION</u>. In the sequel, we will work with the weighted Sobolev space

$$(36.1) \quad W^{k,2}(\Omega;s(d_M))$$

defined as the set of functions $u = u(x)$, $x \in \Omega$, such that

$$(36.2) \quad \|u;W^{k,2}(\Omega;s(d_M))\|^2 = \sum_{|\alpha|\leq k} \int\limits_\Omega |D^\alpha u(x)|^2 \, s(d_M(x)) \, dx < \infty ,$$

while in the case of the Dirichlet problem, we used very substantially the properties of the space

$$(36.3) \quad W_0^{k,2}(\Omega;s(d_M))$$

defined as the closure of the set $C_0^\infty(\Omega)$ with respect to the norm (36.2).

[Let us recall that in the weight function

$$(36.4) \quad s(d_M(x)) ,$$

$s = s(t)$ is a given (continuous) positive function defined on $(0,\infty)$,

$d_M(x) = \text{dist}(x,M)$ and M is an m-dimensional manifold on the boundary $\partial\Omega$ of the domain $\Omega \subset \mathbb{R}^N$, $0 \leq m \leq N - 1$.]

Already in the case of the Dirichlet problem we observed that *imbedding theorems* and estimates for the space $W_0^{k,2}(\Omega;s(d_M))$ played a substantial role; the propositions from [I] already indicate that the assumptions of these theorems are very restrictive *in the case of the spaces* $W^{k,2}(\Omega;s(d_M))$ (compare with Subsection 0.11 for $s(t) = t^\varepsilon$) and consequently, our possibilities will be limited in this case, too.

[Let us recall that in the case $s(t) = t^\varepsilon$, $\dim M = N - 1$ and $k = 1$ – i.e. in the case of power type weights – the main imbedding theorem for the space $W_0^{1,2}(\Omega;s(d_M)) = W_0^{1,2}(\Omega;d_M,\varepsilon)$ holds *for all* $\varepsilon \neq 1$, while for the spaces $W^{1,2}(\Omega;d_M,\varepsilon)$ it holds *only for* $\varepsilon > 1$.]

Since the imbedding theorems just mentioned are very complicated if we consider general weight functions $s(d_M)$, here we restrict ourselves to the case of *power type weights*, i.e., to the case $s(t) = t^\varepsilon$. Nevertheless, the formulation of the Neumann problem will be made using *general weight functions* $s = s(t)$.

36.2. <u>THE DIFFERENTIAL OPERATOR</u>. On a domain $\Omega \subset \mathbb{R}^N$ let us consider a *formal differential operator* $\mathcal{L}$ of order $2k$, given by the formula

$$(36.5) \qquad (\mathcal{L}u)(x) = \sum_{|\alpha|,|\beta| \leq k} (-1)^{|\alpha|} D^\alpha\big(a_{\alpha\beta}(x)D^\beta u(x)\big) .$$

We assume that the coefficients $a_{\alpha\beta}$ of the operator $\mathcal{L}$ satisfy the condition

$$(36.6) \qquad a_{\alpha\beta} \in L^\infty(\Omega) .$$

The operator $\mathcal{L}$ is associated, in the usual way, with a *bilinear form* $a = a(u,v)$ given by the formula

$$(36.7) \qquad a(u,v) = \sum_{|\alpha|,|\beta| \leq k} \int_\Omega a_{\alpha\beta}(x) \, D^\beta u(x) \, D^\alpha v(x) \, dx .$$

We assume that the operator $\mathcal{L}$ is *elliptic* (with respect to the Neumann problem !), i.e., there exists a positive constant c_0 such that for all functions $u \in W^{k,2}(\Omega)$,

$$(36.8) \qquad a(u,u) \geq c_0\|u;W^{k,2}(\Omega)\|^2 .$$

Let us mention that in the sequel we consider *real functions and real spaces only* !

36.3. <u>REMARK</u>. The definition of ellipticity involves classical Sobolev spaces $W^{k,2}(\Omega)$. This reflects the fact that in this chapter we are interested in the

application of weighted spaces to e l l i p t i c operators (as concerns
applications to operators whose ellipticity is in a certain sense perturbed,
see Chapter V).

The fact that we are concerned with the *Neumann problem* is expressed by
our requirement that the estimate (36.8) should hold for functions from
$W^{k,2}(\Omega)$ - in contradistinction to the *Dirichlet problem* where we assumed that
this estimate takes place for functions u from the rather smaller class
$W_0^{k,2}(\Omega)$ (or - which is the same - for $u \in C_0^\infty(\Omega)$: see (35.6) or [I], Section
13.2, formula (13.4)).

Since we are working with domains Ω with a "smooth boundary", we may
assume that the estimate (36.8) takes place for all functions u from the
class $C^\infty(\overline{\Omega})$ which is *dense* in $W^{k,2}(\Omega)$.

36.4. <u>THE NEUMANN PROBLEM</u>. Let $\Omega \subset \mathbb{R}^N$, $M \subset \partial\Omega$ and let $s = s(t)$ be a
weight function. We shall consider the weighted Sobolev spaces

$$W^{k,2}\big(\Omega;s(d_M)\big) \quad \text{and} \quad W^{k,2}\big(\Omega;s^{-1}(d_M)\big)$$

where $s^{-1}(t) = 1/s(t)$ (see Subsection 36.1, formulae (36.1) and (36.2)).
Further, let F be a given continuous linear functional on $W^{k,2}\big(\Omega;s^{-1}(d_M)\big)$,
i.e.,

$$F \in \big[W^{k,2}\big(\Omega;s^{-1}(d_M)\big)\big]^* .$$

We shall say that a function $u \in W^{k,2}\big(\Omega;s(d_M)\big)$ is a *weak solution of
the Neumann problem* (for the differential operator $\mathscr{L}$ from Subsection 36.2)
if for all functions $v \in W^{k,2}\big(\Omega;s^{-1}(d_M)\big)$ we have

(36.9) $a(u,v) = \langle F,v \rangle$

where $a(u,v)$ is the bilinear form (36.7) and the symbol $\langle F,v \rangle$ stands for
the value of the functional F from the dual space $\big[W^{k,2}\big(\Omega;s^{-1}(d_M)\big)\big]^*$ at
the "point" v .

36.5. <u>REMARKS</u>. (i) Again it suffices to require that the "integral identity"
(36.9) be satisfied only for all $v \in C^\infty(\overline{\Omega})$ - compare with the end of Remark
36.3.

(ii) Usually we assume that the functional $F \in \big[W^{k,2}\big(\Omega;s^{-1}(d_M)\big)\big]^*$ has
the form of the sum of two such functionals, the second of which vanishes on
$C_0^\infty(\Omega)$:

$$F = F_1 + F_2 , \quad \langle F_2,v \rangle = 0 \quad \text{for} \quad v \in C_0^\infty(\Omega) .$$

In this case, the functional F_1 describes the right hand side in the *formal*
differential equation $\mathscr{L}u = f$ on Ω while the functional F_2 expresses the
right hand sides in the corresponding Neumann boundary conditions on $\partial\Omega$. We

will not go into details here since the situation corresponds to that commonly
known in the case when the solution is sought in the *classical* Sobolev space
$W^{k,2}(\Omega)$: it suffices to take $s(t) \equiv 1$ here and compare with the literature
– see, e.g., J. NEČAS [1], Chapter 3 ; K. REKTORYS [1], Chapter 32, or S. FU-
ČÍK, A. KUFNER [1], Chapter III. Let us only mention the case $k = 1$, where we
consider second order equations and work with the spaces $W^{1,2}(\Omega;s(d_M))$ and
$W^{1,2}(\Omega;s^{-1}(d_M))$: We can take $F = F_1 + F_2$ with

$$(36.10) \qquad <F_1,v> = \int_\Omega f(x) \; v(x) \; dx \; , \qquad <F_2,v> = \int_{\partial\Omega} g(S) \; v(S) \; dS$$

where the given function f represents the right hand side in the *formal* differential equation

$$\mathscr{L} u = f \quad \text{on} \quad \Omega \; ,$$

and the given function g represents the right hand side in the *formal* Neumann
boundary condition

$$\sum_{|\alpha|,|\beta|=1} a_{\alpha\beta} \; \nu_\alpha \; D^\beta u = g \quad \text{on} \quad \partial\Omega \; .$$

Here $\nu = \{\nu_\alpha, |\alpha| = 1\}$ is the unit vector of the outer normal to the boundary
$\partial\Omega$ of the domain Ω . Of course, the possibility to represent the functionals
F_1 , F_2 in the form (36.10) is closely connected with the properties of *traces*
of functions from $W^{1,2}(\Omega;s(d_M))$ on $\partial\Omega$. The question of a full characteriza-
tion of traces is still open, and therefore we used (36.10) mainly in order to
illustrate the whole situation. We will come back to this question later for
the case of power type weights.

§ 37 . E x i s t e n c e   t h e o r e m s

In what follows, our aim is to decide for *which* weight functions $s = s(t)$
the *existence* of a weak solution of the Neumann problem in the weighted space
$W^{k,2}(\Omega;s(d_M))$ can be guaranteed. As was already said in Remark 36.5 (ii),
there exists at least one such weight function, namely, the function $s(t) \equiv 1$.

The main tool for deriving an existence theorem for a weak solution will
be the following generalization of the Lax–Milgram Theorem (see Subsection
15. 5) due to J. NEČAS [1], Chap. 6, Sec. 3.1 :

<u>37.1. LEMMA</u>. *Let* H_1 , H_2 *be two Hilbert spaces. Let* $b(u,v)$ *be a bilinear
form defined on the cartesian product* $H_1 \times H_2$, *and let there exist positive
constants* c_1 , c_2 , c_3 *such that*

    (i) *for all* $u \in H_1$ *and* $v \in H_2$ *we have*

(37.1) $\qquad |b(u,v)| \leq c_1 \|u;H_1\| \cdot \|v;H_2\|$;

(ii) *for all* $u \in H_1$ *we have*

(37.2) $\qquad \sup_{\|v;H_2\| \leq 1} |b(u,v)| \geq c_2 \|u;H_1\|$;

(iii) *for all* $v \in H_2$ *we have*

(37.3) $\qquad \sup_{\|u;H_1\| \leq 1} |b(u,v)| \geq c_3 \|v;H_2\|$.

Let h *be a continuous linear functional from* H_2^* . *Then there exists one and only one element* $u \in H_1$ *such that for all elements* $v \in H_2$ *we have*

(37.4) $\qquad b(u,v) = \langle h,v \rangle$,

and there is a positive constant c *such that*

(37.5) $\qquad \|u;H_1\| \leq c\|h;H_2^*\|$.

<u>37.2. REMARKS</u>. (i) Condition (i) of Lemma 37.1 states that the bilinear form $b(u,v)$ is *continuous* on $H_1 \times H_2$. A bilinear form satisfying conditions (ii) and (iii) of Lemma 37.1 - i.e., the relations (37.2) and (37.3) - is said to be (H_1,H_2)-*elliptic*.

(ii) Later, we shall need the classical Lax-Milgram Theorem mentioned above. Therefore, let us recall that the *Lax-Milgram Theorem is a special case of Lemma 37.1* where we set
$$H_1 = H_2 = H .$$
Conditions (ii), (iii) of Lemma 37.1 are then replaced by a *single* condition

(37.6) $\qquad b(u,u) \geq c_4 \|u;H\|^2$

which should hold for all $u \in H$ with a constant $c_4 > 0$ independent of u ; in this case we say that the bilinear form $b(u,v)$ is H-*elliptic*.

<u>37.3. THE CONTINUITY OF THE BILINEAR FORM $a(u,v)$</u> . Let us now choose the bilinear form $a(u,v)$ from (36.7) for the form $b(u,v)$ and let us choose the Hilbert spaces H_1 , H_2 in the following way :

(37.7) $\qquad H_1 = W^{k,2}\big(\Omega;s(d_M)\big)$, $\quad H_2 = W^{k,2}\big(\Omega;s^{-1}(d_M)\big)$

with $s^{-1}(t) = 1/s(t)$. Using the boundedness of the coefficients $a_{\alpha\beta}$ - see (36.6) - and the Hölder inequality, we successively obtain

$$|a(u,v)| \leq \sum_{|\alpha|,|\beta| \leq k} \|a_{\alpha\beta};L^\infty(\Omega)\| \int_\Omega |D^\beta u(x)\, D^\alpha v(x)|\ dx =$$

$$= \sum_{|\alpha|,|\beta|\le k} \|a_{\alpha\beta};L^\infty(\Omega)\| \int_\Omega |D^\beta u(x)\ s^{1/2}(d_M(x))|\cdot|D^\alpha v(x)s^{-1/2}(d_M(x))|\ dx$$

$$\le \sum_{|\alpha|,|\beta|\le k} \|a_{\alpha\beta};L^\infty(\Omega)\|\cdot\|D^\beta u;L^2(\Omega;s(d_M))\|\cdot\|D^\alpha v;L^2(\Omega;s^{-1}(d_M))\|$$

$$\le c_1\|u;W^{k,2}(\Omega;s(d_M))\|\cdot\|v;W^{k,2}(\Omega;s^{-1}(d_M))\|$$

with the constant $c_1 = \sum_{|\alpha|,|\beta|\le k}\|a_{\alpha\beta};L^\infty(\Omega)\|$. So we have already proved the following assertion :

The bilinear form $a(u,v)$ *from (36.7) is continuous on* $H_1 \times H_2$ *provided the Hilbert spaces* H_1 , H_2 *are chosen according to (37.7) and the conditions (36.6) are fulfilled.*

37.4. POWER TYPE WEIGHTS.

In the sequel we shall deal mainly with power type weights, i. e., we set $s(t) = t^\varepsilon$, $\varepsilon \in \mathbb{R}$, so that we have

$$(37.8) \qquad H_1 = W^{k,2}(\Omega;d_M,\varepsilon) \ , \quad H_2 = W^{k,2}(\Omega;d_M,-\varepsilon)$$

for (37.7). Let us recall the important fact that by the imbedding theorems mentioned in Subsection 0.11 we have for $u \in W^{k,2}(\Omega;d_M,\eta)$ the inclusion

$$(37.9) \qquad D^\gamma u \in L^2(\Omega;\ d_M,\ \eta - 2(k - |\gamma|)) \quad \text{for} \quad |\gamma| \le k$$

and, moreover,

$$(37.10) \qquad \|D^\gamma u;\ L^2(\Omega;\ d_M,\ \eta - 2(k - |\gamma|))\| \le c_\gamma\|u;W^{k,2}(\Omega;d_M,\eta)\|$$

(with $c_\gamma > 0$ independent of u) provided the following condition is fulfilled :

$$(37.11) \qquad \eta > 2k + m - N$$

(see formulae (0.33) and (0.34) for $p = 2$; also see [I], Chapter 8).

In view of (37.8), we will use this last result for $\eta = \varepsilon$ as well as for $\eta = - \varepsilon$, and consequently, the following two conditions should be fulfilled *simultaneously* :

$$\varepsilon > 2k + m - N \quad \text{and} \quad - \varepsilon > 2k + m - N \ ,$$

i.e.,

$$2k + m - N < \varepsilon < N - m - 2k \ .$$

This leads to the inequality $2k < N - m$ or, since N and m are non negative integers, to the inequality

$$(37.12) \qquad N - m \ge 2k + 1 \ .$$

This inequality will play an important (and very restrictive) role in our

further considerations : It indicates that the dimension m of the manifold
$M \subset \partial\Omega$ (from which we are taking the distance d_M) should be small as compared with the dimension N of the domain Ω itself, and namely, it should be
the smaller the bigger is the order 2k of the differential operator $\mathcal{L}$.
Since we have $k \geq 1$, we should have *at least*

(37.13) $N - m \geq 3$

which excludes a number of important particular cases from our considerations
(as for example vertices of polygonal plane domains − N = 2 , m = 0 ; or
edges and sides of threedimensional cubes − N = 3 , m = 1 and m = 2 ,
respectively).

We can rewrite inequality (37.12) into the form

(37.14) $k < \dfrac{N - m}{2}$

which gives us the "admissible" order of the differential operator $\mathcal{L}$ with
respect to the dimensions of the domain Ω and the set $M \subset \partial\Omega$ considered.

37.5. THE CONTINUITY OF a(u,v) ON THE SPACES (37.8). It follows immediately
from the results of Subsection 37.3 that the bilinear form a(u,v) from (36.7)
is continuous on $H_1 \times H_2$, where the Hilbert spaces H_1 , H_2 are chosen according to (37.8), provided condition (36.6) is fulfilled, i.e., $a_{\alpha\beta} \in L^\infty(\Omega)$.
This last assumption can be w e a k e n e d :

Let us suppose

(37.15) $a_{\alpha\beta} d_M^{2k-|\alpha|-|\beta|} \in L^\infty(\Omega)$, $|\alpha|$, $|\beta| \leq k$.

If, moreover, condition (37.12) is fulfilled then the bilinear form a(u,v)
from (36.7) is continuous on $H_1 \times H_2$ *provided the Hilbert spaces* H_1 , H_2
are chosen according to (37.8) and ε *is such that*

(37.16) $|\varepsilon| < N - m - 2k$.

The p r o o f is simple : In view of (37.15) and of the Hölder inequality we have

$$\left| \int_\Omega a_{\alpha\beta}(x)\, D^\beta u(x)\, D^\alpha v(x)\, dx \right|$$

$$= \left| \int_\Omega a_{\alpha\beta}\, d_M^{2k-|\alpha|-|\beta|}\, D^\beta u\, d_M^{\varepsilon/2 -k+|\beta|}\, D^\alpha v\, d_M^{-\varepsilon/2 -k+|\alpha|}\, dx \right|$$

$$\leq \left\| a_{\alpha\beta}\, d_M^{2k-|\alpha|-|\beta|}\, ;\, L^\infty(\Omega) \right\| \cdot \left\| D^\beta u;\, L^2(\Omega;\, d_M,\, \varepsilon - 2(k - |\beta|)) \right\| \cdot$$

$$\cdot \left\| D^\alpha v;\, L^2(\Omega\,;\, d_M,\, \varepsilon - 2(k - |\alpha|)) \right\|$$

and hence, using the estimate (37.10) for $\gamma = \beta$ and $\eta = \varepsilon$ in the case of the function u , and for $\gamma = \alpha$ and $\eta = -\varepsilon$ in the case of the function v , we immediately obtain, in view of (37.16), the estimate

$$|a(u,v)| \le c_1 \|u; \ W^{k,2}(\Omega;d_M,\varepsilon)\| \cdot \|v; \ W^{k,2}(\Omega;d_M,-\varepsilon)\|$$

with the constant $c_1 = \sum\limits_{|\alpha|,|\beta| \le k} c_\alpha c_\beta \|a_{\alpha\beta} d_M^{2k-|\alpha|-|\beta|} ; \ L^\infty(\Omega)\|$; c_α , c_β are the corresponding constants from (37.10).

<u>37.6.</u> <u>REMARK.</u> Due to conditions (37.15), also such coefficients $a_{\alpha\beta}$ of the operator $\mathscr{L}$ are admissible which can be unbounded in the neighbourhood of the set $M \subset \partial\Omega$, of course under the rather restrictive assumption (37.12).

Considering in [I] the Dirichlet problem, we investigated the bilinear form $a(u,v)$ as a form on the cartesian product $W_0^{k,2}(\Omega;d_M,\varepsilon) \times W_0^{k,2}(\Omega;d_M,-\varepsilon)$, i.e., on the product of W_0-spaces. For functions from these spaces the estimate (37.10) also holds, but under substantially w e a k e r assumptions on η : The number $\eta \in \mathbb{R}$ can be arbitrary if $m < N - 1$ and η has to be different from the numbers $2j + m - N$, $j = 1,2,\ldots,k$, if $m = N - 1$ (see again Subsection 0.11 or [I], Chapter 8). Consequently, *we can weaken conditions (36.6) to the form (37.15) in the case of the Dirichlet problem, too, assuming only that* $\varepsilon \ne \pm (2j + m - N)$, $j = 1,2,\ldots,k$, *for* $m = N - 1$, *and otherwise assuming* $\varepsilon \in \mathbb{R}$. [We have already used these weaker conditions on the coefficients $a_{\alpha\beta}$ in Example 35.3 - see the footnote on p. 142.]

<u>37.7.</u> <u>SEVERAL ESTIMATES.</u> In what follows we shall seek conditions on the number ε guaranteeing that under the choice (37.8) the *bilinear form* $a(u,v)$ *from (36.7) is* (H_1,H_2)-*elliptic.* For this purpose, we shall derive some estimates. During the calculation, we will - among other - differentiate the distance function $d_M(x)$. Therefore, let us agree that in case of ambiguity we shall have in mind the so-called *regularized distance* which belongs to $C^\infty(\Omega)$ and is *equivalent* to d_M . In the sequel, we shall denote this regularized distance again by d_M , making use of the estimate

$$(37.17) \qquad |D^\sigma d_M^\varepsilon(x)| \le |\varepsilon| c_\sigma^{*} d_M^{\varepsilon-|\sigma|}(x) \ ;$$

let us point out that c_σ^{*} depends on ε , $c_\sigma^{*} = c_\sigma^*(\varepsilon)$ - more precisely, c_σ^* is a polynomial in ε . If we restrict our considerations *apriori* to values ε such that $|\varepsilon|$ is bounded, $|\varepsilon| < C_0$ with a suitable constant C_0 , then we can assume that c_σ^* *does not depend on* ε .

(i) For $u \in W^{k,2}(\Omega;d_M,\varepsilon)$ let us set

$$v = u d_M^\varepsilon$$

and let us investigate under what conditions

$$v \in W^{k,2}(\Omega;d_M,-\varepsilon) \ .$$

For simplicity, we shall use the symbol $\|\cdot\|_\eta$ for the norm in $W^{k,2}(\Omega;d_M,\eta)$. We have

$$\|v\|_{-\varepsilon}^2 = \|u d_M^\varepsilon\|_{-\varepsilon} = \sum_{|\alpha|\leq k} \int_\Omega \left|D^\alpha(u d_M^\varepsilon)\right|^2 d_M^{-\varepsilon}\,dx$$

$$= \sum_{|\alpha|\leq k} \int_\Omega \left|D^\alpha u\,d_M^\varepsilon + \sum_{\substack{\gamma+\delta=\alpha\\|\delta|\geq 1}} c_{\gamma\delta}D^\gamma u\,D^\delta d_M^\varepsilon\right|^2 d_M^{-\varepsilon}\,dx$$

$$= \sum_{|\alpha|\leq k} \int_\Omega \Big\{ |D^\alpha u|^2\,d_M^{2\varepsilon} + 2\sum_{\substack{\gamma+\delta=\alpha\\|\delta|\geq 1}} c_{\gamma\delta}D^\alpha u\,d_M^\varepsilon\,D^\gamma u\,D^\delta d_M^\varepsilon$$

$$+ \sum_{\substack{\gamma+\delta=\alpha\\|\delta|\geq 1}}\sum_{\substack{\omega+\tau=\alpha\\|\tau|\geq 1}} c_{\gamma\delta}c_{\omega\tau}D^\gamma u\,D^\delta d_M^\varepsilon\,D^\omega u\,D^\tau d_M^\varepsilon\Big\} d_M^{-\varepsilon}\,dx \ .$$

Using now the estimate (37.17) for $\sigma=\delta$ and $\sigma=\tau$, we obtain

$$(37.18)\quad \|v\|_{-\varepsilon}^2 \leq \sum_{|\alpha|\leq k}\int_\Omega\Big\{|D^\alpha u|^2\,d_M^{2\varepsilon} + |\varepsilon|c_1\sum_{\substack{\gamma+\delta=\alpha\\|\delta|\geq 1}}|D^\alpha u|\,d_M^\varepsilon|D^\gamma u|\,d_M^{\varepsilon-|\delta|}$$

$$+ |\varepsilon|^2 c_2\sum_{\substack{\gamma+\delta=\alpha\\|\delta|\geq 1}}\sum_{\substack{\omega+\tau=\alpha\\|\tau|\geq 1}}|D^\gamma u|\,d_M^{\varepsilon-|\delta|}|D^\omega u|\,d_M^{\varepsilon-|\tau|}\Big\}\,d_M^{-\varepsilon}\,dx$$

$$= \sum_{|\alpha|\leq k}\int_\Omega |D^\alpha u|^2\,d_M^\varepsilon\,dx + c_1|\varepsilon|\sum_{|\alpha|\leq k}\sum_{\substack{\gamma+\delta=\alpha\\|\delta|\geq 1}}\int_\Omega|D^\alpha u|\,|D^\gamma u|\,d_M^{\varepsilon-|\delta|}\,dx$$

$$+ c_2|\varepsilon|^2\sum_{|\alpha|\leq k}\sum_{\substack{\gamma+\delta=\alpha\\|\delta|\geq 1}}\sum_{\substack{\omega+\tau=\alpha\\|\tau|\geq 1}}\int_\Omega|D^\gamma u|\,|D^\omega u|\,d_M^{\varepsilon-|\delta|-|\tau|}\,dx \ .$$

[Here, $c_1 = 2\sum_{\gamma+\delta} c_{\gamma\delta}c_\delta^*$ and $c_2 = \sum_{\gamma+\delta}\sum_{\omega+\tau} c_{\gamma\delta}c_{\omega\tau}c_\delta^*c_\tau^*$ where c_δ^* , c_τ^* are the constants from estimate (37.17).] The first sum in the last expression is equal to $\|u\|_\varepsilon^2$; for the integrals in the second and third sums we have, from the Hölder inequality,

$$J_1 = \int_\Omega |D^\alpha u|\,|D^\gamma u|\,d_M^{\varepsilon-|\delta|}\,dx = \int_\Omega |D^\alpha u|\,d_M^{\varepsilon/2}\,|D^\gamma u|\,d_M^{\varepsilon/2-|\delta|}\,dx$$

$$\leq \|D^\alpha u;\ L^2(\Omega;d_M,\varepsilon)\| \cdot \|D^\gamma u;\ L^2(\Omega;d_M,\varepsilon-2|\delta|)\|$$

and

$$J_2 = \int_\Omega |D^\gamma u|\,|D^\omega u|\,d_M^{\varepsilon-|\delta|-|\tau|}\,dx = \int_\Omega |D^\gamma u|\,d_M^{\varepsilon/2-|\delta|}|D^\omega u|d_M^{\varepsilon/2-|\tau|}\,dx \leq$$

$$\leq \|D^\gamma u;\ L^2(\Omega;d_M,\varepsilon-2|\delta|)\| \cdot \|D^\omega u;\ L^2(\Omega;d_M,\varepsilon-2|\tau|)\|\ .$$

In view of the conditions $\gamma + \delta = \alpha$ and $\omega + \tau = \alpha$ we have $|\delta| = |\alpha| - |\gamma|$ and $|\tau| = |\alpha| - |\omega|$, so that

$$\|D^\gamma u;\ L^2(\Omega;d_M,\varepsilon-2|\delta|)\| = \|D^\gamma u;\ L^2(\Omega;d_M,\varepsilon-2(|\alpha|-|\gamma|))\|$$

and

$$\|D^\omega u;\ L^2(\Omega;d_M,\varepsilon-2|\tau|)\| = \|D^\omega u;\ L^2(\Omega;d_M,\varepsilon-2(|\alpha|-|\omega|))\|\ .$$

Supposing

$$\varepsilon > 2|\alpha| + m - N$$

and using the imbedding theorems mentioned in Subsection 37.4 we obtain upper estimates of the last two norms by the expressions

$$c_\gamma \|u;\ W^{|\alpha|,2}(\Omega;d_M,\varepsilon)\| \quad \text{and} \quad c_\omega \|u;\ W^{|\alpha|,2}(\Omega;d_M,\varepsilon)\|\ ,$$

respectively [see estimate (37.10) for $\eta = \varepsilon$ and $|\alpha|$ instead of k]. However, the condition $|\alpha| \leq k$ implies that

$$\|u;\ W^{|\alpha|,2}(\Omega;d_M,\varepsilon)\| \leq \|u;\ W^{k,2}(\Omega;d_M,\varepsilon)\| = \|u\|_\varepsilon\ ,$$

so that we have shown that

$$J_1 \leq c_\gamma \|u\|_\varepsilon^2\ , \quad J_2 \leq c_\gamma c_\omega \|u\|_\varepsilon^2$$

provided the condition

$$(37.19) \qquad \varepsilon > 2k + m - N$$

is fulfilled.

Using all these estimates in (37.18) we have

$$\|v\|_{-\varepsilon}^2 \leq \left(1 + c_3|\varepsilon| + c_4|\varepsilon|^2\right) \|u\|_\varepsilon^2\ ,$$

i.e., we have shown that *under the condition* (37.19) *the following estimate holds*:

$$(37.20) \qquad \|ud_M^\varepsilon;\ W^{k,2}(\Omega;d_M,-\varepsilon)\|^2 \leq \left(1 + c_3|\varepsilon| + c_4|\varepsilon|^2\right) \|u;\ W^{k,2}(\Omega;d_M,\varepsilon)\|^2$$

with positive constants c_3 , c_4 *independent of* u .

(ii) For $u \in W^{k,2}(\Omega;d_M,\varepsilon)$, let us now estimate the norm of $ud_M^{\varepsilon/2}$ in the *classical* Sobolev space $W^{k,2}(\Omega)$. We obtain

$$(37.21) \qquad \|ud_M^{\varepsilon/2};\ W^{k,2}(\Omega)\|^2 = \sum_{|\alpha|\leq k} \int_\Omega |D^\alpha(ud_M^{\varepsilon/2})|^2\ dx$$

$$= \sum_{|\alpha|\leq k} \int_\Omega \Big[D^\alpha u\, d_M^{\varepsilon/2} + \sum_{\substack{\gamma+\delta=\alpha \\ |\delta|\geq 1}} c_{\gamma\delta} D^\gamma u\, D^\delta d_M^{\varepsilon/2}\Big]^2\ dx \geq$$

$$\geq \sum_{|\alpha|\leq k} \int_\Omega |D^\alpha u|^2 \, d_M^\varepsilon \, dx - 2 \sum_{\substack{|\alpha|\leq k \\ }} \sum_{\substack{\gamma+\delta=\alpha \\ |\delta|\geq 1}} |c_{\gamma\delta}| \int_\Omega |D^\alpha u| \, d_M^{\varepsilon/2} |D^\gamma u| \, |D^\delta d_M^{\varepsilon/2}| dx$$

$$- \sum_{|\alpha|\leq k} \sum_{\substack{\gamma+\delta=\alpha \\ |\delta|\geq 1}} \sum_{\substack{\omega+\tau=\alpha \\ |\tau|\geq 1}} |c_{\gamma\delta} c_{\omega\tau}| \int_\Omega |D^\gamma u| \, |D^\delta d_M^{\varepsilon/2}| \, |D^\omega u| \, |D^\tau d_M^{\varepsilon/2}| \, dx \; .$$

Using again — as in point (i) above, see (37.18) — estimate (37.17) for $\sigma = \delta$ and $\sigma = \tau$, we obtain

$$J_3 = \int_\Omega |D^\alpha u| \, d_M^{\varepsilon/2} |D^\gamma u| \, |D^\delta d_M^{\varepsilon/2}| dx \leq |\tfrac{\varepsilon}{2}| c_\delta^* \int_\Omega |D^\alpha u| d_M^{\varepsilon/2} |D^\gamma u| d_M^{\varepsilon/2-|\delta|} dx$$

and

$$J_4 = \int_\Omega |D^\gamma u| \, |D^\delta d_M^{\varepsilon/2}| \, |D^\omega u| \, |D^\tau d_M^{\varepsilon/2}| dx$$

$$\leq |\tfrac{\varepsilon}{2}|^2 c_\delta^* c_\tau^* \int_\Omega |D^\gamma u| d_M^{\varepsilon/2-|\delta|} |D^\omega u| d_M^{\varepsilon/2-|\tau|} \, dx \; .$$

The integrals on the right hand sides of the last two inequalities appeared already in point (i), and so we obtain in the same way as above the estimates

$$J_3 \leq c_5 |\varepsilon| \; \|u; \; W^{k,2}(\Omega;d_M,\varepsilon)\|^2$$

and

$$J_4 \leq c_6 |\varepsilon|^2 \; \|u; \; W^{k,2}(\Omega;d_M,\varepsilon)\|^2 \; .$$

Using these estimates in (37.21), we come eventually to the following result : *Under the condition* (37.19) *the following lower estimate holds:*

$$(37.22) \qquad \|ud_M^{\varepsilon/2}; \; W^{k,2}(\Omega)\|^2 \geq \left(1 - c_7 |\varepsilon| - c_8 |\varepsilon|^2\right) \|u; \; W^{k,2}(\Omega;d_M,\varepsilon)\|^2$$

with positive constants c_7 , c_8 *independent of* u .

(iii) Let $a(u,v)$ be the bilinear form from (36.7). Setting $v = ud_M^\varepsilon$ with $u \in W^{k,2}(\Omega;d_M,\varepsilon)$, we have $v \in W^{k,2}(\Omega;d_M,-\varepsilon)$ by (37.20). Consequently, the expression $a(u,ud_M^\varepsilon)$ is meaningful and can be rewritten in the form

$$(37.23) \qquad a(u,ud_M^\varepsilon) = a(ud_M^{\varepsilon/2},ud_M^{\varepsilon/2}) + J(\varepsilon)$$

where

$$(37.24) \qquad J(\varepsilon) = \sum_{|\alpha|,|\beta|\leq k} \int_\Omega a_{\alpha\beta} \left[D^\alpha(ud_M^\varepsilon)D^\beta u - D^\alpha(ud_M^{\varepsilon/2})D^\beta(ud_M^{\varepsilon/2})\right] \, dx \; .$$

Let us now estimate the expression $J(\varepsilon)$. Using — analogously as in points (i), (ii) above — the fact that

$$D^{\alpha}(ud_M^{\lambda}) = D^{\alpha}u \, d_M^{\lambda} + \sum_{\substack{\gamma+\delta=\alpha \\ |\delta| \geq 1}} c_{\gamma\delta} D^{\gamma}u \, D^{\delta}d_M^{\lambda} \, ,$$

$$D^{\beta}(ud_M^{\lambda}) = D^{\beta}u \, d_M^{\lambda} + \sum_{\substack{\omega+\tau=\beta \\ |\tau| \geq 1}} c_{\omega\tau} D^{\omega}u \, D^{\tau}d_M^{\lambda}$$

($\lambda = \varepsilon$ and $\lambda = \varepsilon/2$) we easily obtain that $J(\varepsilon)$ is a linear combination of integrals of the following four types :

$$
\begin{aligned}
\tilde{J}_1 &= \int_{\Omega} a_{\alpha\beta} D^{\gamma}u \, D^{\delta}d_M^{\varepsilon} \, D^{\beta}u \, dx \, , & \gamma + \delta = \alpha \, , \quad |\delta| \geq 1 \, , \\[4pt]
\tilde{J}_2 &= \int_{\Omega} a_{\alpha\beta} D^{\gamma}u \, D^{\delta}d_M^{\varepsilon/2} \, D^{\beta}u \, d_M^{\varepsilon/2} \, dx \, , & \gamma + \delta = \alpha \, , \quad |\delta| \geq 1 \, , \\[4pt]
\tilde{J}_3 &= \int_{\Omega} a_{\alpha\beta} D^{\alpha}u \, d_M^{\varepsilon/2} \, D^{\omega}u \, D^{\tau}d_M^{\varepsilon/2} \, dx \, , & \omega + \tau = \beta \, , \quad |\tau| \geq 1 \, , \\[4pt]
\tilde{J}_4 &= \int_{\Omega} a_{\alpha\beta} D^{\gamma}u \, D^{\delta}d_M^{\varepsilon/2} \, D^{\omega}u \, D^{\tau}d_M^{\varepsilon/2} \, dx \, , & \gamma + \delta = \alpha \, , \quad |\delta| \geq 1 \, , \\
& & \omega + \tau = \beta \, , \quad |\tau| \geq 1 \, .
\end{aligned}
$$

(37.25)

(iii-1) Let us deal with the integral $\tilde{J}_1$. An application of the formula (37.17) to the term $D^{\delta}d_M^{\varepsilon}$ leads to the estimate

$$
\begin{aligned}
|\tilde{J}_1| &\leq |\varepsilon| c_{\delta}^{*} \int_{\Omega} |a_{\alpha\beta}| \, |D^{\gamma}u| \, |D^{\beta}u| \, d_M^{\varepsilon-|\delta|} \, dx \\[4pt]
&= |\varepsilon| c_{\delta}^{*} \int_{\Omega} |a_{\alpha\beta}| \, d_M^{2k-|\alpha|-|\beta|} \, |D^{\gamma}u| \, d_M^{\varepsilon/2-(k-|\gamma|)} |D^{\beta}u| \, d_M^{\varepsilon/2-(k-|\beta|)} \, dx \, ;
\end{aligned}
$$

(37.26)

here the last equality follows from the fact that $\gamma + \delta = \alpha$ and consequently, $|\delta| = |\alpha| - |\gamma|$: for the exponents of d_M we have $2k - |\alpha| - |\beta| + \varepsilon/2 - (k - |\gamma|) + \varepsilon/2 - (k - |\beta|) = \varepsilon - |\alpha| + |\gamma| = \varepsilon - |\delta|$. Using the fact that $|a_{\alpha\beta}| d_M^{2k-|\alpha|-|\beta|}$ is bounded – see condition (37.15) – and then estimating the last integral by the Hölder inequality we immediately obtain

$$|\tilde{J}_1| \leq |\varepsilon| c_{\delta}^{*} \, \| a_{\alpha\beta} d_M^{2k-|\alpha|-|\beta|} ; \, L^{\infty}(\Omega) \| \cdot \| D^{\gamma}u; \, L^2(\Omega; \, d_M, \, \varepsilon-2(k-|\gamma|)) \|$$

$$\cdot \| D^{\beta}u; \, L^2(\Omega; \, d_M, \, \varepsilon-2(k-|\beta|)) \| \, .$$

The last two norms can be estimated by $c_{\gamma} \|u\|_{\varepsilon}$ and $c_{\beta} \|u\|_{\varepsilon}$ – similarly as in point (i) above – provided condition (37.19) is fulfilled. So we have obtained the estimate

$$(37.27) \qquad |\tilde{J}_1| \leq |\varepsilon| \tilde{c}_1 \, \| u; \, W^{k,2}(\Omega; d_M, \varepsilon) \|^2$$

with $\tilde{c}_1 = \| a_{\alpha\beta} d_M^{2k-|\alpha|-|\beta|} ; \, L^{\infty}(\Omega) \| \, c_{\delta}^{*} c_{\gamma} c_{\beta}$.

(iii-2) Similarly we proceed in the case of the integral $\tilde{J}_2$: An application of formula (37.17) to the term $D^{\delta}d_M^{\varepsilon/2}$ yields the estimate

$$|\tilde{J}_2| \leq \left|\frac{\varepsilon}{2}\right| c_{\delta}^* \int_{\Omega} |a_{\alpha\beta}| \ |D^{\gamma}u| \ d_M^{\varepsilon/2 - |\delta|} \ |D^{\beta}u| \ d_M^{\varepsilon/2} \ dx$$

$$= \left|\frac{\varepsilon}{2}\right| c_{\delta}^* \int_{\Omega} |a_{\alpha\beta}| \ |D^{\gamma}u| \ |D^{\beta}u| \ d_M^{\varepsilon - |\delta|} \ dx \ ,$$

and since the last integral is the same as the first integral in (37.26), we obtain by the same argument and assumptions as in (iii-1) the estimate

$$(37.28) \quad |\tilde{J}_2| \leq |\varepsilon| \ \tilde{c}_2 \ \|u; \ W^{k,2}(\Omega;d_M,\varepsilon)\|^2$$

with $\tilde{c}_2 = \tilde{c}_1/2$.

(iii-3) Since the integral $\tilde{J}_3$ can be obtained from the integral $\tilde{J}_2$ by exchanging the role of multiindices α and β and by replacing γ , δ by ω , τ , respectively, we immediately have the estimate

$$(37.29) \quad |\tilde{J}_3| \leq |\varepsilon| \ \tilde{c}_3 \ \|u; \ W^{k,2}(\Omega;d_M,\varepsilon)\|^2$$

with $\tilde{c}_3 = \|a_{\alpha\beta}d_M^{2k-|\alpha|-|\beta|}; \ L^{\infty}(\Omega)\| \ c_{\tau}^* \ c_{\omega} \ c_{\alpha}/2$.

(iii-4) A repeated application of formula (37.17) — to $D^{\delta}d_M^{\varepsilon/2}$ and to $D^{\tau}d_M^{\varepsilon/2}$ — yields in the case of the integral $\tilde{J}_4$ the estimate

$$|\tilde{J}_4| \leq \left|\frac{\varepsilon}{2}\right|^2 c_{\delta}^* c_{\tau}^* \int_{\Omega} |a_{\alpha\beta}| \ |D^{\gamma}u| \ d_M^{\varepsilon/2 - |\delta|} \ |D^{\omega}u| \ d_M^{\varepsilon/2 - |\tau|} \ dx$$

$$= \left|\frac{\varepsilon}{2}\right|^2 c_{\delta}^* c_{\tau}^* \int_{\Omega} |a_{\alpha\beta}| d_M^{2k-|\alpha|-|\beta|} |D^{\gamma}u| d_M^{\varepsilon/2 - (k-|\gamma|)} |D^{\omega}u| d_M^{\varepsilon/2 - (k-|\omega|)} \ dx$$

where we have used the fact that $|\delta| = |\alpha| - |\gamma|$ and $|\tau| = |\beta| - |\omega|$. An application of condition (37.15), the Hölder inequality and the imbedding theorems lead to the inequality

$$|\tilde{J}_4| \leq \left|\frac{\varepsilon}{2}\right|^2 c_{\delta}^* c_{\tau}^* \|a_{\alpha\beta}d_M^{2k-|\alpha|-|\beta|}; \ L^{\infty}(\Omega)\|$$

$$(37.30) \qquad \cdot \|D^{\gamma}u; \ L^2(\Omega; \ d_M, \ \varepsilon-2(k-|\gamma|))\| \cdot \|D^{\omega}u; \ L^2(\Omega; \ d_M, \ \varepsilon-2(k-|\omega|))\|$$

$$\leq |\varepsilon|^2 \ \tilde{c}_4 \ \|u; \ W^{k,2}(\Omega;d_M,\varepsilon)\|^2$$

with $\tilde{c}_4 = \|a_{\alpha\beta}d_M^{2k-|\alpha|-|\beta|}; \ L^{\infty}(\Omega)\| \ c_{\delta}^* \ c_{\tau}^* \ c_{\gamma} \ c_{\omega} \ /4$.

Summarizing the estimates (37.27) - (37.30), we can state that *under the conditions (37.15) and (37.19) the following estimate for the expression* $J(\varepsilon)$ *from (37.24) holds :*

$$(37.31) \qquad |J(\varepsilon)| \leq \left(c_9 |\varepsilon| + c_{10} |\varepsilon|^2 \right) \| u; \, W^{k,2}(\Omega; d_M, \varepsilon) \|^2$$

with positive constants c_9 , c_{10} *independent of* u .

Now we are able to prove the following assertion :

37.8. LEMMA. *Let* $a(u,v)$ *be the bilinear form from (36.7) with coefficients* $a_{\alpha\beta}$ *fulfilling the condition*

$$(37.15) \qquad a_{\alpha\beta} d_M^{2k-|\alpha|-|\beta|} \in L^\infty(\Omega) \, , \quad |\alpha|, \, |\beta| \leq k \, .$$

Let $\varepsilon \in \mathbb{R}$ *be such that*

$$(37.19) \qquad \varepsilon > 2k + m - N \, .$$

Then for $u \in W^{k,2}(\Omega; d_M, \varepsilon)$ *and* $v = u d_M^\varepsilon$ *we have*

$$(37.32) \qquad \frac{a(u,v)}{\| v; \, W^{k,2}(\Omega; d_M, -\varepsilon) \|} \geq c(\varepsilon) \| u; \, W^{k,2}(\Omega; d_M, \varepsilon) \|$$

where

$$(37.33) \qquad c(\varepsilon) = \frac{c_{11} - c_{12}|\varepsilon| - c_{13}|\varepsilon|^2}{\sqrt{1 + c_{14}|\varepsilon| + c_{15}|\varepsilon|^2}}$$

with positive constants c_{11} *to* c_{15} *independent of* u .

P r o o f . From (37.23) it follows that

$$(37.34) \qquad a(u, u d_M^\varepsilon) \geq a(u d_M^{\varepsilon/2}, \, u d_M^{\varepsilon/2}) - |J(\varepsilon)| \, .$$

The ellipticity condition (36.8) yields

$$a(u d_M^{\varepsilon/2}, u d_M^{\varepsilon/2}) \geq c_0 \| u d_M^{\varepsilon/2}; \, W^{k,2}(\Omega) \|^2$$

and from (37.22) we have

$$a(u d_M^{\varepsilon/2}, \, u d_M^{\varepsilon/2}) \geq \left(c_0 - c_0 c_7 |\varepsilon| - c_0 c_8 |\varepsilon|^2 \right) \| u; \, W^{k,2}(\Omega; d_M, \varepsilon) \|^2 \, .$$

In view of (37.34) and (37.31) we obtain

$$(37.35) \qquad \begin{aligned} &a(u, u d_M^\varepsilon) \\ &\geq \left[\left(c_0 - c_0 c_7 |\varepsilon| - c_0 c_8 |\varepsilon|^2 \right) - \left(c_9 |\varepsilon| + c_{10} |\varepsilon|^2 \right) \right] \| u; \, W^{k,2}(\Omega; d_M, \varepsilon) \|^2 \, , \end{aligned}$$

and this inequality together with (37.20), which can be rewritten in the form

$$\frac{1}{\| u d_M^\varepsilon; \, W^{k,2}(\Omega; d_M, -\varepsilon) \|} \geq \frac{1}{\sqrt{1 + c_3 |\varepsilon| + c_4 |\varepsilon|^2}} \| u; \, W^{k,2}(\Omega; d_M, \varepsilon) \|^{-1} \, ,$$

yields the desired lower estimate (37.32) with $c(\varepsilon)$ given by (37.33), where

$$c_{11} = c_0 \; , \quad c_{12} = c_0 c_7 + c_9 \; , \quad c_{13} = c_0 c_8 + c_{10} \; , \quad c_{14} = c_3 \quad \text{and} \quad c_{15} = c_4 \; .$$

37.9. REMARK. The exponent ε in the foregoing considerations was almost *arbitrary* - except for condition (37.19). Therefore, it is evident that we can repeat our foregoing considerations in order to estimate the expression $a(ud_M^{-\varepsilon}, u)$ with $u \in W^{k,2}(\Omega; d_M, -\varepsilon)$: it suffices to replace ε by $-\varepsilon$ and use a certain "symmetry" of the differential operator $\mathcal{L}$ from (36.5). Condition (37.19) then has the form $-\varepsilon > 2k + m - N$, i.e.

$$(37.36) \qquad \varepsilon < -2k - m + N \; ,$$

and we immediately obtain the following analogue of Lemma 37.8 :

Let $a(u,v)$ *be the bilinear form from* (36.7) *with coefficients* $a_{\alpha\beta}$ *fulfilling condition* (37.15). *Let* $\varepsilon \in \mathbb{R}$ *satisfy condition* (37.36). *Then for* $\tilde{v} \in W^{k,2}(\Omega; d_M, -\varepsilon)$ *and* $\tilde{u} = \tilde{v} d_M^{-\varepsilon}$ *we have*

$$(37.37) \qquad \frac{a(\tilde{u}, \tilde{v})}{\| \tilde{u}; \, W^{k,2}(\Omega; d_M, \varepsilon) \|} \geq \tilde{c}(\varepsilon) \, \| \tilde{v}; \, W^{k,2}(\Omega; d_M, -\varepsilon) \|$$

where

$$(37.38) \qquad \tilde{c}(\varepsilon) = \frac{\tilde{c}_{11} - \tilde{c}_{12}|\varepsilon| - \tilde{c}_{13}|\varepsilon|^2}{\sqrt{1 + \tilde{c}_{14}|\varepsilon| + \tilde{c}_{15}|\varepsilon|^2}}$$

with positive constants $\tilde{c}_{11}$ *to* $\tilde{c}_{15}$ *independent of* v .

37.10. THE (H_1, H_2)-ELLIPTICITY OF $a(u,v)$. Now we are able to formulate conditions under which the bilinear form $a(u,v)$ from (36.7) is (H_1, H_2)-elliptic :

Let $\mathcal{L}$ *be the differential operator of order* $2k$ *from* (36.5). *Let us suppose that* $\mathcal{L}$ *is elliptic* (with respect to the Neumann problem) *in the sense of* (36.8) *and that its coefficients* $a_{\alpha\beta}$ *satisfy condition* (37.15). *Let* H_1 *and* H_2 *be the Hilbert spaces from* (37.8) *and let us suppose that the dimension* N *and* m *of* Ω *and* $M \subset \partial\Omega$, *respectively, satisfy the condition*

$$(37.12) \qquad N - m \geq 2k + 1 \; .$$

Then there exists an open interval I *containing the origin and such that for* $\varepsilon \in I$ *the bilinear form* $a(u,v)$ *from* (36.7) *is* (H_1, H_2)-*elliptic.*

The p r o o f is again simple : Using the notation from (37.8), we can rewrite inequalities (37.32) and (37.37) in the form

$$(37.39) \qquad \begin{aligned} \frac{a(u,v)}{\|v; H_2\|} &\geq c(\varepsilon) \|u; H_1\| \quad \text{with} \quad v = u d_M^{\varepsilon} \; , \\[1.5ex] \frac{a(\tilde{u}, \tilde{v})}{\|\tilde{u}; H_1\|} &\geq \tilde{c}(\varepsilon) \|\tilde{v}; H_2\| \quad \text{with} \quad \tilde{u} = \tilde{v} d_M^{-\varepsilon} \; . \end{aligned}$$

Moreover, it follows from (37.39) that for every $u \in H_1$ there is a function $w \in H_2$ with $\|w;H_2\| = 1$ - namely, the function $w = ud_M^\varepsilon / \|ud_M^\varepsilon;H_2\|$ - such that

$$a(u,w) \geq c(\varepsilon)\|u;H_1\| .$$

Then, *a fortiori*,

$$(37.40) \qquad \sup_{\|v;H_2\| \leq 1} a(u,v) \geq c(\varepsilon)\|u;H_1\|$$

holds for every $u \in H_1$ with $c(\varepsilon)$ independent of u . Analogously, from the second inequality in (37.39) we can derive that

$$(37.41) \qquad \sup_{\|u;H_1\| \leq 1} a(u,v) \geq \tilde{c}(\varepsilon)\|v;H_2\|$$

holds for every $v \in H_2$.

The function $g(\varepsilon) = c_{11} - c_{12}|\varepsilon| - c_{13}|\varepsilon|^2$ - see (37.33) - has two symmetric zero points ε_1 and $-\varepsilon_1$ ($\varepsilon_1 > 0$); if we denote $I_1 = (-\varepsilon_1,\varepsilon_1)$ then $g(\varepsilon) > 0$ for $\varepsilon \in I_1$ and consequently also $c(\varepsilon) > 0$ for $\varepsilon \in I_1$. If we analogously denote by $I_2 = (-\varepsilon_2,\varepsilon_2)$ ($\varepsilon_2 > 0$) the interval on which the function $\tilde{g}(\varepsilon) = \tilde{c}_{11} - \tilde{c}_{12}|\varepsilon| - \tilde{c}_{13}|\varepsilon|^2$ is positive, then $\tilde{c}(\varepsilon) > 0$ for $\varepsilon \in I_2$.

The estimates (37.39) take place if ε satisfies *simultaneously* conditions (37.19) and (37.36), i.e., if

$$\varepsilon \in I_3 = (2k + m - N, N - m - 2k) .$$

Condition (37.12) guarantees that the set I_3 is not empty.

Consequently, for $\varepsilon \in I$ where

$$(37.42) \qquad I = I_1 \cap I_2 \cap I_3$$

the estimates (37.40) and (37.41) hold with *positive* constants $c(\varepsilon)$ and $\tilde{c}(\varepsilon)$, respectively. But in view of formulas (37.2) and (37.3) this fact expresses the (H_1,H_2)-ellipticity of the form $a(u,v)$ - see Remark 37.2 (i). Consequently, I from (37.42) is the desired interval.

Now, we are able to formulate the main (existence and uniqueness) theorem of § 37.

<u>37.11.</u> THEOREM. *Let* $\mathcal{L}$ *be the linear differential operator from Subsection 36.2 with coefficients* $a_{\alpha\beta}$ *such that*

$$(37.15) \qquad a_{\alpha\beta}d_M^{2k-|\alpha|-|\beta|} \in L^\infty(\Omega) , \quad |\alpha|, |\beta| \leq k ,$$

and elliptic in the sense of (36.8) Let I *be the interval (37.42),* $\varepsilon \in I$ *,*

F   *a continuous linear functional on* $W^{k,2}(\Omega;d_M,-\varepsilon)$.

Then there is one and only one weak solution $u \in W^{k,2}(\Omega;d_M,\varepsilon)$ *of the Neumann problem; moreover, there is a positive constant*   c   *such that*

$$(37.43) \qquad \|u; W^{k,2}(\Omega;d_M,\varepsilon)\| \leq c \, \|F; [W^{k,2}(\Omega;d_M,-\varepsilon)]^*\| \ .$$

P r o o f .   If we use the notation (37.8), then we have to show – in view of Subsection 36.4 – that there is one and only one $u \in H_1$ such that

$$a(u,v) = <F,v> \quad \text{for every} \quad v \in H_2 \ .$$

But this follows directly from the Lax–Milgram–Nečas Lemma 37.1, where we take for $b(u,v)$ our bilinear form $a(u,v)$ from (36.7) : In Subsection 37.5 it was shown that $a(u,v)$ is continuous and thus fulfils condition (i) of Lemma 37.1, in Subsection 37.10 it was shown that $a(u,v)$ is (H_1,H_2)-elliptic for $\varepsilon \in I$ and thus fulfils conditions (ii), (iii) of Lemma 37.1. Therefore, the existence of a uniquely determined weak solution $u \in H_1$ is guaranteed and inequality (37.43) is a direct consequence of (37.5).

37.12. CONCLUDING REMARKS. (i) We have dealt with the *Neumann problem* only, but it is evident that the foregoing considerations   apply   also to other boundary value problems, in which the space $W^{k,2}(\Omega;d_m,\varepsilon)$ is replaced by its appropriate subspace V such that

$$W_0^{k,2}(\Omega;d_M,\varepsilon) \subset V \subset W^{k,2}(\Omega;d_M,\varepsilon) \ .$$

In particular, the *Dirichlet problem* can be considered; in this case $V = W_0^{k,2}(\Omega;d_M,\varepsilon)$ and some assumptions can be substantially weakened or completely removed. This concerns mainly the (very restrictive) assumptions (37.19) and (37.36) which lead to the condition $N - m \geq 2k + 1$. Since the Dirichlet problem was thoroughly investigated in [I], Chapter 14, the considerations presented here represent an extension as concerns the assumption on $a_{\alpha\beta}$: we can derive existence theorems for the case of the Dirichlet problem under the assumptions (37.15), while in [I] only $a_{\alpha\beta} \in L^\infty(\Omega)$ have been considered.

(ii) An important role is played by the interval I from (37.42). The assertion in Subsection 37.10 guarantees its *existence* and the calculations from Subsection 37.7 give some ideas how to describe *actual bounds* for I . Nevertheless, it should be pointed out that the calculations mentioned are rather rough and therefore sometimes a little restrictive.

§ 38 . **T h e   c a s e $N - m \leq 2k$**

<u>38.1. INTRODUCTION</u>. The condition

(38.1) $N - m \geq 2k + 1$

which appeared in § 37 - see (37.12) - is very restrictive and excludes,
as was pointed out in Subsection 37.4, a number of important particular cases
even for *second order* equations, where condition (38.1) reads

$$N - m \geq 3 .$$

This condition arises from our method of deriving the (H_1, H_2)-ellipticity :
In the course of estimating the expression $a(u,v)$ we used "test" functions
v of the form $ud_M^{\pm\varepsilon}$ and we needed imbedding theorems of the type
$W^{k,2}(\Omega; d_M, \pm\varepsilon) \subsetneq L^2(\Omega; d_M, \pm\varepsilon - 2k)$ which are valid under the assumption
$\pm \varepsilon > 2k + m - N$. Obviously , the "H_1-ellipticity" [i.e., the lower estimate
(37.40)] can be also derived provided we succeed in finding for
$u \in W^{k,2}(\Omega; d_M, \varepsilon)$ a certain function v (not necessarily of the form ud_M^ε)
such that the following estimates hold :

(38.2) $\| v;\ W^{k,2}(\Omega; d_M, -\varepsilon) \| \leq c_1 \| u;\ W^{k,2}(\Omega; d_M, \varepsilon) \|$

(i.e., $\| v; H_2 \| \leq c_1 \| u; H_1 \|$) and

(38.3) $a(u,v) \geq c_2 \| u;\ W^{k,2}(\Omega; d_M, \varepsilon) \|^2$

(i.e., $a(u,v) \geq c_2 \| u; H_1 \|^2$) with positive constants c_1 , c_2 independent of
u : from these two inequalities we can easily derive - in the same way as
in Subsection 37.10 - the desired estimate

$$\sup_{\| v; H_2 \| \leq 1} a(u,v) \geq \frac{c_2}{c_1} \| u; H_1 \| .$$

Similarly we can proceed when deriving the "H_2-ellipticity" [i.e., the estimate
(37.41)]. Therefore we will deal here only with the H_1-case and show how inequalities (38.2) and (38.3) can be derived if $N - m \leq 2k$.

[Note that (38.2) and (38.3) contain, as special cases, inequalities
(37.20) and (37.35).]

In what follows we will show how to construct the "test" function $v \in H_2$
such that inequalities (38.2) and (38.3) hold even if condition (38.1) is *not
fulfilled* and the approach of Subsections 37.7 and 37.8 *cannot be used*. We will
deal with second order operators only, i.e. with

$$k = 1 ,$$

so that we will have in mind the case $N - m < 3$; we will investigate the cases $N - m = 1$ and $N - m = 2$ separately. We restrict ourselves to $k = 1$ not only for the sake of simplicity; it will be seen that for extending the construction of v to $k > 1$ a little more sophisticated procedure will probably be required, at least in some cases.

Therefore, let us assume than $\mathscr{L}$ is an elliptic differential operator of second order :

$$(\mathscr{L}u)(x) = \sum_{|\alpha|,|\beta| \leq 1} (-1)^{|\alpha|} D^{\alpha}\left(a_{\alpha\beta}(x)D^{\beta}u(x)\right) ;$$

we assume $a_{\alpha\beta} \in L^{\infty}(\Omega)$ and express the ellipticity condition in the following form :

$$(38.4) \qquad \sum_{|\alpha|,|\beta| \leq 1} a_{\alpha\beta}(x) \, \xi_{\alpha} \, \xi_{\beta} \geq c_0 \sum_{|\alpha| \leq 1} |\xi_{\alpha}|^2$$

for almost all $x \in \Omega$ and for all $\xi = \{\xi_{\alpha}, |\alpha| \leq 1\} \in \mathbb{R}^{N-1}$ with a suitable positive constant c_0 .

We are interested in *under what conditions on* ε *there exists* - *for a given function* $u \in W^{1,2}(\Omega;d_M,\varepsilon) = H_1$ - *a function* $v \in W^{1,2}(\Omega;d_M,-\varepsilon) = H_2$ *such that the following inequalities hold :*

$$(38.5) \qquad \|v;H_2\| \leq c_1 \|u;H_1\| ,$$

$$(38.6) \qquad a(u,v) \geq c_2 \|u;H_1\|^2$$

with positive constants c_1 , c_2 *independent of* u .

<u>38.2. THE CASE $N - m = 1$, Ω A CUBE</u>. Let Ω be the cube $(0,1)^N$ and let us take its base for M : $M = \{x = (x_1,\ldots,x_{N-1},x_N) \in \partial\Omega , \; x_N = 0\}$. In this case $d_M(x) = x_N$.

Let us denote $x = (x',x_N)$, $x' = (x_1,\ldots,x_{N-1})$. For a given $u \in W^{1,2}(\Omega;d_M,\varepsilon)$ we define the function v by the formula

$$(38.7) \qquad v(x) = u(x',1) - \int_{x_N}^{1} \frac{\partial u}{\partial x_N} (x',t) \, t^{\varepsilon} \, dt .$$

Then we have

$$(38.8) \qquad \frac{\partial v}{\partial x_N} (x) = \frac{\partial u}{\partial x_N} (x) \, x_N^{\varepsilon} .$$

Integration by parts yields

$$v(x) = u(x)x_N^{\varepsilon} + \varepsilon \int_{x_N}^{1} u(x',t) \, t^{\varepsilon-1} \, dt ,$$

so that

$$\frac{\partial v}{\partial x_k}(x) = \frac{\partial u}{\partial x_k}(x) \, x_N^\varepsilon + \varepsilon \int_{x_N}^1 \frac{\partial u}{\partial x_k}(x',t) \, t^{\varepsilon-1} \, dt \, , \quad k = 1,\ldots,N-1 \, .$$

The last two formulas can be expressed in a unified form :

$$(38.9) \qquad D^\alpha v(x) = D^\alpha u(x) x_N^\varepsilon + \varepsilon \int_{x_N}^1 D^\alpha u(x',t) t^{\varepsilon-1} \, dt, \quad |\alpha| \le 1, \ \alpha \ne (0,\ldots,0,1) \, .$$

    (i) *For $\varepsilon < 1$, inequality (38.5) holds with the number c_1 given by*

$$c_1 = \left(1 + \frac{2|\varepsilon|}{1-\varepsilon}\right)$$

For the p r o o f , we need a special case of the *Hilbert inequalities*

$$\int_0^\infty \int_0^\infty K(s,t) \, f(s) \, g(t) \, ds \, dt \le c \|f; \, L^p(0,\infty)\| \cdot \|g; \, L^q(0,\infty)\| \, ,$$

$$\int_0^\infty \left(\int_0^\infty K(s,t) \, f(s) \, ds \right)^p dt \le c^p \|f; \, L^p(0,\infty)\|^p$$

which hold with $p > 1$, $q = p/(p-1)$, provided the kernel $K(s,t)$ is non-negative, homogeneous of degree -1 and such that $\int_0^\infty K(s,1)s^{-1/p} \, ds$ $= \int_0^\infty K(1,t)t^{-1/q} \, dt = c$ (see G. H. HARDY, J. E. LITTLEWOOD, G. PÓLYA [1], Theorem 319).

    The special choice $p = 2$,

$$K(s,t) = s^{\varepsilon/2-1} t^{-\varepsilon/2} \quad \text{for} \quad s > t \, , \quad K(s,t) = 0 \quad \text{for} \quad s \le t \, ,$$

$$f(s) = g(s) = w(s)s^{\varepsilon/2} \quad \text{for} \quad s \in (a,b) \, , \quad f(s) = g(s) = 0 \quad \text{otherwise}$$

with $0 \le a < b \le \infty$ and with $\varepsilon < 1$ leads to the inequalities

$$(38.10) \qquad \int_a^b \left(\int_t^b w(s) \, s^{\varepsilon-1} \, ds \right) w(t) \, dt \le \frac{2}{1-\varepsilon} \int_a^b |w(s)|^2 \, s^\varepsilon \, ds \, ,$$

$$(38.11) \qquad \int_a^b \left(\int_t^b w(s) \, s^{\varepsilon-1} \, ds \right)^2 t^{-\varepsilon} \, dt \le \left(\frac{2}{1-\varepsilon} \right)^2 \int_a^b |w(s)|^2 \, s^\varepsilon \, ds \, .$$

    Now it follows from (38.8) and (38.9) that

$$\|v; \; W^{1,2}(\Omega;d_M,-\varepsilon)\|^2 \; = \; \sum_{|\alpha|\le 1} \int\limits_{(0,1)^N} |D^\alpha v(x)|^2 \; x_N^{-\varepsilon} \; dx$$

$$= \; \sum_{|\alpha|\le 1} \int\limits_{(0,1)^N} |D^\alpha u(x)|^2 \; x_N^{2\varepsilon} \; x_N^{-\varepsilon} \; dx$$

$$(38.12) \qquad + \sum_\alpha{}' \Big[2\varepsilon \int\limits_{(0,1)^{N-1}} \int\limits_0^1 D^\alpha u(x) \int\limits_{x_N}^1 D^\alpha u(x',t) \; t^{\varepsilon-1} \; dt \; dx_N \; dx'$$

$$+ \; \varepsilon^2 \int\limits_{(0,1)^{N-1}} \int\limits_0^1 x_N^{-\varepsilon} \Big(\int\limits_{x_N}^1 D^\alpha u(x',t) \; t^{\varepsilon-1} \; dt \Big)^2 dx_N \; dx' \Big]$$

where $\sum_\alpha{}'$ means summation over $|\alpha| \le 1$ such that $\alpha \ne (0,\dots,0,1)$. The first
term after the last equality sign in (38.12) is exactly the square
$\|u; \; W^{1,2}(\Omega;d_M,\varepsilon)\|^2$; the last two integrals can be estimated with the help of
formulas (38.10), (38.11) where we take $a = 0$, $b = 1$, $w(t) = D^\alpha u(x',t)$.
In this way we obtain

$$\|v; \; W^{1,2}(\Omega;d_M,-\varepsilon)\|^2 \le \|u; \; W^{1,2}(\Omega;d_M,\varepsilon)\|^2$$

$$+ \; \Big[2|\varepsilon| \frac{2}{1-\varepsilon} + \varepsilon^2 \frac{4}{(1-\varepsilon)^2} \Big] \sum_\alpha{}' \int\limits_{(0,1)^{N-1}} \int\limits_0^1 |D^\alpha u(x',t)|^2 \; t^\varepsilon \; dt \; dx'$$

$$\le \; \Big[1 + \frac{4|\varepsilon|}{1-\varepsilon} + \frac{4\varepsilon^2}{(1-\varepsilon)^2} \Big] \|u; \; W^{1,2}(\Omega;d_M,\varepsilon)\|^2 \; ,$$

and this is assertion (i).

(ii) *For* $\varepsilon < 1$, *inequality (38.6) holds with* $c_2 = c_0 - 2|\varepsilon|(1-\varepsilon)^{-1}k_0$
where c_0 *is the "ellipticity" constant from (38.4) and*

$$(38.13) \qquad k_0 = \Big(\sum_\alpha{}' \sum_{|\beta|\le 1} \|a_{\alpha\beta}; L^\infty(\Omega)\|^2 \Big)^{1/2} \; .$$

P r o o f : From (38.8) and (38.9) we have

$$(38.14) \qquad a(u,v) = \sum_{|\alpha|,|\beta|\le 1} \int\limits_{(0,1)^N} a_{\alpha\beta} \; D^\beta u \; D^\alpha v \; dx$$

$$= \sum_{|\alpha|,|\beta|\le 1} \int\limits_{(0,1)^N} a_{\alpha\beta} \; D^\beta u \; D^\alpha u \; x_N^\varepsilon \; dx$$

$$+ \; \varepsilon \sum_{|\beta|\le 1} \sum_\alpha{}' \int\limits_{(0,1)^N} a_{\alpha\beta}(x) \; D^\beta u(x) \int\limits_{x_N}^1 D^\alpha u(x',t) \; t^{\varepsilon-1} \; dt \; dx \; .$$

From the ellipticity condition (38.4) it follows that

$$\sum_{|\alpha|,|\beta|\leq 1} \int_{(0,1)^N} a_{\alpha\beta}\, D^\beta u\, D^\alpha u\, x_N^\varepsilon\, dx$$

$$= \sum_{|\alpha|,|\beta|\leq 1} \int_{(0,1)^N} a_{\alpha\beta}\, D^\beta u\, x_N^{\varepsilon/2}\, D^\alpha u\, x_N^{\varepsilon/2}\, dx$$

$$\geq c_0 \sum_{|\alpha|\leq 1} \int_{(0,1)^N} |D^\alpha u|^2\, x_N^\varepsilon\, dx = c_0 \|u;\ W^{1,2}(\Omega;d_M,\varepsilon)\|^2 .$$

Using the Hölder inequality and formula (38.11) similarly as in point (i) above
we obtain

$$\left| \int_{(0,1)^N} a_{\alpha\beta}\, D^\beta u(x) \int_{x_N}^1 D^\alpha u(x\,,t)\, t^{\varepsilon-1}\, dt\, dx \right|$$

$$\leq \|a_{\alpha\beta};L^\infty(\Omega)\| \int_{(0,1)^N} \left| D^\beta u(x)\, x_N^{\varepsilon/2}\, x_N^{-\varepsilon/2} \int_{x_N}^1 D^\alpha u(x',t)\, t^{\varepsilon-1}\, dt \right| dx$$

$$\leq \|a_{\alpha\beta};L^\infty(\Omega)\| \left(\int_{(0,1)^N} |D^\beta u|^2\, x_N^\varepsilon\, dx \right)^{1/2}$$

$$\cdot \left(\int_{(0,1)^N} x_N^{-\varepsilon} \left(\int_{x_N}^1 D^\alpha u(x',t)\, t^{\varepsilon-1}\, dt \right)^2 dx \right)^{1/2}$$

$$\leq \|a_{\alpha\beta};L^\infty(\Omega)\| \left(\int_{(0,1)^N} |D^\beta u|^2\, x_N^\varepsilon\, dx \right)^{1/2} \frac{2}{1-\varepsilon} \left(\int_{(0,1)^N} |D^\alpha u|^2\, x_N^\varepsilon\, dx \right)^{1/2} .$$

Using the last estimates in (38.14) we have

$$a(u,v) \geq c_0 \|u;\ W^{1,2}(\Omega;d_M,\varepsilon)\|^2$$

$$- \frac{2|\varepsilon|}{1-\varepsilon} \sum_\alpha{}' \sum_{|\beta|\leq 1} \|a_{\alpha\beta};L^\infty(\Omega)\|\ \|D^\alpha u;\ L^2(\Omega;d_M,\varepsilon)\|\ \|D^\beta u;\ L^2(\Omega;d_M,\varepsilon)\|$$

$$\geq \left[c_0 - \frac{2|\varepsilon|}{1-\varepsilon} k_0 \right] \|u;\ W^{1,2}(\Omega;d_M,\varepsilon)\|^2$$

with k_0 from (38.13). Consequently, assertion (ii) is proved.

(iii) *There exists an open interval* I *containing the origin and such
that for* $\varepsilon \in I$ *the bilinear form* $a(u,v)$ *from (38.14) is* (H_1,H_2)*-elliptic
with* $H_1 = W^{1,2}(\Omega;d_M,\varepsilon)$, $H_2 = W^{1,2}(\Omega;d_M,-\varepsilon)$.

P r o o f : Let us denote by I_1 the open interval $\left(\dfrac{-c_0}{2k_0 - c_0} \; , \; \dfrac{c_0}{2k_0 + c_0}\right)$ with c_0 from (38.4) and k_0 from (38.13). The constant $c_2 =$
$= c_0 - 2|\varepsilon|(1 - \varepsilon)^{-1}k_0$ appearing in (38.6) is then positive for $\varepsilon \in I_1$.

Analogously as in the foregoing considerations of this Subsection we can show that for every function $\tilde{v} \in H_2$ there is a function $\tilde{u} \in H_1$ such that

$$(38.15) \quad \|\tilde{u};H_1\| \le \tilde{c}_1\|\tilde{v};H_2\| \; , \quad a(\tilde{u},\tilde{v}) \ge \tilde{c}_2\|\tilde{v};H_2\|^2$$

with $\tilde{c}_1$, $\tilde{c}_2$ independent of $\tilde{v}$, $\tilde{c}_2 = c_0 - 2|\varepsilon|(1 - \varepsilon)^{-1} \tilde{k}_0$ where $\tilde{k}_0 =$
$= \left(\sum\limits_{|\alpha|\le 1} {\sum\limits_{\beta}}' \|a_{\alpha\beta};L^\infty(\Omega)\|^2\right)^{1/2}$. If we denote by I_2 the open interval
$\left(-\dfrac{c_0}{2\tilde{k}_0 - c_0} \; , \; \dfrac{c_0}{2\tilde{k}_0 + c_0}\right)$, then $\tilde{c}_2 > 0$ for $\varepsilon \in I_2$.

Consequently, for $\varepsilon \in I$, where

$$I = I_1 \cap I_2 \; ,$$

both constants c_2 and $\tilde{c}_2$ are *positive* and we can now derive the (H_1,H_2)-
-ellipticity of $a(u,v)$ for $\varepsilon \in I$ from (38.5), (38.6) and (38.15) in the
same way as in Subsection 37.10 .

The just proved (H_1,H_2)-ellipticity together with the continuity of the
bilinear form $a(u,v)$, which obviously takes place by virtue of the results
of Subsection 37.3, allows us to formulate the following existence result, which
is completely analogous to Theorem 37.11 :

__38.3.__ __LEMMA.__ *Let $\mathcal{L}$ be the second order linear differential operator from
Subsection 38.1 with coefficients $a_{\alpha\beta} \in L^\infty(\Omega)$, and elliptic in the sense of
(38.4). Let Ω be the cube $(0,1)^N$, $M = \{x \in \overline{\Omega}, \; x_N = 0\}$, I the interval
from assertion 38.2 (iii), and $\varepsilon \in I$. Let F be a continuous linear functional on $W^{1,2}(\Omega;d_M,-\varepsilon)$.*

*Then there exists one and only one weak solution $u \in W^{1,2}(\Omega;d_M,\varepsilon)$ of the
Neumann problem; moreover, there is a positive constant c such that*

$$\|u; \; W^{1,2}(\Omega;d_M,\varepsilon)\| \le c\|F; \; [W^{1,2}(\Omega;d_M,-\varepsilon)]^*\| \; .$$

__38.4.__ __EXAMPLE.__ Let us consider the very special differential operator

$$\mathcal{L}u = - \Delta u + u \; ;$$

here $a_{\alpha\beta}(x) \equiv 1$ for $\alpha = \beta$ and $a_{\alpha\beta}(x) \equiv 0$ for $\alpha \ne \beta$, $|\alpha|, |\beta| \le 1$.
Thus evidently $a_{\alpha\beta} \in L^\infty(\Omega)$ and the ellipticity condition (38.4) is fulfilled

with $c_0 = 1$. For this operator, the interval of admissible values of ε is
at least the interval $(-1/3, 1/3)$, and we can assert that the (formal) Neumann problem

$$- \Delta u + u = f \quad \text{on} \quad \Omega = (0,1)^N \ ,$$
$$\partial u / \partial \nu = g \quad \text{on} \quad \partial \Omega$$

has a uniquely determined weak solution $u \in W^{1,2}(\Omega; d_M, \varepsilon)$ for $|\varepsilon| < 1/3$,
provided the functional F , which has in this case the form

$$<F,v> = \int_\Omega f(x) \ v(x) \ dx + \int_{\partial \Omega} g(S) \ v(S) \ dS$$

belongs to $\left[W^{1,2}(\Omega; d_M, -\varepsilon) \right]^*$ – see Remark 36.5 (ii), formula (36.10). The
condition $F \in \left[W^{1,2}(\Omega; d_M, -\varepsilon) \right]^*$ is satisfied if e.g. $f \in L^2(\Omega; d_M, \varepsilon)$ and
$g \in L^2(\partial \Omega)$ (see [I], Chap. 9).

38.5. THE CASE $N - m = 1$; MORE GENERAL DOMAINS Ω . The choice of the
"test" function v according to (38.7) was suggested by J. VOLDŘICH [3]. The
same author proposed also methods how to extend the construction of v from
cubes to more general domains in $\mathbb{R}^N$. We will not go into details here since
the description of v is very technical; let us only mention the case of a
domain Ω with "sufficiently smooth" (e.g. C^2) boundary $\partial \Omega$; we take for M
the whole boundary, $M = \partial \Omega$, and consequently $m = N - 1$: Using – roughly
speaking – a covering of $\overline{\Omega}$ by some open neighbourhoods and local coordinates
which map the intersection of $\overline{\Omega}$ with the corresponding neighbourhood U_i
onto a cube $Q = (0,1)^N$ and particularly the intersection of $\partial \Omega$ with U_i
onto the base of Q , it is possible to describe v *locally* in the form (38.7),
of course in the *local* coordinates.

The corresponding result about the existence of a weak solution
$u \in W^{1,2}(\Omega; d_M, \varepsilon)$ is completely analogous to the result formulated in Lemma
38.3.

J. VOLDŘICH [3] succeeded also in extending these results to certain
domains Ω with "non-smooth" boundaries, in particular to plane domains with
convex corners and convex cusps (with smooth sides). The extension to the case

$$N - m = 2$$

described below is due to the same author.

38.6. THE CASE $N - m = 2$. A SPECIAL PLANE DOMAIN. Again we have to find
for every $u \in W^{1,2}(\Omega; d_M, \varepsilon)$ a function $v \in W^{1,2}(\Omega; d_M, -\varepsilon)$ such that the inequalities (38.5) and (38.6) hold.

Let us take for M the origin $(0,0)$ in $\mathbb{R}^2$ (i.e., $m = 0$ and $N = 2$)
and consider a special *plane* domain Ω which is described in the *polar co-*

ordinates r , ϕ in the form

(38.16) $\quad \Omega = \{(r,\phi); \ 0 < r < a, \ \ \psi_1(r) < \phi < \psi_2(r)\}$

where a is a positive constant and ψ_1 , ψ_2 are continuous and monotone functions on $[0,a]$ and map the interval $[0,a]$ onto $[0,b_1]$, $[b_2,b_3]$, respectively (b_1 , b_2 , b_3 are non-negative constants with $b_2 \leq b_3 \leq 2\pi$). Further we suppose that there are numbers R_1 , $R_2 > 0$ and $\lambda \geq 0$ such that

(38.17) $\quad R_1 r^\lambda \leq \psi_2(r) - \psi_1(r) \leq R_2 r^\lambda$ for $r \in (0,a)$.

Such domains will be called *generalized cones*. For $b_1 = 0$, $b_2 = b_3$ (i.e., $\psi_1(r) \equiv 0$, $\psi_2(r) \equiv b_2$), Ω is the usual cone with its vertex at the origin and with one side coinciding with the interval $[0,a]$ on the x-axis.

In what follows, we will assume — without loss of generality — that $a = 1$.

Notice that in our case

(38.18) $\quad d_M(x) = |x| = r$.

38.7. SOME AUXILIARY RESULTS. Let Ω be a generalized cone and let us consider a function $u \in W^{1,2}(\Omega;d_M,\varepsilon)$ with $d_M = |x|$; if we restrict ourselves to ε such that $|\varepsilon| < 1$ then we can consider functions from $C^\infty(\overline{\Omega})$ since such functions are then dense in $W^{1,2}(\Omega;d_M,\varepsilon)$.

Now we extend the function u from Ω to 2Ω , where

$$2\Omega = \{(r,\phi); \ 0 < r < 1, \ \ \psi_1(r) < \phi < 2\psi_2(r) - \psi_1(r)\} \ ,$$

defining the extension u^* by the formula

$$u^*(r,\phi) = \begin{cases} u(r,\phi) & \text{for } (r,\phi) \in \Omega \ , \\ u(r, \ 2\psi_2(r) - \phi) & \text{for } (r,\phi) \in 2\Omega \setminus \Omega \ . \end{cases}$$

Obviously, u^* is smooth in Ω as well as in $2\Omega \setminus \overline{\Omega}$ and has the same trace from "both sides" on the set $\{(r, \ \psi_2(r)), \ 0 < r < 1\}$. Since

$$\frac{\partial u^*}{\partial \phi}(r,\vartheta) = - \frac{\partial u}{\partial \phi}(r, \ 2\psi_2(r) - \vartheta) \ ,$$

the function $u^*(r,\cdot)$ is absolutely continuous on the interval

$$I(r) = [\psi_1(r), \ 2\psi_2(r) - \psi_1(r)]$$

and the functions $u^*(r,\cdot)$ and $\frac{\partial u^*}{\partial \phi}(r,\cdot)$ belong to $L^2(I(r))$ for every $r \in (0,1)$. Moreover, due to the symmetry of the extension, we have

$$(38.19) \qquad \int\limits_{I(r)} |u^*(r,\phi)|^2 \, d\phi = 2 \int\limits_{\psi_1(r)}^{\psi_2(r)} |u(r,\phi)|^2 \, d\phi$$

and

$$(38.20) \qquad \int\limits_{I(r)} \left|\frac{\partial u^*}{\partial \phi}(r,\phi)\right|^2 \, d\phi = 2 \int\limits_{\psi_1(r)}^{\psi_2(r)} \left|\frac{\partial u}{\partial \phi}(r,\phi)\right|^2 \, d\phi \; .$$

Therefore, the Fourier expansion of $u^*(r,\phi)$ with respect to ϕ makes sense for every $r \in (0,1)$:

$$(38.21) \qquad u^*(r,\phi) = A(r) + \sum_{n=1}^{\infty} A_n(r) \, \cos \frac{n\pi}{\psi_2(r) - \psi_1(r)} \, \phi \; ,$$

and we have

$$(38.22) \qquad \frac{\partial u^*}{\partial \phi}(r,\phi) = - \frac{\pi}{\psi_2(r) - \psi_1(r)} \sum_{n=1}^{\infty} nA_n(r) \, \sin \frac{n\pi}{\psi_2(r) - \psi_1(r)} \, \phi \; .$$

The "test" function v corresponding to u will be defined by the formula

$$(38.23) \qquad v(r,\phi) = r^\varepsilon u(r,\phi) + \varepsilon \int\limits_r^1 A(t) \, t^{\varepsilon-1} \, dt$$

where $A(t)$ is the first "Fourier coefficient" in the expansion (38.21) [compare formula (38.23) with (38.9) for $|\alpha| = 0$]. From (38.23) we have

$$(38.24) \qquad \frac{\partial v}{\partial r}(r,\phi) = r^\varepsilon \frac{\partial u}{\partial r}(r,\phi) + \varepsilon r^{\varepsilon-1}\left[u(r,\phi) - A(r)\right] \; ,$$

$$(38.25) \qquad \frac{\partial v}{\partial \phi}(r,\phi) = r^\varepsilon \frac{\partial u}{\partial \phi}(r,\phi) \; .$$

To derive estimates (38.5) and (38.6), we need some inequalities.

(i) *The inequality*

$$(38.26) \qquad \|u(r,\phi) - A(r); \; L^2(\Omega; \, d_M, \, \varepsilon-2)\|^2 \leq C\|u; \; W^{1,2}(\Omega;d_M,\varepsilon)\|^2$$

holds with the constant

$$(38.27) \qquad C = \max_{0 \leq r \leq 1} \left[\psi_2(r) - \psi_1(r)\right]\pi^{-2} \; .$$

P r o o f : Using the polar coordinates we immediately obtain

$$\|u(r,\phi) - A(r); \ L^2(\Omega; \ d_M, \ \varepsilon-2)\|^2$$

$$= \int_0^1 r^{\varepsilon-1} \left(\int_{\psi_1(r)}^{\psi_2(r)} |u(r,\phi) - A(r)|^2 \ d\phi \right) dr \ .$$

The *Parseval identity* for the function $u^*(r,\phi) - A(r)$ in $L^2\big(I(r)\big)$ yields, in view of (38.19) and (38.21),

$$\int_{\psi_1(r)}^{\psi_2(r)} |u(r,\phi) - A(r)|^2 \ d\phi = \frac{1}{2} \int_{I(r)} |u^*(r,\phi) - A(r)|^2 \ d\phi$$

$$= \frac{\psi_2(r) - \psi_1(r)}{2} \sum_{n=1}^{\infty} A_n^2(r) \ .$$

The Parseval identity for the function $\frac{\partial u^*}{\partial\phi}(r,\phi)$ yields, in view of (38.20) and (38.22),

$$\int_{\psi_1(r)}^{\psi_2(r)} \left|\frac{\partial u}{\partial\phi}(r,\phi)\right|^2 \ d\phi = \frac{1}{2} \int_{I(r)} \left|\frac{\partial u^*}{\partial\phi}(r,\phi)\right|^2 \ d\phi$$

$$= \frac{1}{2} \frac{\pi^2}{\psi_2(r) - \psi_1(r)} \sum_{n=1}^{\infty} n^2 A_n^2(r) \ .$$

Comparing the right hand sides of the last two formulas and using the definition of C from (38.27) we obtain that

$$\int_{\psi_1(r)}^{\psi_2(r)} |u(r,\phi) - A(r)|^2 \ d\phi \leq C \int_{\psi_1(r)}^{\psi_2(r)} \left|\frac{\partial u}{\partial\phi}(r,\phi)\right|^2 \ d\phi \ .$$

Multiplying both sides of this inequality by $r^{\varepsilon-1}$ and integrating by r over the interval $(0,1)$, we immediately obtain inequality (38.26) in view of the fact that

$$\|u; \ W^{1,2}(\Omega; d_M, \varepsilon)\|^2$$

(38.28)
$$= \int_0^1 \int_{\psi_1(r)}^{\psi_2(r)} \left[|u(r,\phi)|^2 \ r^{\varepsilon+1} + \left|\frac{\partial u}{\partial r}(r,\phi)\right|^2 r^{\varepsilon+1} + \left|\frac{\partial u}{\partial\phi}(r,\phi)\right|^2 r^{\varepsilon-1}\right] d\phi \ dr \ .$$

  (ii) *Let* R_1 , λ *be the constants from* (38.17). *Then*

(38.29)
$$\int_0^1 A^2(r) \ r^{\varepsilon+\lambda+1} \ dr \leq \frac{1}{R_1} \|u; \ W^{1,2}(\Omega; d_M, \varepsilon)\|^2 \ .$$

P r o o f : The Parseval identity for the function $u^*(r,\phi)$ yields, in view of (38.19), (38.21) and (38.17),

$$\int_{\psi_1(r)}^{\psi_2(r)} |u(r,\phi)|^2 \, d\phi = \frac{1}{2} \int_{I(r)} |u^*(r,\phi)|^2 \, d\phi$$

$$= [\psi_2(r) - \psi_1(r)]\left[A^2(r) + \frac{1}{2}\sum_{n=1}^{\infty} A_n^2(r)\right] \geq R_1 \, r^\lambda \, A^2(r) \ .$$

Multiplying this inequality by $r^{\varepsilon+1}$ and integrating by r over the interval $(0,1)$, we obtain inequality (38.29) in view of formula (38.28).

38.8.   THE CASE $N - m = 2$. (H_1,H_2)-ELLIPTICITY; EXISTENCE THEOREM.   Let Ω be a generalized cone in $\mathbb{R}^2$ and suppose that – for $u \in H_1 = W^{1,2}(\Omega;d_M,\varepsilon)$ – the "test" function v is defined according to (38.23). Then we have

$$\|v; \, H_2\|^2 = \|v; \, W^{1,2}(\Omega;d_M,-\varepsilon)\|^2$$

$$= \int_0^1 \int_{\psi_1(r)}^{\psi_2(r)} \left[|v(r,\phi)|^2 r^{-\varepsilon+1} + |\tfrac{\partial v}{\partial r}(r,\phi)|^2 r^{-\varepsilon+1} + |\tfrac{\partial v}{\partial \phi}(r,\phi)|^2 r^{-\varepsilon-1}\right] d\phi \, dr \ .$$

(i) Using formulas (38.23) – (38.25) and the inequality $(a + b)^2 \leq 2a^2 + 2b^2$, we obtain

$$\|v;H_2\|^2 \leq \int_0^1 \int_{\psi_1(r)}^{\psi_2(r)} \left[2|u(r,\phi)|^2 \, r^{\varepsilon+1} + 2\varepsilon^2\left(\int_r^1 A(t)t^{\varepsilon-1} \, dt\right)^2 r^{-\varepsilon+1}\right.$$

$$+ 2|\tfrac{\partial u}{\partial r}(r,\phi)|^2 \, r^{\varepsilon+1} + 2\varepsilon^2 r^{\varepsilon-1}\left(u(r,\phi) - A(r)\right)^2$$

$$\left. + |\tfrac{\partial u}{\partial \phi}(r,\phi)|^2 \, r^{\varepsilon-1}\right] d\phi \, dr$$

$$\leq 2\|u;H_1\|^2 + 2\varepsilon^2 \int_0^1 \int_{\psi_1(r)}^{\psi_2(r)} r^{-\varepsilon+1}\left(\int_r^1 A(t) \, t^{\varepsilon-1} \, dt\right)^2 d\phi \, dr$$

$$+ 2\varepsilon^2 \|u(r,\phi) - A(r); \, L^2(\Omega; \, d_M, \, \varepsilon-2)\|^2 \ .$$

The integral J in the last expression can be estimated – by virtue of (38.17) and the Hilbert inequality (38.11) with $a = 0$, $b = 1$, $w(s) = A(s)s^{\lambda+1}$ and with $\varepsilon - \lambda - 1$ instead of ε – as follows :

$$(38.30) \quad J = \int_0^1 [\psi_2(r) - \psi_1(r)]r^{-\varepsilon+1}\left(\int_r^1 A(t) \, t^{\varepsilon-1} \, dt\right)^2 dr \leq$$

$$\leq R_2 \int\limits_0^1 r^{\lambda-\varepsilon+1}\Big(\int\limits_r^1 A(t)\ t^{\varepsilon-1}\ dt\Big)^2 dr \leq \frac{4R_2}{(2-\varepsilon+\lambda)^2} \int\limits_0^1 A^2(t)\ t^{\lambda+\varepsilon+1}\ dt$$

provided $\varepsilon < 2 + \lambda$. But this condition is fulfilled since we have supposed $|\varepsilon| < 1$ and $\lambda \geq 0$. Consequently, it follows from the assertions 38.7 (i), (ii), i.e. from (38.26) and (38.29), that

$$\|v;H_2\|^2 \leq 2\Big(1 + \frac{R_2}{R_1}\ \frac{4\varepsilon^2}{(2-\varepsilon+\lambda)^2} + C\varepsilon^2\Big)\|u;H_1\|^2 \ .$$

Thus we have proved inequality (38.5).

(ii) Let us assume that the linear elliptic operator $\mathcal{L}$ of the second order is elliptic in the sense of (38.4) and that v is given by formula (38.23). Then it follows from (38.24) and (38.25) that for $|\alpha| = 1$ we have

$$D^\alpha v = r^\varepsilon\ D^\alpha u + \varepsilon r^{\varepsilon-1}\ D^\alpha r\big[u(r,\phi) - A(r)\big] \ ,$$

and consequently,

$$
\begin{aligned}
a(u,v) = &\sum_{|\alpha|,|\beta|\leq 1} \int\limits_0^1 \int\limits_{\psi_1(r)}^{\psi_2(r)} a_{\alpha\beta}\ D^\beta u\ D^\alpha u\ r^{\varepsilon+1}\ d\phi\ dr \\[2mm]
&+ \varepsilon \sum_{\substack{|\alpha|=1 \\ |\beta|\leq 1}} \int\limits_0^1 \int\limits_{\psi_1(r)}^{\psi_2(r)} a_{\alpha\beta}\ D^\beta u\ r^\varepsilon\ D^\alpha r\big[u(r,\phi) - A(r)\big]\ d\phi\ dr \\[2mm]
&+ \varepsilon \sum_{|\beta|\leq 1} \int\limits_0^1 \int\limits_{\psi_1(r)}^{\psi_2(r)} a_{\Theta\beta}\ D^\beta u\Big(\int\limits_r^1 A(t)\ t^{\varepsilon-1}\ dt\Big)r\ d\phi\ dr
\end{aligned}
$$

with $\Theta = (0,0,\ldots,0)$. The first sum on the right hand side can be, in view of the ellipticity condition (38.4), estimated from below by

$$c_0 \sum_{|\alpha|\leq 1} \int\limits_0^1 \int\limits_{\psi_1(r)}^{\psi_2(r)} |D^\alpha u|^2\ r^{\varepsilon+1}\ d\phi\ dr = c_0\|u;H_1\|^2 \ .$$

If we denote the second and the third sum by J_2 and J_3 , respectively, we obtain — using the boundedness of $a_{\alpha\beta}$, the Hölder inequality, the assertion 38.7 (i) [i.e., inequality (38.26)] and the estimate for J [see (38.30)] and the assertion 38.7 (ii) [i.e., inequality (38.29)], respectively — the following estimates :

$$|J_2| \leq |\varepsilon| \sum_{\substack{|\alpha|=1 \\ |\beta| \leq 1}} \|a_{\alpha\beta}; \, L^\infty(\Omega)\| \left(\int_0^1 \int_{\psi_1(r)}^{\psi_2(r)} |D^\beta u|^2 \, r^{\varepsilon+1} \, d\phi \, dr \right)^{1/2} \cdot$$

$$\cdot \; \|u(r,\phi) - A(r); \, L^2(\Omega; \, d_M, \, \varepsilon-2)\|$$

$$\leq |\varepsilon| \sqrt{C} \sum_{\substack{|\alpha|=1 \\ |\beta| \leq 1}} \|a_{\alpha\beta}; \, L^\infty(\Omega)\| \cdot \|u; \, H_1\|^2$$

[notice that $|D^\alpha r| \leq 1$; C is the constant from (38.27)], and

$$|J_3| \leq |\varepsilon| \sum_{|\beta| \leq 1} \|a_{0\beta}; \, L^\infty(\Omega)\| \left(\int_0^1 \int_{\psi_1(r)}^{\psi_2(r)} |D^\beta u|^2 \, r^{\varepsilon+1} \, d\phi \, dr \right)^{1/2} \cdot$$

$$\cdot \left(\int_0^1 \int_{\psi_1(r)}^{\psi_2(r)} r^{-\varepsilon+1} \left(\int_r^1 A(t) \, t^{\varepsilon-1} \, dt \right)^2 d\phi \, dr \right)^{1/2}$$

$$\leq |\varepsilon| \sqrt{\frac{R_2}{R_1}} \, \frac{2}{(2 - \varepsilon + \lambda)} \sum_{|\beta| \leq 1} \|a_{0\beta}; \, L^\infty(\Omega)\| \; \|u; H_1\|^2$$

provided $2 - \varepsilon + \lambda > 0$. Thus, we have shown that, for such ε ,

$$a(u,v) \geq c_0 \|u; H_1\|^2 - |J_2| - |J_3| \geq c^* \|u; H_1\|^2$$

where

$$c^* = c_0 - |\varepsilon| \sqrt{C} \sum_{\substack{|\alpha|=1 \\ |\beta| \leq 1}} \|a_{\alpha\beta}; L^\infty(\Omega)\| - \frac{2|\varepsilon|}{(2 - \varepsilon + \lambda)} \sqrt{\frac{R_2}{R_1}} \sum_{|\beta| \leq 1} \|a_{0\beta}; L^\infty(\Omega)\| \; .$$

Since c^* is obviously positive for $|\varepsilon|$ sufficiently small, we have shown that *there is an open interval* I *containing the origin and such that for* $\varepsilon \in I$ *inequality* (38.6) *holds with a positive constant* c_2 .

(iii) The assertions from points (i) and (ii) above guarantee that a formula analogous to (37.40) holds, i.e., that the form $a(u,v)$ is "H_1-elliptic". We can proceed analogously in order to prove the "H_2-ellipticity" (37.41), and consequently, we have shown that *an existence and uniqueness assertion completely analogous to that of* Lemma 38.3 *holds for second order elliptic operators* $\mathcal{L}$ *also if* Ω *is a g e n e r a l i z e d c o n e in* $\mathbb{R}^2$.

<u>38.9.</u> <u>THE CASE $N - m = 2$; A LITTLE MORE GENERAL DOMAIN</u> Ω . (i) Let Ω be a plane domain (i.e., $N = 2$) and x_0 a point on $\partial\Omega$. Let us assume that *the set*

$$\Omega_1 = \Omega \cap \{x; \, |x - x_0| < a\}$$

with a suitable positive constant a *is a generalized cone,* i.e.

$$\Omega_1 = \{(r,\phi); \ r = |x - x_0| \in (0,a), \ \ \phi \in (\psi_1(r),\psi_2(r))\}$$

where r , ϕ are polar coordinates centered at x_0 and ψ_1 , ψ_2 are the functions from Subsection 38.6. If we denote

$$\Omega_2 = \Omega \smallsetminus \Omega_1$$

and

$$a_i(u,v) = \sum_{|\alpha|,|\beta|\leq 1} \int_{\Omega_1} a_{\alpha\beta} \ D^\beta u \ D^\alpha v \ dx \ , \quad i = 1,2 \ ,$$

where $a_{\alpha\beta}$ are the coefficients of an elliptic second order differential operator $\mathscr{L}$ defined on Ω , then obviously

$$a(u,v) = a_1(u,v) + a_2(u,v) \ .$$

For a given $u \in W^{1,2}(\Omega;d_M,\epsilon)$ with $M = \{x_0\}$ and thus $d_M(x) = |x - x_0|$, we define the corresponding "test" function v by the formula

$$v(r,\phi) = \begin{cases} u(r,\phi)r^\epsilon + \epsilon \displaystyle\int_r^a A(t) \ t^{\epsilon-1} \ dt & \text{for} \quad (r,\phi) \in \Omega_1 \ , \\[2mm] u(r,\phi)r^\epsilon & \text{for} \quad (r,\phi) \in \Omega_2 \ . \end{cases}$$

Now, we can easily show that

$$(38.31) \qquad \|v; \ W^{1,2}(\Omega_i;d_M,\epsilon)\| \leq c_{1i}\|u; \ W^{1,2}(\Omega_i;d_M,\epsilon)\|$$

and

$$(38.32) \qquad a_i(u,v) \geq c_{2i}\|u; \ W^{1,2}(\Omega_i;d_M,\epsilon)\|^2$$

($i = 1,2$) with c_{11} , c_{12} positive for $|\epsilon|$ sufficiently small : For $i = 1$ this follows from the considerations of Subsections 38.6 - 38.8, since Ω_1 is a generalized cone; for $i = 2$ this is obvious since the weight function d_M^ϵ is in Ω_2 separated from zero and bounded : there are K_1 , K_2 such that

$$0 < K_1 \leq d_M^\epsilon(x) = |x - x_0| \leq K_2 < \infty \quad \text{for} \quad x \in \Omega_2 \ ,$$

and consequently we can estimate c_{2i} with the help of K_1 , K_2 and (in the case of c_{22}) of the ellipticity constant c_0 .

It follows from (38.31) and (38.32) that inequalities (38.5) and (38.6) are fulfilled, too. Consequently, we   deduce the existence and uniqueness theorem for this case again as in 38.8 (iii) .

(ii) For $\Omega \subset \mathbb{R}^N$ with $N > 2$ and $M \subset \partial\Omega$ with $\dim M = m = N - 2$, we can sometimes reduce our problem with the help of some special curvilinear coordinates to the case of a plane domain. If this "reduced plane domain" is a generalized cone, we can use our foregoing considerations.

Section 13 . *A m o d i f i e d   c o n c e p t   o f   t h e   w e a k
        s o l u t i o n*

§ 39 . **F o r m u l a t i o n   o f   t h e   p r o b l e m**

39.1. <u>INTRODUCTION</u>. In [I] and in the foregoing paragraphs 35 - 38 we have
dealt with weak solutions of certain boundary value problems; that is, we asked
whether there is - roughly speaking - a function $u \in V$ (V being a certain Banach space) such that the identity

(39.1) $a(u,v) = \langle F,v \rangle$

holds for every $v \in V$. Here F was a continuous linear functional on V ,
$f \in V^{*}$, and $a(u,v)$ was the bilinear form

$$(39.2) \qquad \sum_{|\alpha|, |\beta| \leq k} \int_{\Omega} a_{\alpha\beta}(x) \, D^{\beta}u(x) \, D^{\alpha}v(x) \, dx \, ,$$

defined and continuous on the cartesian product $V \times V$ and generated by a
linear differential operator of order 2k ,

$$(39.3) \qquad (\mathcal{L}u)(x) = \sum_{|\alpha|, |\beta| \leq k} (-1)^{|\alpha|} D^{\alpha}\left(a_{\alpha\beta}(x)D^{\beta}u(x)\right) \, , \quad x \in \Omega \subset \mathbb{R}^{N} \, .$$

We have assumed that the operator is *elliptic* in a sense which allows us to use
for V   some subspace of the classical Sobolev space $W^{k,2}(\Omega)$, e.g. in the
sense that there is a positive constant c_0 such that

$$(39.4) \qquad a(u,u) \geq c_0 \|u;V\|^2$$

holds for all $u \in V$.

   Our aim was to extend the results about the existence and uniqueness of a
weak solution of the corresponding boundary value problem for $\mathcal{L}$ from *classical* Sobolev spaces to *weighted* Sobolev spaces $W^{k,2}(\Omega;s(d_M))$ [see Subsection
 36.1]. This extension makes it possible to consider wider classes of "right
hand sides", represented here by the functional F in (39.1), and thus to
obtain existence results - in terms of w e a k   s o l u t i o n s - for
a substantially richer set of data in our boundary value problems.

   The use of weighted spaces made it necessary to input weight functions
into the bilinear form $a(u,v)$, even if *formally*. One way how to carry out
this step was described in Subsection 37.3 : We *rewrite* the bilinear form
$a(u,v)$ from (39.2) in the form

$$(39.5) \qquad a(u,v) = \sum_{|\alpha|,|\beta|\leq k} \int_\Omega a_{\alpha\beta}\, D^\beta u\, s^{1/2}(d_M)\, D^\alpha v\, s^{-1/2}(d_M)\, dx \; ;$$

this change is very formal, but eventually it requires to consider a(u,v) on a *pair* of Hilbert spaces $H_1 \times H_2$ – see (37.7) – and to use more sophisticated tools – namely the generalized Lax–Milgram–Nečas Lemma 37.1 instead of the simpler Lax–Milgram Theorem (see Subsection 15.5 or Subsection 39.5 below) – if we desire to derive some existence and uniqueness assertion.

Let us summarize : The drawback of the approach of §§ 35 – 38 consists in the fact that we have to work with *two different spaces* and to use *more complicated tools* as Lemma 37.1 etc.

In this Section we want to suggest another approach which again enables us to consider elliptic boundary value problems in w e i g h t e d Sobolev spaces but which needs only o n e space and for which the classical Lax– –Milgram Theorem is sufficient. We again introduce the weight function *formally,* changing a little the concept of a weak solution.

39.2. <u>A NEW BILINEAR FORM</u>. Let us recall how the bilinear form a(u,v) from (39.2) could be derived from the differential operator $\mathscr{L}$ in (39.3) : The (formal) differential equation

$$(39.6) \qquad \mathscr{L}\,u = f \quad \text{on}\ \Omega$$

should be *multiplied* by a function v , say from $C_0^\infty(\Omega)$, and the resulting equality $\mathscr{L}\,u\cdot v = f\cdot v$ should be *integrated* over Ω . We obtain the "integral identity"

$$(39.7) \qquad \int_\Omega \mathscr{L}\,u\ v\ dx = \int_\Omega f\ v\ dx$$

and an application of *Green's formula* to the left hand side leads to the identity

$$(39.8) \qquad a(u,v) = \int_\Omega f\ v\ dx \; ;$$

here finally the bilinear form from (39.2) appears and the last formula can be viewed as a certain "special case" of identity (39.1).

Now, let us repeat this procedure with only one slight change : instead of the Lebesgue measure dx , let us consider in the *integration step* a *more general measure* $d\mu$. More precisely, let us consider the measure μ given by

$$d\mu = w\ dx$$

where w is a *weight function,* i.e., a function measurable and positive almost everywhere on Ω . Then we have, instead of (39.7), the "integral identity"

$$(39.9) \qquad \int_\Omega \mathcal{L} u \ v \ w \ dx = \int_\Omega f \ v \ w \ dx$$

and since we are in fact working with the product vw instead of just with
the function v , we obtain by Green's formula, applied to the left hand side
of (39.9), the identity

$$(39.10) \qquad \sum_{|\alpha|,|\beta| \leq k} \int_\Omega a_{\alpha\beta} D^\beta u \ D^\alpha(vw) \ dx = \int_\Omega f \ v \ w \ dx \ ,$$

i.e., the identity

$$(39.11) \qquad a(u,vw) = \int_\Omega f \ v \ w \ dx \ .$$

It is this identity which will be the starting point of our considerations:
Instead of investigating the identity (39.1) with the bilinear form $a(u,v)$
we will investigate the analogous identity with the *new* (modified) *bilinear
form* $a(u,vw)$ with a given (sufficiently smooth) weight function w .

39.3. <u>DEFINITION</u>. Let $\mathcal{L}$ be the linear differential operator of order $2k$
from (39.3) with coefficients $a_{\alpha\beta}$ satisfying the condition

$$(39.12) \qquad a_{\alpha\beta} \in L^\infty(\Omega) \ , \quad |\alpha|, \ |\beta| \leq k \ .$$

Further, let w be a weight function on Ω of class $C^k(\Omega)$ (among other, w
is positive a.e. on Ω). Then we denote by $a_w(u,v)$ the bilinear form defined
by the formula

$$(39.13) \qquad a_w(u,v) = \sum_{|\alpha|,|\beta| \leq k} \int_\Omega a_{\alpha\beta}(x) \ D^\beta u(x) \ D^\alpha\big(v(x)w(x)\big) \ dx \ .$$

Provided $a_w(u,v)$ is defined on the cartesian product $V \times V$ with V
a suitable Banach (Hilbert) space and F is a given functional from V^* ,
a function $u \in V$ for which

$$(39.14) \qquad a_w(u,v) = <F,v> \quad \text{for every} \quad v \in V$$

will be called a *w-weak solution* (of "$\mathcal{L}u = F$ in Ω ").

39.4. <u>REMARKS</u>. (i) As was pointed out in Subsection 39.2, our new bilinear
form a_w is closely connected with the "usual" bilinear form a from (39.2) :

$$(39.15) \qquad a_w(u,v) = a(u,vw) \ .$$

Nevertheless, this small formal change will enable us to consider the form a_w
on the cartesian product $V \times V$ with a single suitable chosen weighted space
V , and to use the simpler Lax–Milgram Theorem. We will recall this theorem in
the following Subsection 39.5.

178

(ii) We will establish conditions under which there exists a w-weak
solution, i.e., a function $u \in V$ such that (39.14) holds. The concept of
a w-weak solution represents a small change of the concept of the *weak* solution,
which made use of (39.1) instead of (39.14). However, this change seems not to
be very substantial from the point of view of applications : While the *weak*
solution generalizes (or - for smooth data - coincides with) the solution
of the differential equation $\mathcal{L}u = f$ on Ω [provided $\langle F,v \rangle$ can be expressed in the form $\int_\Omega f\, v\, dx$], the w-*weak* solution generalizes - roughly speaking - the solution of the differential equation

$$w \, \mathcal{L} \, u = f \quad \text{on} \quad \Omega \; .$$

However, since w is positive a.e. on Ω , the difference between this equation and equation (39.6) is not substantial.

(iii) The concept of the w-weak solution is a natural extension of the
concept of the weak solution : If we take for the weight function w the
trivial weight $w(x) \equiv 1$, then the w-weak (i.e., 1-weak) solution is precisely
the weak solution. Consequently, our approach contains as a special case (for
$w \equiv 1$) the "classical" approach, which works with classical Sobolev spaces
$W^{k,2}(\Omega)$.

The main tool for deriving existence theorems for w-weak solutions will
be the following assertion :

39.5. LAX-MILGRAM THEOREM (see, e.g., J. NEČAS [1], Chap. 1, Lemma 3.1).
*Let H be a Hilbert space. Let $b(u,v)$ be a bilinear form defined on the
cartesian product $H \times H$ and let there exist positive constants c_1 , c_2
such that*

(i) *for all $u, v \in H$ we have*

(39.16) $\left| b(u,v) \right| \leq c_1 \|u;H\| \cdot \|v;H\|$;

(ii) *for all $u \in H$ we have*

(39.17) $b(u,u) \geq c_2 \|u;H\|^2$.

*Let h be a continuous linear functional from H^* . Then there exists
one and only one element $u \in H$ such that for all elements $v \in H$ we have*

(39.18) $b(u,v) = \langle h,v \rangle$

and, moreover,

(39.19) $\|u;H\| \leq \dfrac{1}{c_2} \|h;H^*\|$.

<u>39.6.</u> <u>REMARK.</u> Condition (i) of Theorem 39.5 states that the bilinear form $b(u,v)$ is *continuous* on $H \times H$.

A bilinear form satisfying condition (ii) of Theorem 39.5 is said to be H-*elliptic*.

<u>39.7.</u> <u>THE CONTINUITY OF THE BILINEAR FORM $a_w(u,v)$. THE WEIGHTED SPACE.</u>

(i) Let w be the given weight function which determines the bilinear form $a_w(u,v)$ from (39.13). We introduce the *weighted Sobolev space*

$$(39.20) \quad W^{k,2}(\Omega;w)$$

as the set of functions $u = u(x)$, $x \in \Omega$, such that

$$(39.21) \quad \|u;W^{k,2}(\Omega;w)\|^2 = \sum_{|\alpha| \leq k} \int_\Omega |D^\alpha u(x)|^2 \, w(x) \, dx < \infty \ ;$$

thus all derivatives of u of order $|\alpha| \leq k$ belong to $L^2(\Omega;w)$ with the *same* weight function w .

We have supposed that $w \in C^k(\Omega)$; moreover, let us assume that the weight function w has the following property, called *property* (P_1) : there are positive constants c_α such that

$$(39.22) \quad |D^\alpha w(x)| \leq c_\alpha w(x) \quad \text{for} \quad x \in \Omega \quad \text{and} \quad |\alpha| \leq k .$$

(ii) We assume that $u, v \in W^{k,2}(\Omega)$ and investigate the expression $a_w(u,v)$: Using similarly as in § 37 the identity

$$D^\alpha(vw) = \sum_{\gamma+\delta=\alpha} c_{\gamma\delta} \, D^\gamma v \, D^\delta w \ ,$$

we have

$$a_w(u,v) = \sum_{|\alpha|,|\beta| \leq k} \int_\Omega a_{\alpha\beta} \, D^\beta u \, D^\alpha(vw)$$

$$= \sum_{|\alpha|,|\beta| \leq k} \sum_{\gamma+\delta=\alpha} c_{\gamma\delta} \int_\Omega a_{\alpha\beta} \, D^\beta u \, D^\gamma v \, D^\delta w \, dx \ .$$

Using succesively the property (P_1) of the weight w [i.e., the estimate (39.22)], the condition (39.12) of $a_{\alpha\beta}$ and the Hölder inequality, we obtain

$$\left| \int_\Omega a_{\alpha\beta} \, D^\beta u \, D^\gamma v \, D^\delta w \, dx \right| \leq c_\delta \int_\Omega |a_{\alpha\beta} \, D^\beta u \, D^\gamma v \, w| \, dx$$

$$\leq c_\delta \|a_{\alpha\beta};L^\infty(\Omega)\| \int_\Omega |D^\beta u| \, w^{1/2} \, |D^\gamma v| \, w^{1/2} \, dx$$

$$\leq c_\delta \|a_{\alpha\beta};L^\infty(\Omega)\| \cdot \|D^\beta u;L^2(\Omega;w)\| \cdot \|D^\gamma v;L^2(\Omega;w)\| \leq$$

$$\leq c_\delta \|a_{\alpha\beta};L^\infty(\Omega)\| \cdot \|u; \ W^{k,2}(\Omega;w)\| \cdot \|v; \ W^{k,2}(\Omega;w)\| \ .$$

Consequently, we have

$$(39.23) \qquad |a_w(u,v)| \ \leq \ c_1 \|u; \ W^{k,2}(\Omega;w)\| \cdot \|v; \ W^{k,2}(\Omega;w)\|$$

with the constant $c_1 = \displaystyle\sum_{|\alpha|,|\beta|\leq k} \ \sum_{\gamma+\delta=\alpha} c_\gamma c_\delta \|a_{\alpha\beta};L^\infty(\Omega)\|$. Thus we have already proved the following assertion :

The bilinear form $a_w(u,v)$ *from* (39.13) *is continuous on* $H \times H'$ *provided the Hilbert space* H *is chosen as*

$$(39.24) \qquad H = W^{k,2}(\Omega;w) \ ,$$

conditions (39.12) *are fulfilled and* w *has the property* (P_1) .

39.8. <u>REMARKS</u>. (i) The property (P_1) is very restrictive, but inequality (39.22) is fulfilled e.g. for weight functions of the type

$$w(x) = \exp\left(\lambda d_M(x)\right)$$

with $\lambda \in \mathbb{R}$ and $d_M(x) = \mathrm{dist}(x,M)$, $M \subset \partial\Omega$, in virtue of the fact that the first derivatives of $d_M(x)$ are bounded [see (37.17)].

Later we will give some other conditions on w which again guarantee the continuity of the bilinear form a_w [see Subsection 40.2 (ii), property (P_2)].

(ii) The estimates established in the course of calculations in Subsection 39.7 (ii) have been rather rough; evidently, the constant c_1 in (39.23) can be diminished provided we derive some finer estimates.

39.9. <u>CONVENTION</u>. To avoid *technical* difficulties, we will in the sequel consider the case $k = 1$ only, that is, we will deal only with *second order* differential operators $\mathscr{L}$. The reader will certainly understand that our investigations can be extended to the case $k > 1$ but at the cost of complicated and intricate hypotheses. The case $k = 1$ is instructive enough to point out the *idea* of our approach.

§ 40 . T h e   D i r i c h l e t   p r o b l e m

40.1. <u>THE WEIGHTED SOBOLEV SPACE</u>. (i) Let w_0 , w_1 be two weight functions on Ω , i.e., functions measurable and positive a.e. on Ω . Further, we suppose that

$$(40.1) \qquad w_i \in L^1_{loc}(\Omega) \ , \qquad w_i^{-1} \in L^1_{loc}(\Omega) \ , \qquad i = 0,1 \ .$$

(ii) We denote by

$$(40.2) \qquad W^{1,2}(\Omega;w_0,w_1)$$

the set of functions $u \in L^2(\Omega;w_0)$ such that $D^\alpha u \in L^2(\Omega;w_1)$ for $|\alpha| = 1$.
$W^{1,2}(\Omega;w_0,w_1)$ is a *Banach (Hilbert) space* if equipped with the norm

$$(40.3) \qquad \|u;\ W^{1,2}(\Omega;w_0,w_1)\| = \left(\|u;\ L^2(\Omega;w_0)\|^2 + \sum_{|\alpha|=1} \|D^\alpha u;\ L^2(\Omega;w_1)\|^2 \right)^{1/2}$$

$$= \left(\int_\Omega |u(x)|^2\ w_0(x)\ dx + \sum_{|\alpha|=1} \int_\Omega |D^\alpha u(x)|^2\ w_1(x)\ dx \right)^{1/2} .$$

(iii) Further, we denote by

$$(40.4) \qquad W_0^{1,2}(\Omega;w_0,w_1)$$

the closure of the set $C_0^\infty(\Omega)$ with respect to the norm (40.3).

(iv) If $w_0 = w_1 = w$, we shall write

$$W^{1,2}(\Omega;w) \quad \text{and} \quad W_0^{1,2}(\Omega;w)$$

instead of $W^{1,2}(\Omega;w,w)$ and $W_0^{1,2}(\Omega;w,w)$, respectively. The space $W^{1,2}(\Omega;w)$
just defined obviously coincides with the space $W^{k,2}(\Omega;w)$ from (39.20) for
$k = 1$.

For $w_0(x) = w_1(x) \equiv 1$ we obtain the *classical* Sobolev spaces $W^{1,2}(\Omega)$
and $W_0^{1,2}(\Omega)$.

40.2. THE WEIGHT FUNCTIONS. (i) We shall say that the weight function w
has the *property* (P_1) if there exists a positive constant c_1^* such that

$$(40.5) \qquad |\nabla w(x)| \leq c_1^*\, w(x) \qquad \text{for a.e. } x \in \Omega .$$

[Here $|\nabla w| = |\text{grad } w| = \left(\sum_{|\alpha|=1} |D^\alpha w|^2 \right)^{1/2}$. For $k = 1$ the property (P_1) just
introduced coincides with the property (P_1) introduced in Subsection 39.7 (i).]

(ii) We shall say that the weight function w has the *property* (P_2) if
there exists a weight function w_0 and positive constants c_2^* and c_3^* such
that

$$(40.6) \qquad \|u;\ L^2(\Omega;w_0)\| \leq c_2^* \left(\sum_{|\gamma|=1} \|D^\gamma u;\ L^2(\Omega;w)\|^2 \right)^{1/2}$$

$$\text{for every } u \in W_0^{1,2}(\Omega;w)$$

and

$$(40.7) \qquad |\nabla w(x)|^2\, w^{-1}(x) \leq c_3^{*2}\, w_0(x) \qquad \text{for a.e. } x \in \Omega .$$

40.3. __REMARKS. EXAMPLES.__ (i) Conditions (40.1) are very important. The first condition guarantees that $C_0^\infty(\Omega) \subset W^{1,2}(\Omega;w_0,w_1)$ and makes the definition of $W_0^{1,2}(\Omega;w_0,w_1)$ meaningful; the second condition implies that the spaces $W^{1,2}(\Omega;w_0,w_1)$ and $W_0^{1,2}(\Omega;w_0,w_1)$ are *complete* normed linear spaces, i.e. Hilbert (Banach) spaces. For details see A. KUFNER, B. OPIC [4], [6].

(ii) Inequality (40.5) is a special case of inequality (40.7): we obtain (40.5) by taking $w_0 = w$ and $c_3^* = c_1^*$ in (40.7).

(iii) Important examples of weight functions are the power type weights

$$(40.8) \qquad w(x) = d_M^\varepsilon(x) \ , \qquad M \subset \Omega \ , \qquad \varepsilon \in \mathbf{R}$$

[recall that $d_M(x) = \mathrm{dist}(x,M)$] and the "exponential" weights of the type

$$(40.9) \qquad w(x) = \exp\left(\varepsilon d_M(x)\right) \ , \qquad \varepsilon \in \mathbf{R} \ .$$

These special weights obviously fulfil conditions (40.1).

Moreover, as was mentioned in Remark 39.8 (i), the weight (40.9) has property (P_1), and we can take $c_1^* = |\varepsilon|$ for $\varepsilon \neq 0$.

Power type weights (40.8) have property (P_2) : Taking $w_0(x) = d_M^{\varepsilon-2}(x)$ for $w(x) = d_M^\varepsilon(x)$, we obviously have inequality (40.7) with $c_3^* = |\varepsilon|$ (we use the fact that $|\nabla d_M(x)| \leq 1$); inequality (40.6) then follows from the imbedding theorems for weighted Sobolev spaces mentioned in Subsection 0.11, namely from the imbedding

$$(40.10) \qquad W_0^{1,2}(\Omega;d_M,\varepsilon) \subsetneq L^2(\Omega;d_M,\varepsilon-2)$$

which holds for $\varepsilon \neq 1$ if $m = \dim M = N - 1$ and for every $\varepsilon \in \mathbf{R}$ if $m < N - 1$. Since the main tool for deriving this imbedding is the Hardy inequality (0.32) with $p = 2$ [see also formula (35.11)], we can easily show that the constant c_2^* in (40.6) can be expressed as follows :

$$(40.11) \qquad c_2^* = \begin{cases} \dfrac{c}{|\varepsilon - 1|} & \text{for } \varepsilon \neq 1 \ , \\[3mm] \dfrac{c}{|\varepsilon + N - m - 2|} & \text{for } \varepsilon \neq m + 2 - N \end{cases}$$

where $m = \dim M$ and c is a positive constant depending only on Ω and M (and thus independent of ε). For details see e.g. [I] (in particular for $m = N - 1$ and $m = 0$) and J. RÁKOSNÍK [1] (for general m , $0 \leq m \leq N-1$).

40.4. __DEFINITION : w-WEAK SOLUTIONS OF THE DIRICHLET PROBLEM.__ (i) Let be a linear elliptic differential operator of the *second* order

$$(40.12) \qquad (\mathcal{L}u)(x) = \sum_{|\alpha|,|\beta|\leq 1} (-1)^{|\alpha|} D^\alpha\left(a_{\alpha\beta}(x)\, D^\beta u(x)\right)$$

with coefficients $a_{\alpha\beta} \in L^{\infty}(\Omega)$. Let $\mathscr{L}$ be *elliptic with respect to the Dirichlet problem*, i.e., let the corresponding bilinear form

$$(40.13) \qquad a(u,v) = \sum_{|\alpha|,|\beta| \leq 1} \int_{\Omega} a_{\alpha\beta}(x) \, D^{\beta}u(x) \, D^{\alpha}v(x) \, dx$$

be $W_0^{1,2}(\Omega)$-elliptic   (in the terminology of Remark 39.6) :

$$(40.14) \qquad a(u,u) \geq c_0 \|u;W^{1,2}(\Omega)\|^2 \qquad \text{for every } u \in W_0^{1,2}(\Omega)$$

with a positive constant c_0 independent of u .

Further, let w be a weight function on Ω and let the bilinear form $a_w(u,v)$ be defined by

$$(40.15) \qquad a_w(u,v) = a(u,vw) = \sum_{|\alpha|,|\beta| \leq 1} \int_{\Omega} a_{\alpha\beta} \, D^{\beta}u \, D^{\alpha}(vw) \, dx \ .$$

Finally, let the function $u_0 \in W^{1,2}(\Omega;w)$ and the functional $F \in \left[W_0^{1,2}(\Omega;w)\right]^*$ be given.

We shall say that the function $u \in W^{1,2}(\Omega;w)$ is a *w-weak solution of the Dirichlet problem* for the operator $\mathscr{L}$ if

$$(40.16) \qquad u - u_0 \in W_0^{1,2}(\Omega;w)$$

and

$$(40.17) \qquad a_w(u,v) = \langle F,v \rangle \qquad \text{for every } v \in W_0^{1,2}(\Omega;w) \ .$$

<u>40.5.</u>   <u>REMARKS</u>.   (i)   The classical formulation of the Dirichlet problem, for instance in the form

$$\mathscr{L} u = f \quad \text{on } \Omega \ , \quad u = g \quad \text{on } \partial\Omega$$

with f and g given functions defined on Ω and $\partial\Omega$, respectively, is closely connected with the "w-weak formulation" : The function u_0 in (40.16) represents the boundary condition, and it is assumed that g is in fact the *trace* of $u_0 \in W^{1,2}(\Omega;w)$ [or - vice versa - that g can be extended from $\partial\Omega$ to Ω in such a way that the extension u_0 belongs to $W^{1,2}(\Omega;w)$]. The functional F in (40.17) represents the right hand side of the differential equation and the usual assumption is $\langle F,v \rangle = \int_{\Omega} f v \, dx$.

(ii)   For the case $w(x) \equiv 1$ the bilinear form $a_w(u,v)$ coincides with $a(u,v)$, $W_0^{1,2}(\Omega;w)$ is the classical Sobolev space $W_0^{1,2}(\Omega)$ and the w-weak formulation coincides with the usual weak formulation. Consequently, our approach contains the usual weak approach. In this last case, the existence of a weak solution is ensured, for $w(x) \equiv 1$, by the Lax-Milgram Theorem 39.5, since the assumptions of this theorem are obviously fulfilled: the continuity follows by the Hölder inequality directly from the assumptions $a_{\alpha\beta} \in L^{\infty}(\Omega)$, and the $W_0^{1,2}(\Omega)$-ellipticity is supposed [see (40.14)].

If $w(x) \neq 1$, then the difference from the foregoing case consists in the fact that we are working with the bilinear form a_w and with the weighted space $W^{1,2}(\Omega;w)$. Therefore, we will try to use again Theorem 39.5; this implies that we have to find conditions under which the form a_w is *bounded* and $W_0^{1,2}(\Omega;w)$-*elliptic*.

<u>40.6. THE CONTINUITY OF THE BILINEAR FORM $a_w(u,v)$</u> . Let us denote by V and V_0 the following (weighted) Hilbert spaces

$$(40.18) \quad V = W^{1,2}(\Omega;w) , \quad V_0 = W_0^{1,2}(\Omega;w) .$$

It follows from the definition of V_0 that V_0 is a subspace of V .

Using the identity $D^{\alpha}(vw) = D^{\alpha}vw + vD^{\alpha}w$ for $|\alpha| = 1$, we can write

$$(40.19) \quad a_w(u,v) = a_1(u,v) + a_2(u,v)$$

where

$$a_1(u,v) = \sum_{|\alpha|,|\beta| \leq 1} \int_{\Omega} a_{\alpha\beta} \, D^{\beta}u \, D^{\alpha}v \, w \, dx ,$$

$$a_2(u,v) = \sum_{|\alpha|=1, |\beta| \leq 1} \int_{\Omega} a_{\alpha\beta} \, D^{\beta}u \, v \, D^{\alpha}w \, dx .$$

The Hölder inequality yields – completely analogously as in Subsection 39.7 – that

$$(40.20) \quad \begin{aligned} |a_1(u,v)| &\leq \sum_{|\alpha|,|\beta| \leq 1} \int_{\Omega} |a_{\alpha\beta}| \, |D^{\beta}u| \, w^{1/2} \, |D^{\alpha}v| \, w^{1/2} \, dx \\ &\leq \tilde{c}_1 \|u;V\| \cdot \|v;V\| \end{aligned}$$

with $\tilde{c}_1 = \sum_{|\alpha|,|\beta| \leq 1} \|a_{\alpha\beta} ; L^{\infty}(\Omega)\|$.

(i) If w has property (P_1), then $|D^{\alpha}w| \leq c_1^* w$ for $|\alpha| = 1$. Using this inequality in $a_2(u,v)$, we obtain by the Hölder inequality that

$$|a_2(u,v)| \leq c_1^* \sum_{|\alpha|=1, |\beta| \leq 1} \int_{\Omega} |a_{\alpha\beta}| \, |D^{\beta}u| \, w^{1/2} \, |v| \, w^{1/2} \, dx$$

$$(40.21) \quad \leq c_1^* \sum_{|\alpha|=1, |\beta| \leq 1} \|a_{\alpha\beta}; L^{\infty}(\Omega)\| \cdot \|D^{\beta}u; L^2(\Omega;w)\| \cdot \|v; L^2(\Omega;w)\|$$

$$\leq \tilde{c}_2 \, c_1^* \, \|u;V\| \cdot \|v;V\|$$

with $\tilde{c}_2 = \sum_{|\alpha|=1, |\beta| \leq 1} \|a_{\alpha\beta} ; L^{\infty}(\Omega)\|$.

(ii) If w has property (P_2), then inequality (40.7) yields $|D^\alpha w| \le c_3^* w^{1/2} w_0^{1/2}$. Using this inequality in $a_2(u,v)$ and then applying the Hölder inequality and the imbedding inequality (40.6), we successively obtain

$$
\begin{aligned}
a_2(u,v) &\le c_3^* \sum_{|\alpha|=1,\,|\beta|\le 1} \int_\Omega |a_{\alpha\beta}| \; |D^\beta u| \; w^{1/2} \; |v| \; w_0^{1/2} \; dx \\[2mm]
&\le c_3^* \sum_{|\alpha|=1,\,|\beta|\le 1} \|a_{\alpha\beta}; \; L^\infty(\Omega)\| \cdot \|D^\beta u; \; L^2(\Omega;w)\| \cdot \|v; \; L^2(\Omega;w_0)\| \\[2mm]
&\le c_3^* c_2^* \sum_{|\alpha|=1,\,|\beta|\le 1} \|a_{\alpha\beta}; L^\infty(\Omega)\| \cdot \|u;V\| \Big(\sum_{|\gamma|=1} \|D^\gamma v; \; L^2(\Omega;w)\|^2 \Big)^{1/2} \\[2mm]
&\le \tilde{c}_2 \; c_2^* \; c_3^* \; \|u;V\| \cdot \|v;V\|
\end{aligned}
$$

(40.22)

with $\tilde{c}_2$ as in (40.21) provided $v \in V_0$.

Consequently, (40.19) – (40.22) imply that for $u,\, v \in V_0$

(40.23) $|a_w(u,v)| \le (\tilde{c}_1 + \tilde{c}_2 c^*)\|u;V\| \cdot \|v;V\|$

with $\tilde{c}_1$, $\tilde{c}_2$ positive constants depending only on the coefficients of the operator $\mathcal{L}$, and with

$$
(40.24) \qquad c^* = \begin{cases} c_1^* & \text{if } w \text{ has property } (P_1) \text{ ,} \\ c_2^* c_3^* & \text{if } w \text{ has property } (P_2) \text{ .} \end{cases}
$$

So we have already proved the following assertion :

The bilinear form $a_w(u,v)$ *from (40.15) is continuous on* $V_0 \times V_0$ *provided the Hilbert space* V_0 *is chosen according to (40.18), the weight* w *has property* (P_1) *or* (P_2) *and the condition* $a_{\alpha\beta} \in L^\infty(\Omega)$ *is fulfilled.*

<u>40.7. REMARK.</u> The only point in our foregoing consideration in which we needed the assumption $u,\, v \in V_0$ was in (40.22) when using inequality (40.6) from property (P_2). Consequently, we can say that *the bilinear form* $a_w(u,v)$ *from* (40.15) *is continuous* not only on $V_0 \times V_0$ but *on* $V \times V$ *as well, provided*

(i) w *has property* (P_1) *or*

(ii) w *has property* (P_2^*) which differs from property (P_2) by the inequality

(40.25) $\|u; \; L^2(\Omega;w_0)\| \le c_2^*\|u; \; W^{1,2}(\Omega;w)\|$ for every $u \in W^{1,2}(\Omega;w)$

instead of the inequality (40.6).

This last condition is more restrictive than condition (40.6) since it expresses the requirement that an imbedding of the form

$$
W^{1,2}(\Omega;w) \hookrightarrow L^2(\Omega;w_0)
$$

should hold while (40.6) expresses in a certain sense the imbedding

$$W_0^{1,2}(\Omega;w) \subsetneq L^2(\Omega;w_0) \ .$$

Property (P_2^*) and thus the continuity of a_w on $V \times V$ will be useful if we consider e.g. the *Neumann problem* for the differential operator $\mathscr{L}$.

40.8. THE V_0-ELLIPTICITY OF THE BILINEAR FORM $a_w(u,v)$. Let V_0 be the weighted space from (40.18) and let us consider the expression $a_w(u,u)$. We can write

$$(40.26) \qquad a_w(u,u) = a(uw^{1/2}, uw^{1/2}) + J_w$$

where a is the bilinear form (40.13).

(i) The ellipticity of $\mathscr{L}$ with respect to the Dirichlet problem, i.e. the $W_0^{1,2}(\Omega)$-ellipticity of a — see (40.14) — implies

$$(40.27) \qquad a(uw^{1/2}, uw^{1/2}) \geq c_0 \| uw^{1/2}; \ W^{1,2}(\Omega) \|^2 \ .$$

We have

$$\| uw^{1/2}; \ W^{1,2}(\Omega) \|^2 = \sum_{|\alpha| \leq 1} \int_\Omega |D^\alpha(uw^{1/2})|^2 \ dx$$

$$= \int_\Omega |u|^2 \ w \ dx + \sum_{|\alpha|=1} \int_\Omega |D^\alpha u w^{1/2} + \tfrac{1}{2} u w^{-1/2} D^\alpha w|^2 \ dx$$

$$(40.28) \qquad \geq \int_\Omega |u|^2 \ w \ dx$$

$$+ \sum_{|\alpha|=1} \int_\Omega \left(|D^\alpha u|^2 \ w - |u| \ |D^\alpha u| \ |D^\alpha w| - \tfrac{1}{4} |u|^2 \ |D^\alpha w|^2 \ w^{-1} \right) \ dx$$

$$= \|u;V\|^2 - S_1 - \tfrac{1}{4} S_2$$

where

$$(40.29) \qquad
\begin{aligned}
S_1 &= \sum_{|\alpha|=1} S_{1\alpha} \ , \qquad & S_{1\alpha} &= \int_\Omega |u| \ |D^\alpha u| \ |D^\alpha w| \ dx \\
S_2 &= \sum_{|\alpha|=1} S_{2\alpha} \ , \qquad & S_{2\alpha} &= \int_\Omega |u|^2 \ |D^\alpha w|^2 \ w^{-1} \ dx \ .
\end{aligned}$$

(i-1) If we assume that the weight w has property (P_1) , then we can use inequality (40.5) in the form $|D^\alpha w| \leq c_1^* w$ and have — using in the case of $S_{1\alpha}$ the Hölder inequality, too — the estimates

$$S_{1\alpha} \leq c_1^* \int_\Omega |u| \ |D^\alpha u| \ w \ dx = c_1^* \int_\Omega |u| \ w^{1/2} \ |D^\alpha u| \ w^{1/2} \ dx$$

$$\leq c_1^* \| u; \ L^2(\Omega;w) \| \cdot \| D^\alpha u; \ L^2(\Omega;w) \| \leq c_1^* \| u;V \|^2$$

and

$$S_{2\alpha} \le (c_1^*)^2 \int_\Omega |u|^2 \, w \, dx = (c_1^*)^2 \|u; \, L^2(\Omega;w)\|^2 \le (c_1^*)^2 \|u;V\|^2 \ .$$

(i-2) If we assume that the weight w has property (P_2), then we can use inequality (40.7) in the form $|D^\alpha w| \le c_3^* \, w_0^{1/2} \, w^{1/2}$ and obtain similarly as in point (i-1) — using additionally inequality (40.6) — the estimates

$$S_{1\alpha} \le c_3^* \int_\Omega |u| \, |D^\alpha u| \, w_0^{1/2} \, w^{1/2} \, dx = c_3^* \int_\Omega |u| \, w_0^{1/2} \, |D^\alpha u| \, w^{1/2} \, dx$$

$$\le c_3^* \|u; \, L^2(\Omega;w_0)\| \cdot \|D^\alpha u; \, L^2(\Omega;w)\|$$

$$\le c_3^* \, c_2^* \Big(\sum_{|\gamma|=1} \|D^\gamma u; \, L^2(\Omega;\,w)\|^2 \Big)^{1/2} \|u;V\| \le c_3^* \, c_2^* \, \|u;V\|^2$$

and

$$S_{2\alpha} \le (c_3^*)^2 \int_\Omega |u|^2 \, w_0 \, w \, w^{-1} \, dx = (c_3^*)^2 \|u; \, L^2(\Omega;w_0)\|^2$$

$$\le (c_3^* c_2^*)^2 \sum_{|\gamma|=1} \|D^\gamma u; \, L^2(\Omega;w)\|^2 \le (c_3^* c_2^*)^2 \|u;V\|^2 \ .$$

(i-3) Using the estimates just established in (40.28) we have together with (40.27)

$$(40.30) \qquad a(uw^{1/2}, \, uw^{1/2}) \ge c_0 \big(1 - c^*N - \tfrac{1}{4} \, (c^*N)^2 \big) \|u; \, W^{1,2}(\Omega;w)\|^2$$

for every $u \in V_0$ with the constant c^* given by (40.24).

(ii) For J_w from (40.26) we have

$$
\begin{aligned}
J_w = \ &\tfrac{1}{2} \sum_{|\alpha|=1,\,|\beta|\le 1} \int_\Omega a_{\alpha\beta} \, D^\beta u \, u \, D^\alpha w \, dx \\[4pt]
(40.31) \qquad &- \tfrac{1}{2} \sum_{|\alpha|\le 1,\,|\beta|=1} \int_\Omega a_{\alpha\beta} \, u \, D^\alpha u \, D^\beta w \, dx \\[4pt]
&- \tfrac{1}{4} \sum_{|\alpha|,\,|\beta|=1} \int_\Omega a_{\alpha\beta} \, u^2 \, w^{-1} \, D^\alpha w \, D^\beta w \, dx \ .
\end{aligned}
$$

The three integrals in J_w can be estimated from above by an argument completely analogous to that in part (i). We obtain

$$\Big| \int_\Omega a_{\alpha\beta} \, D^\beta u \, u \, D^\alpha w \, dx \Big| \le \|a_{\alpha\beta}; \, L^\infty(\Omega)\| \int_\Omega |u| \, |D^\beta u| \, |D^\alpha w| \, dx$$

$$\le c^* \, \|a_{\alpha\beta}; \, L^\infty(\Omega)\| \cdot \|u;V\|^2 \ ,$$

$$\Big| \int_\Omega a_{\alpha\beta} \, u \, D^\alpha u \, D^\beta w \, dx \Big| \le \|a_{\alpha\beta}; \, L^\infty(\Omega)\| \int_\Omega |u| \, |D^\alpha u| \, |D^\beta w| \, dx \le$$

$$\leq c^* \, \|a_{\alpha\beta}; \, L^\infty(\Omega)\| \cdot \|u;V\|^2 \, ,$$

$$\left| \int_\Omega a_{\alpha\beta} \, u^2 \, w^{-1} \, D^\alpha w \, D^\beta w \, dx \right| \leq \|a_{\alpha\beta}; \, L^\infty(\Omega)\| \int_\Omega |u|^2 \, w^{-1} \, |D^\alpha w| \, |D^\beta w| \, dx$$

$$\leq (c^*)^2 \, \|a_{\alpha\beta}; \, L^\infty(\Omega)\| \cdot \|u;V\|^2 \, ,$$

and consequently

$$(40.32) \qquad |J_w| \leq \tilde{c}\left(c^* + \tfrac{1}{4}(c^*)^2\right) \, \|u; \, W^{1,2}(\Omega;w)\|^2$$

with

$$(40.33) \qquad \tilde{c} = \sum_{|\alpha|,|\beta| \leq 1} \|a_{\alpha\beta}; \, L^\infty(\Omega)\|$$

and with c^* from (40.24).

    (iii) Since (40.26) implies

$$a_w(u,u) \geq a(uw^{1/2}, \, uw^{1/2}) - |J_w| \, ,$$

we have shown, using the estimates (40.30) and (40.32), that

$$(40.34) \qquad a_w(u,u) \geq \left[c_0 - (Nc_0 + \tilde{c})c^* - \tfrac{1}{4}(Nc_0 + \tilde{c})(c^*)^2\right] \, \|u; \, W^{1,2}(\Omega;w)\|^2$$

for every $u \in V_0 = W_0^{1,2}(\Omega;w)$ with c_0 the ellipticity constant from (40.14), c^* given by (40.24) and $\tilde{c}$ depending on the operator $\mathcal{L}$.

    The multiplicative constant in the square brackets in (40.34) is positive if and only if

$$(40.35) \qquad c^* < 2\left[\sqrt{1 + \frac{c_0}{Nc_0 + \tilde{c}}} - 1\right]$$

So we have already proved the following assertion :

The bilinear form $a_w(u,v)$ *from (40.15) is* $W_0^{1,2}(\Omega;w)$-*e l l i p t i c provided the bilinear form* a(u,v) *from (40.13) is* $W_0^{1,2}(\Omega)$-*elliptic [with the ellipticity constant* c_0 , *see (40.14)], the conditions* $a_{\alpha\beta} \in L^\infty(\Omega)$ *are fulfilled and the weight* w *has property* (P_1) *or* (P_2) *with the constant* c^* *from (40.24) sufficiently small [so that (40.35) holds].*

<u>40.9.</u> <u>REMARKS.</u> (i) The estimates established in the foregoing Subsections for the constants in (40.30) and (40.32) are again very rough and can be improved by making the calculations more carefully and in a little more sophisticated way. For example, the important constant $\tilde{c}$ from (40.33) was for simplicity chosen in the form

$$\sum_{|\alpha|,|\beta| \leq 1} \|a_{\alpha\beta}; L^\infty(\Omega)\| \, ,$$

but a little better result can be obtained if we work with the constant

$$(40.36) \qquad \left(\sum_{|\alpha|,|\beta|\le 1} \| a_{\alpha\beta}; L^{\infty}(\Omega) \|^2 \right)^{1/2}$$

or if we take, moreover, into account the fact that the particular sums in the expression for J_w [see Subsection 40.8 (ii), formula (40.31)] are taken over different sets of multiindices α , β , namely $\{|\alpha|=1, |\beta|\le 1\}$, $\{|\alpha|\le 1, |\beta|=1\}$ and $\{|\alpha|=1, |\beta|=1\}$, while we used the summation over α , β such that $|\alpha|$, $|\beta| \le 1$.

(ii) As a further example, let us show how the constant appearing in (40.30) can be improved to the form

$$(40.37) \qquad c_0 \left(1 - c^* - \tfrac{1}{4}(c^*)^2 \right)$$

(i.e., with N replaced by 1): Instead of estimating the integrals $S_{1\alpha}$ in (40.29) we estimate directly their sum S_1 using the Hölder inequality for sums

$$\sum_{|\alpha|=1} |D^{\alpha}u| \; |D^{\alpha}w| \le \left(\sum_{|\alpha|=1} |D^{\alpha}u|^2 \right)^{1/2} \left(\sum_{|\alpha|=1} |D^{\alpha}w|^2 \right)^{1/2} = |\nabla u| \; |\nabla w| \; .$$

If w has property (P_1), then we estimate $|\nabla w|$ by inequality (40.5) and obtain that

$$S_1 \le c_1^* \int_{\Omega} |u| \; |\nabla u| \; w \; dx \le c_1^* \| u; L^2(\Omega;w) \| \; \| \nabla u; L^2(\Omega;w) \| \le c_1^* \| u;V \|^2 \; ;$$

since $S_2 = \int_{\Omega} |u|^2 \left(\sum_{|\alpha|=1} |D^{\alpha}w|^2 \right) w^{-1} \, dx = \int_{\Omega} |u|^2 \; |\nabla w|^2 \; w^{-1} \, dx$, we have, again by (40.5),

$$S_1 \le (c_1^*)^2 \int_{\Omega} |u|^2 \; w \; dx \le (c_1^*)^2 \| u;V \|^2 \; .$$

Thus finally $S_1 + \tfrac{1}{4} S_2 \le (c^* + \tfrac{1}{4} c^*) \| u;V \|^2$ with $c^* = c_1^*$ and this estimate leads to the estimate of the form (40.30) with the constant from (40.37). Analogously we proceed if w has property (P_2).

In this case, the condition of $W_0^{1,2}(\Omega;w)$-ellipticity reads as follows :

$$(40.38) \qquad c^* < 2 \left(\sqrt{ \frac{2c_0 + \tilde{c}}{c_0 + \tilde{c}} } - 1 \right) \; .$$

(iii) A more detailed investigation of the ellipticity condition for the bilinear form a offers another possibility of improving the constants just mentioned. For instance, let us use — instead of the "integral" ellipticity condition (40.14) — the *algebraic* ellipticity condition

$$(40.39) \qquad \sum_{|\alpha|,|\beta|\le 1} a_{\alpha\beta}(x) \; \xi_{\alpha} \, \xi_{\beta} \ge c_0 |\xi|^2$$

with $c_0 > 0$ and $\xi = \{\xi_{\alpha}, |\alpha| \le 1\} \in \mathbb{R}^{N+1}$ arbitrary. Condition (40.39) is

stronger than condition (40.14) since (40.14) is a consequence of (40.39). In
this case, instead of (40.34) we derive the estimate

$$(40.40) \quad a_w(u,u) \geq (c_0 - \tilde{c}_2 c^*) \| u; W^{1,2}(\Omega) \|^2$$

with $\tilde{c}_2 = \sum\limits_{|\alpha|=1, |\beta| \leq 1} \| a_{\alpha\beta}; L^\infty(\Omega) \|$ and c^* from (40.24).

Indeed, it follows from (40.19) that

$$a_w(u,u) \geq a_1(u,u) - |a_2(u,u)| \ .$$

Condition (40.39) implies — taking $\xi_\alpha = D^\alpha u(x) \, w^{1/2}(x)$ and then integrating
over Ω — that

$$a_1(u,u) \geq c_0 \| u; \ W^{1,2}(\Omega;w) \|^2 \ ;$$

further we have from (40.21) or (40.22) that $\quad |a_2(u,u)| \leq \tilde{c}_2 c^* \| u; \ W^{1,2}(\Omega;w) \|^2 \ ,$
and finally we obtain for $u \in W_0^{1,2}(\Omega;w)$ the desired estimate

$$a_w(u,u) \geq (c_0 - \tilde{c}_2 c^*) \ \| u; \ W^{1,2}(\Omega;w) \|^2$$

which says that the bilinear form a_w is $W_0^{1,2}(\Omega;w)$-elliptic provided the corresponding ellipticity constant $c_0 - \tilde{c}_2 c^*$ is positive. But this is true for

$$(40.41) \quad c^* < \frac{c_0}{\tilde{c}_2}$$

and this estimate is evidently better then the estimates (40.35) or (40.38).

Thus, the stronger ellipticity condition (40.39) enables us to deal with
a generally larger scale of weight functions w .

<u>40.10.</u>   <u>EXAMPLE.</u>   Let us consider the special operator

$$\mathscr{L} u = - \Delta u + u \ .$$

Here $a_{\alpha\alpha}(x) \equiv 1$, $a_{\alpha\beta}(x) \equiv 0$ for $\alpha \neq \beta$, $|\alpha|, |\beta| \leq 1$, and consequently,
formula (40.33) says that the number $\tilde{c}$ is $N + 1$, while formula (40.36)
yields for $\tilde{c}$ the value $\sqrt{N + 1}$. Further we have $c_0 = 1$ and $\tilde{c}_2 = N$ — see
(40.40). According to the estimates established in the foregoing Subsections,
such weights w are admissible for which the corresponding constant c^* from
properties (P_1) and/or (P_2) is — in the best possible case, expressed by
formula (40.41) — such that

$$(40.42) \quad c^* < \frac{1}{N} \ .$$

Nevertheless, even this estimate can be improved : For our operator $\mathscr{L}$, we
have

$$a_w(u,u) = \|u; W^{1,2}(\Omega;w)\|^2 + \sum_{|\alpha|=1} \int_\Omega D^\alpha u \; u \; D^\alpha w \; dx \; .$$

The last term is precisely the expression S_1 from (40.29); using the Hölder inequality for sums as indicated in Remark 40.9 (ii), we have

$$S_1 \leq c_1^* \int_\Omega u \; |\nabla u| \; w \; dx \leq \frac{c_1^*}{2} \int_\Omega \left(|u|^2 + |\nabla u|^2\right) w \; dx = \frac{c_1^*}{2} \|u; W^{1,2}(\Omega;w)\|^2$$

[we have assumed that the weight function w has property (P_1) and used the inequality $ab \leq \frac{1}{2}(a^2 + b^2)$] . Thus

$$a_w(u,u) \geq \left(1 - \frac{c_1^*}{2}\right) \|u; W^{1,2}(\Omega;w)\|^2 .$$

and the ellipticity constant $1 - c_1^*/2$ is positive if

$$c^* < 2 \; .$$

This result is substantially better than the estimate (40.42).

Now, we can prove the main existence and uniqueness assertion of this Section.

40.11. THEOREM. *Let $\mathcal{L}$ be the linear differential operator of the second order from (40.12) with coefficients $a_{\alpha\beta} \in L^\infty(\Omega)$ and elliptic in the sense of (40.14). Then there exists a positive number C with the following property : If w is a weight function which has property (P_1) or (P_2) with the corresponding constant c^* from (40.24) such that*

(40.43) $c^* < C$

and if a function $u_0 \in W^{1,2}(\Omega;w)$ and a functional $F \in \left[W_0^{1,2}(\Omega;w)\right]^*$ are given, then there exists one and only one w-weak solution $u \in W^{1,2}(\Omega;w)$ of the Dirichlet problem for the operator $\mathcal{L}$ (in the sense of Definition 40.4). Moreover, there is a positive constant c independent of u_0 and F and such that

(40.44) $\|u; W^{1,2}(\Omega;w)\| \leq c\left(\|u_0; W^{1,2}(\Omega;w)\| + \|F; \left[W_0^{1,2}(\Omega;w)\right]^*\|\right) .$

P r o o f : Let us use the notation (40.18) for the spaces involved. Further, let us denote by $\tilde{F}$ the functional defined by the formula

(40.45) $\langle\tilde{F},v\rangle = \langle F,v\rangle - a_w(u_0,v)$

with a_w the bilinear form from (40.15). It follows from the continuity of a_w — see (40.23) — that $\tilde{F} \in V^*$ with $\|\tilde{F};V^*\| \leq \|F;V^*\| + c_1\|u_0;V\|$, c_1 being a suitable positive constant.

In view of the results of Subsections 40.6 and 40.8, the bilinear form a_w is continuous and V_0-elliptic provided the number C in (40.43) is appropriately chosen - e.g. as the right hand side in some of the estimates (40.35), (40.38) or (40.41). Therefore we can use Theorem 39.5 taking for $b(u,v)$ the form $a_w(u,v)$, for H the weighted space V_0 and for h the functional $\tilde{F}$ just defined.

According to this theorem, there exists one and only one function $\tilde{u} \in V_0$ such that

$$(40.46) \qquad a_w(\tilde{u},v) = \langle \tilde{F},v \rangle \qquad \text{for every} \ \ v \in V_0$$

and

$$(40.47) \qquad \|\tilde{u};V_0\| \leq \frac{1}{c_2} \|\tilde{F};V_0^*\| \ .$$

Now, the function
$$u = \tilde{u} + u_0$$

is the desired w-weak solution of the Dirichlet problem for $\mathcal{L}$. Indeed, we have $u - u_0 = \tilde{u} \in V_0$ so that condition (40.16) from Definition 40.4 is fulfilled, and further, in view of the linearity of $a_w(u,v)$ and of (40.46) and (40.45),

$$a_w(u,v) = a_w(\tilde{u} + u_0, \ v) = a_w(\tilde{u},v) + a_w(u_0,v)$$
$$= \langle \tilde{F},v \rangle + a_w(u_0,v) = \langle F,v \rangle$$

for every $v \in V$, so that (40.17) is fulfilled, too.

Since $\|u;V\| \leq \|\tilde{u};V\| + \|u_0;V\|$, the estimate (40.44) follows from (40.47):

$$\|u;V\| \leq \frac{1}{c_2}\big(\|F;V_0^*\| + c_1\|u_0;V\|\big) + \|u_0;V\|$$

and we can take $c = \max\big(1/c_2, \ 1 + c_1/c_2\big)$.

<u>40.12. AN EXTENSION : (w_0,w)-WEAK SOLUTIONS.</u> Let us suppose that *the weight function w has property* (P_2). Then it follows from the estimate (40.6) that for every $u \in C_0^\infty(\Omega)$,

$$\|u; \ W^{1,2}(\Omega;w_0,w)\|^2 \leq \big[1 + (c_2^*)^2\big] \ \|u; \ W^{1,2}(\Omega;w)\|^2 \ .$$

This means that the imbedding

$$W_0^{1,2}(\Omega;w) \subsetneqq W_0^{1,2}(\Omega;w_0,w)$$

holds. Using the *larger* class $W_0^{1,2}(\Omega;w_0,w)$, it seems to be meaningful to consider (w_0,w)-*weak solutions of the Dirichlet problem* for the operator $\mathcal{L}$ from (40.12).

The definition of such a solution, which follows Definition 40.4 word for word, only replacing the spaces $W^{1,2}(\Omega;w)$ and $W_0^{1,2}(\Omega;w)$ by the spaces $W^{1,2}(\Omega;w_0,w)$ and $W_0^{1,2}(\Omega;w_0,w)$, respectively, is left to the reader, as well

as the formulation and proof of an existence and uniqueness theorem for (w_0,w)-
-weak solutions analogous to Theorem 40.11.

§ 41 . P o w e r   t y p e   w e i g h t s .   O t h e r   b o u n d a r y
v a l u e   p r o b l e m s

41.1. <u>INTRODUCTION</u>. The main tool we have used in our foregoing considerations have been the properties (P_1) and/or (P_2) of the weight function w .
The method described can be used if the constant c^* from (40.24) is sufficiently small - see condition (40.43). The number C from condition (40.43)
gives in a certain sense an idea about the scale of the addmissible weights.
Therefore, let us now look what is the situation with the particular weights
mentioned in Example 40.3 (iii).

(i) As was mentioned in this example, the "exponential" weight

$$w(x) = \exp\left(\varepsilon\, d_M(x)\right)$$

with $\varepsilon \in \mathbf{R}$ and $d_M(x) = \mathrm{dist}(x,M)$, M being an m-dimensional manifold on
$\partial\Omega$, *has property* (P_1) *with* $c_1^* = c^* = |\varepsilon|$. Therefore, exponential weights
are admissible and existence results about a w-weak solution in the space
$W^{1,2}(\Omega;e^{\varepsilon d_M})$ can be obtained provided $|\varepsilon|$ *is sufficiently small*. The bounus
for the admissible values of ε are given mainly in terms of the coefficients
of the differential operator $\mathscr{L}$, more precisely, in terms of the L^∞-norm of
$a_{\alpha\beta}$.

(ii) More important and more interesting are power type weights. Therefore, let us consider the weight function

(41.1) $w(x) = \left[d_M(x)\right]^\varepsilon$, $\varepsilon \in \mathbf{R}$;

for the corresponding weighted spaces $W^{1,2}(\Omega;w)$ and $W_0^{1,2}(\Omega;w)$ we use the
notation $W^{1,2}(\Omega;d_M,\varepsilon)$ and $W_0^{1,2}(\Omega;d_M,\varepsilon)$, respectively.

It was already mentioned in Example 40.3 (iii) that the weight $d_M^\varepsilon(x)$ *has*
property (P_2) , if we take for $w_0(x)$ the weight $d_M^{\varepsilon-2}(x)$; for the constant
c_3^* from (40.7) we have $c_3^* = |\varepsilon|$ while the constant c_2^* from (40.6) is given
by formula (40.11). Consequently, for the constant c^* from (40.24) we have the
expression

(41.2) $c^* = |\varepsilon| \min \left(\dfrac{c_1}{|\varepsilon - 1|} \,,\, \dfrac{c_2}{|\varepsilon + N - m - 2|}\right)$

where $N = \dim\Omega$, $m = \dim M$, $M \subset \partial\Omega$, and c_1 , c_2 are positive constants
depending only on Ω and M .

The analogue of Theorem 40.11 reads now as follows :

41.2. THEOREM. *Let* $\mathcal{L}$ *be the differential operator of the second order from (40.12) with coefficients* $a_{\alpha\beta} \in L^{\infty}(\Omega)$ *, and elliptic in the sense of (40.14). Then there exists an open interval* I *containing the origin which has the following property :*

For every $\varepsilon \in I$ *and for* $u_0 \in W^{1,2}(\Omega;d_M,\varepsilon)$ *and* $F \in \left[W_0^{1,2}(\Omega;d_M,\varepsilon)\right]^*$ *, there is one and only one* d_M^{ε}*-weak solution* $u \in W^{1,2}(\Omega;d_M,\varepsilon)$ *of the Dirichlet problem for the operator* $\mathcal{L}$ *(in the sense of Definition 40.4); moreover, there is a positive constant* c *independent of* u_0 *and* F *and such that*

$$\|u;\ W^{1,2}(\Omega;d_M,\varepsilon)\| \leq c\left(\|u_0;\ W^{1,2}(\Omega;d_M,\varepsilon)\| + \|F;\ \left[W_0^{1,2}(\Omega;d_M,\varepsilon)\right]^*\|\right) .$$

P r o o f : Since the weight function d_M^{ε} considered has property (P_2), we only have to show that there exists an open interval I containing the origin and such that for $\varepsilon \in I$, condition (40.43) of Theorem 40.11 is satisfied. But this follows immediately from formula (41.2) : If we regard c^* as a function of ε , then obviously $c^* = c^*(\varepsilon) > 0$ for $\varepsilon \neq 0$, $c^*(0) = 0$ and $c^*(\varepsilon)$ is continuous in an open neighbourhood of the origin (the only possible point of discontinuity is $\varepsilon = 1$ if $m = N - 1$). Consequently, there exists a neighbourhood I of the origin such that

$$c^*(\varepsilon) < C \quad \text{for} \quad \varepsilon \in I ,$$

where C is, e.g., the constant from the right hand side in (40.38) or - if the algebraic ellipticity condition (40.39) is fulfilled - from the right hand side in (40.41). The assertion then follows from Theorem 40.11.

41.3. REMARK. A little more detailed investigation of the interval I of the admissible powers ε in the weight function d_M^{ε} in Theorem 40.11 shows that we can put

$$I = I_1 \cup I_2$$

where

$$I_1 = (C,+\infty) \cap \left(\frac{-C(N - m - 2)}{c_2 + C} ,\ \frac{C(N - m - 2)}{c_2 - C}\right) ,$$

$$I_2 = (-\infty,C) \cap \left(-\frac{C}{c_1 - C} ,\ \frac{C}{c_1 + C}\right)$$

and $C = 2\left(\sqrt{(2c_0 + \tilde{c})/(c_0 + \tilde{c})} - 1\right)$ with c_0 the ellipticity constant from (40.14) and $\tilde{c} = \sum\limits_{|\alpha|,\,|\beta| \leq 1} \|a_{\alpha\beta};L^{\infty}(\Omega)\|$, or $C = c_0/\tilde{c}_2$ with c_0 the ellipticity constant from (40.39) and $\tilde{c}_2 = \sum\limits_{|\alpha|=1,\,|\beta|\leq 1} \|a_{\alpha\beta};L^{\infty}(\Omega)\|$.

Indeed, if we denote $t = \dfrac{c_2 - c_1(N - m - 2)}{c_2 + c_1}$, then (41.2) implies

$$c^* = \left\{ \begin{array}{ll} \dfrac{c_1|\varepsilon|}{|\varepsilon - 1|} & \text{for } \varepsilon \le t , \\[3mm] \dfrac{c_2|\varepsilon|}{|\varepsilon + N - m - 2|} & \text{for } \varepsilon > t , \end{array} \right.$$

and the above estimates follow from the inequality $c^* < C$ which guarantees the $W_0^{1,2}(\Omega;d_M,\varepsilon)$-ellipticity of the bilinear form a_w (with $w = d_M^\varepsilon$) — see Remarks 40.9 (ii), (iii).

The constant C was chosen according to inequalities (40.38) and (40.41), respectively. Obviously , we can take for C any other constant which guarantees that for $c^* < C$ the bilinear form a_w with $w = d_M^\varepsilon$ is $W_0^{1,2}(\Omega;d_M,\varepsilon)$-
-elliptic.

41.4. __EXAMPLE.__ Let Ω be the cube $(0,1)^N$ and $M = \{x = (x_1,x_2,\ldots,x_N) \in \overline{\Omega};$ $x_i = 0$ for $i = m+1,m+2,\ldots,N\}$ with $0 \le m \le N - 1$. Then $M \subset \partial\Omega$, $\dim M = m$ and

$$(41.3) \qquad d_M^\varepsilon(x) = \left(\sum_{i=m+1}^{N} x_i^2 \right)^{\varepsilon/2} .$$

Further, let us consider the operator

$$(41.4) \qquad \mathscr{L}u = - \Delta u + u ;$$

then $a_{\alpha\alpha}(x) \equiv 1$ and $a_{\alpha\beta}(x) \equiv 0$ for $\alpha \ne \beta$, $|\alpha|, |\beta| \le 1$; the operator satisfies the ellipticity condition (40.39) with $c_0 = 1$ and is $W_0^{1,2}(\Omega;w)$-
-elliptic for $c^* = c_2^* c_3^* < 1$ _provided the weight_ w _has property_ (P_2).

Indeed, proceeding as in Example 40.10 we obtain

$$a_w(u,u) = \|u; W^{1,2}(\Omega;w)\|^2 + S_1$$

where — see Remark 40.9 (ii) —

$$|S_1| = \left| \sum_{|\alpha|=1} \int_\Omega D^\alpha u \, u \, D^\alpha w \, dx \right| \le \int_\Omega |\nabla u| \cdot |u| \cdot |\nabla w| \, dx ,$$

and by inequalities (40.7) and (40.6) we have

$$|S_1| \le c_3^* \int_\Omega |\nabla u| \cdot |u| \cdot w^{1/2} w_0^{1/2} \, dx \le c_3^* \|\nabla u; L^2(\Omega;w)\| \cdot \|u; L^2(\Omega;w_0)\|$$

$$\le c_3^* c_2^* \|\nabla u; L^2(\Omega;w)\|^2 \le c_3^* c_2^* \|u; W^{1,2}(\Omega;w)\|^2 .$$

Consequently,

$$|a_w(u,u)| \ge (1 - c_3^* c_2^*)\|u; W^{1,2}(\Omega;w)\|^2$$

and the operator $\mathscr{L}$ from (41.4) is $W_0^{1,2}(\Omega;w)$-elliptic for $c_3^* c_2^* = c^* < 1$.

In particular, the weight d_M^ε from (41.3) has property (P_2). In view of the foregoing considerations, we can take $C = 1$ in the estimates from Remark 41.3. In order to obtain the exact values for the constants c_1 , c_2 in (41.2) we proceed as follows :

We extend the function $u \in C_0^\infty(\Omega)$ by zero for $x_i \geq 1$, $i = m+1, m+2, \ldots, N$, use the generalized cylindrical coordinates $(x_1, \ldots, x_m, \vartheta_1, \ldots, \vartheta_{N-m-1}, r) = (\tilde{x}, \vartheta, r)$, so that $d_M(x) = r$, and express the $L^2(\Omega; d_M, \varepsilon-2)$-norm in the following form :

$$
\| u; \ L^2(\Omega; d_M, \varepsilon-2) \|^2
$$

$$
= \int\limits_{(0,1)^m} \left[\int\limits_{(0,\pi/2)^{N-m-1}} \left(\int\limits_0^\infty |u(\tilde{x}, \vartheta, r)|^2 \ r^{\varepsilon-2} \ r^{N-m-1} \ dr \right) d\vartheta \right] d\tilde{x} \ .
$$

Applying the Hardy inequality – see Subsection 0.11, formula (0.32) with $p = 2$ and with $\varepsilon + N - m - 1$ instead of ε – to the inner integral under the assumption $\varepsilon \neq m + 2 - N$ and passing again to the Cartesian coordinates we obtain

$$
\| u; \ L^2(\Omega; d_M, \varepsilon-2) \|^2 \leq \frac{4}{|\varepsilon + N - m - 2|^2} \int\limits_\Omega \left| \frac{\partial u}{\partial r} \right|^2 d_M^\varepsilon \ dx
$$

$$
\leq \frac{4}{|\varepsilon + N - m - 2|^2} \| u; \ W^{1,2}(\Omega; d_M, \varepsilon) \|^2 \ ;
$$

this means that $c_2 = 2$ in (41.2) – see (40.11). – Analogously we obtain for $\varepsilon \neq -1$ the estimate

$$
\| u; \ L^2(\Omega; d_M, \varepsilon-2) \| \leq \frac{c_1^2}{|\varepsilon - 1|^2} \| u; \ W^{1,2}(\Omega; d_M, \varepsilon) \|
$$

where

$$
c_1 = \left\{ \begin{array}{ll} 2^{(6-\varepsilon)/4} \ (N - m)^{-1/2} & \text{for } \varepsilon \leq 0 \ , \\[2mm] 2^{3/2} \ (N - m)^{-1/2} & \text{for } 0 < \varepsilon \leq 2 \ , \\[2mm] 2^{(4-\varepsilon)/4} \ (N - m)^{-1/2} & \text{for } \varepsilon > 2 \end{array} \right.
$$

(see J. RÁKOSNÍK [1]).

A more detailed discussion carried out by J. RÁKOSNÍK has led to the following estimates of the interval I of the admissible values of ε for the operator $\mathcal{L}$ from (40.4) :

$$
\begin{array}{llll}
\text{for} & N - m = 1 & \text{we have} & I = (-0.48, \ 0.26) \ , \\
\text{for} & N - m = 2 & \text{we have} & I = (-0.78, \ 0.33) \ , \\
\text{for} & N - m = 3 & \text{we have} & I = (-1.04, \ 1) \ , \\
\text{for} & N - m = 4 & \text{we have} & I = (-1.30, \ 2) \ , \\
\text{for} & N - m = 5 & \text{we have} & I = (-1.54, \ 3) \ .
\end{array}
$$

41.5. REMARKS. (i) In this Section, we follow very closely the ideas developed in the paper A. KUFNER, J. RÁKOSNÍK [1].

(ii) This paper deals also with (w_0,w)-weak solutions mentioned in Subsection 40.12. In particular, for the operator $\mathcal{L}$ from (40.4), it is shown for what values of ε we can guarantee existence and uniqueness of a $(d_M^{\varepsilon-2}, d_M^{\varepsilon})$ -weak solution of the Dirichlet problem. Here let us only mention that the corresponding intervals I again contain the origin, but they are substantially smaller than the intervals mentioned at the end of Example 41.4.

(iii) It should be pointed out that all the intervals just mentioned give only a rough picture of the admissible values of ε : the condition $\varepsilon \in I$ is only a *sufficient condition* for the existence of a d_M^{ε}-weak [or $(d_M^{\varepsilon-2}, d_M^{\varepsilon})$-weak] solution of the Dirichlet problem for the operator $\mathcal{L}$ in question.

41.6. A COMPARISON WITH THE APPROACH FROM [I]. In Section 12 we proposed a method for treating the Neumann problem in weighted Sobolev spaces; this method imitated the method proposed in [I] for the Dirichlet problem. A comparison with the method proposed in this Section 13 shows that the last described approach gives generally *a larger class of admissible weights*. In fact, in Section 13 the main problem was the investigation of the $W_0^{1,2}(\Omega;w)$-ellipticity, i.e. the derivation of the inequality

(41.5) $\quad a_w(u,u) = a(u,uw) \geq c\|u;\ W^{1,2}(\Omega;w)\|$,

and we obtained certain restrictive conditions on the weight w (in particular, certain restrictive conditions on the power ε for the case of the weight d_M^{ε}).

However, the approach from [I] required the investigation of *two* inequalities of the type (41.5), namely, the inequalities

$$a(u,uw) \geq c_1\|u;\ W^{1,2}(\Omega;w^{-1})\| ,$$
$$a(vw^{-1},v) \geq c_2\|v;\ W^{1,2}(\Omega;w)\|$$

[see, e.g., formulas (37.39) where $w = d_M^{\varepsilon}$]. This eventually generates additional conditions on w and can cause a further restriction of the class of admissible weights.

41.6*. OTHER BOUNDARY VALUE PROBLEMS. The main tools for establishing the existence and uniqueness of a w-weak solution of the Dirichlet problem for the operator $\mathcal{L}$ were the boundedness (= continuity) and the $W_0^{1,2}(\Omega;w)$-ellipticity of the bilinear form $a_w(u,v)$, i.e., the validity of estimates of the type

(41.6)
$$|a_w(u,v)| \leq c_1\|u;V_0\| \cdot \|v;V_0\|$$
$$|a_w(u,u)| \geq c_2\|u;V_0\|^2$$

198

for every pair $u, v \in V_0 = W_0^{1,2}(\Omega;w)$ [we use the notation from (40.18), so
that $V = W^{1,2}(\Omega;w)$].

For other boundary value problems we have to derive analogous estimates,
but now for functions $u, v \in \tilde{V}$, where $\tilde{V}$ is a larger space,

(41.7) $$V_0 \subset \tilde{V} \subset V$$

(e.g., we have $\tilde{V} = V$ for the *Neumann problem*). As was shown in Remark 40.7,
the bilinear form $a_w(u,v)$ is continuous on $V \times V$ provided the weight w
has property (P_1) or property (P_2^*), which differs from property (P_2) by the
requirement that inequality (40.6) is replaced by inequality (40.25). So we
have derived the f i r s t inequality in (41.6) [with V instead of V_0].
A detailed analysis of the evaluations made in Subsections 40.8 and 40.9 shows
that the V-ellipticity of the bilinear form $a_w(u,v)$ – i.e., the s e c o n d
inequality in (41.6) with V instead of V_0 – can be also derived by replacing property (P_2) by property (P_2^*).

The main problem which arises here is caused by the more restrictive conditions on the weights w , w_0 for which (40.25) holds, i.e. for which the
imbedding

(41.8) $$W^{1,2}(\Omega;w) \subsetneqq L^2(\Omega;w_0)$$

[or more precisely, the imbedding $\tilde{V} \subsetneqq L^2(\Omega;w_0)$ where $\tilde{V}$ is the space from
(41.7)] takes place. The situation is similar to that described in Section 12;
even for the case of power type weights, the imbedding

(41.9) $$W^{1,2}(\Omega;d_M,\varepsilon) \subsetneqq L^2(\Omega;d_M,\varepsilon-2)$$

[i.e., the imbedding (41.8) with $w = d_M^\varepsilon$ and $w_0 = d_M^{\varepsilon-2}$] *holds only for*
$\varepsilon > 2 + m - N$ while the corresponding imbedding for V_0 – see (40.10) –
holds for all real numbes ε (for $\varepsilon \neq 1$ if $m = N - 1$).

For example, if we consider the *Neumann problem* for the operator $\mathscr{L}$ from
(40.12) and are interested in a d_M^ε*-weak solution*, we can proceed completely
analogously as in the foregoing Subsections but with the imbedding (41.9)
[i.e., with inequality (40.25) from property (P_2^*)] instead of the imbedding
(40.10) [i.e., the inequality (40.7) from property (P_2)]. Since the constant
c_2^* is in both cases the same and has the form (40.11), we have only to compare the interval of those ε's for which the imbedding (41.9) holds, with
the intervals I mentioned at the end of Subsection 41.4. This comparison
yields the following result :

(i) For $N - m = 1$, i.e. for $m = N - 1$, these intervals are *disjoint*
and we are not able to guarantee the existence of a d_M^ε-weak solution
$u \in W^{1,2}(\Omega;d_M,\varepsilon)$ of the Neumann problem for $\mathscr{L}$ for any value of ε .

(ii) For $N - m = 2$, i.e. for $m = N - 2$, the Neumann problem is

d_M^ε-weakly solvable for $\varepsilon \in I \cap {<}0,\infty)$, i.e., for $\varepsilon \in [0, 0.33)$.

(iii) For $N - m \geq 3$, i.e. for $m \leq N - 3$, the Neumann problem is d_M^ε-weakly solvable in $W^{1,2}(\Omega; d_M, \varepsilon)$ for the same values of ε as the Dirichlet problem.

41.7. REMARK. The last example shows that the case of Ω and $M \subset \partial\Omega$ such that

$$N - m \geq 3$$

causes no troubles in the case of the Neumann problem, similarly as in Section 12 [compare with condition (38.1) for $k = 1$, i.e., for second order equations]. On the other hand, again similarly as in Section 12, the case of Ω and M such that $N - m < 3$ is less satisfactory.

This similarity of the conditions of Section 12 and Section 13 indicates that for higher order equations (i.e. of order $2k$), we will again meet with the restrictive condition $N - m \geq 2k + 1$. We will say a few words about higher order equations in Subsection 41.9, but first let us give an example which is more satisfactory than the results for the Neumann problem.

41.8. EXAMPLE. Let us consider a d_M^ε-weak analogue of the *mixed boundary value problem*

$$(41.10) \qquad \begin{aligned} &\mathscr{L}\,u = f \quad \text{on} \quad \Omega \\ &u = g_1 \quad \text{on} \quad \Gamma \subset \partial\Omega , \quad \frac{\partial u}{\partial \nu} = g_2 \quad \text{on} \quad \partial\Omega \setminus \Gamma \end{aligned}$$

for the second order elliptic differential operator $\mathscr{L}$ of the form (40.12) with $\frac{\partial u}{\partial \nu} = \sum_{|\alpha|,|\beta|=1} a_{\alpha\beta} \nu_\alpha D^\beta u$, $\nu = \{\nu_\alpha,\ |\alpha| = 1\}$ being the unit vector of the outer normal to the boundary $\partial\Omega$ of the domain Ω .

We introduce the space $\tilde{V}$ as the closure of the subset of all functions $v \in C^\infty(\overline{\Omega})$ such that $\operatorname{supp} v \cap \overline{\Gamma} = \emptyset$ in the norm of $W^{1,2}(\Omega; d_M, \varepsilon)$ with $M \subset \Gamma$, and call $u \in W^{1,2}(\Omega; d_M, \varepsilon)$ a d_M^ε-weak solution of the mixed problem (41.10) if

$$u - u_0 \in \tilde{V}$$

and

$$a_w(u,v) = {<}F,v{>} \quad \text{for every} \quad v \in \tilde{V}$$

where u_0 is a given function from $W^{1,2}(\Omega; d_M, \varepsilon)$ such that its trace on Γ concides with g_1 from the first boundary condition in (41.10) and F is a functional from $\tilde{V}^*$ defined by

$$\langle F,v\rangle = \int_\Omega f(x)\, v(x)\, dx + \int_{\partial\Omega \setminus \Gamma} g_2(S)\, v(S)\, dS$$

with f the right hand side of the equation $\mathscr{L}\,u = f$ and g_2 the function from the second boundary condition in (41.10). [Naturally, we suppose that the given functions f , g_1 , g_2 are so behaved that the trace $g_1|_\Gamma$ and the functional F above make sense.]

Thanks to the fact that on Γ a Dirichlet-type boundary condition appears, it is possible to show that a uniquely determined d_M^ε-weak solution $u \in W^{1,2}(\Omega;d_M,\varepsilon)$ of the mixed boundary value problem exists exactly for the same values of ε for which the existence of a d_M^ε-weak solution of the Dirichlet problem is guaranteed [in particular, for $\varepsilon \in I$ with I the intervals from the end of Subsection 41.4 if $\mathcal{L}u = -\Delta u + u$].

41.9. <u>DIFFERENTIAL EQUATIONS OF HIGHER ORDERS</u>. Now, let us consider the linear differential operator $\mathcal{L}$ of order $2k$ from (39.3); let us, for simplicity, suppose that $\mathcal{L}$ is *elliptic in the algebraic sense*, i.e. that there exists a constant $c_0 > 0$ such that

$$(41.11) \qquad \sum_{|\alpha|,|\beta|\le k} a_{\alpha\beta}(x)\,\xi_\alpha\,\xi_\beta \ge c_0 \sum_{|\gamma|\le k} |\xi_\gamma|^2$$

for a.e. $x \in \Omega$ and for all real vectors $\xi = \{\xi_\gamma,\ |\gamma| \le k\}$. Further, let $a_w(u,v)$ be the corresponding bilinear form defined by the formula (39.13).

(i) In Subsection 39.7, we have shown that the bilinear form $a_w(u,v)$ is *continuous on* $H \times H$ with

$$H = W^{k,2}(\Omega;w)$$

provided the weight w has *property* (P_1) *in the sense of* (39.22) [i.e., for $k \ge 1$; property (P_1) from Subsection 40.2 (i) is a special case]. It can be easily shown that the form $a_w(u,v)$ is H-elliptic provided the constants c_α in (39.22) are sufficiently small :

Indeed, using the identity $D^\alpha(vw) = D^\alpha v\, w + \sum_{\substack{\gamma+\delta=\alpha \\ |\delta|\ge 1}} c_{\gamma\delta}\, D^\gamma v\, D^\delta w$, we have

$$\begin{aligned}
(41.12) \qquad a_w(u,v) &= \sum_{|\alpha|,|\beta|\le k} \int_\Omega a_{\alpha\beta}\, D^\beta u\, D^\alpha v\, w \\
&+ \sum_{\substack{|\alpha|,|\beta|\le k \\ |\alpha|\ge 1}} \sum_{\substack{\gamma+\delta=\alpha \\ |\delta|\ge 1}} c_{\gamma\delta} \int_\Omega a_{\alpha\beta}\, D^\beta u\, D^\gamma v\, D^\delta w \ .
\end{aligned}$$

Denoting the first and the second sum on the right hand side by $S_1(u,v)$ and $S_2(u,v)$, respectively, we immediately have from (41.11) that

$$\begin{aligned}
S_1(u,u) &= \sum_{|\alpha|,|\beta|\le k} \int_\Omega a_{\alpha\beta}\, D^\beta u\, w^{1/2}\, D^\alpha u\, w^{1/2}\, dx \\
&\ge c_0 \sum_{|\gamma|\le k} \int_\Omega |D^\gamma u|^2\, w\, dx = c_0 \|u;\, W^{k,2}(\Omega;w)\|^2 = c_0 \|u;H\|^2
\end{aligned}$$

while (P_1)-property - see (39.22) - implies that

$$|S_2(u,u)| \le \sum_{\substack{|\alpha|,|\beta| \le k \\ |\alpha| \ge 1}} \sum_{\substack{\gamma+\delta=\alpha \\ |\delta| \ge 1}} c_{\gamma\delta} \|a_{\alpha\beta} ; L^\infty(\Omega)\| \, c_\delta \int_\Omega |D^\beta u| \, |D^\gamma u| \, w \, dx$$

$$\le \sum_{\alpha,\beta} \sum_{\gamma+\delta=\alpha} c_{\gamma\delta} \, \tilde{c}_{\alpha\beta} \, c_\delta \, \|D^\beta u; L^2(\Omega;w)\| \cdot \|D^\gamma u; L^2(\Omega;w)\|$$

$$\le c_1 \max_{|\delta| \le k} c_\delta \, \|u; W^{k,2}(\Omega;w)\|^2 = c_1 \max_{|\delta| \le k} c_\delta \, \|u;H\|^2$$

with c_1 depending only on $\mathscr{L}$ and Ω and c_δ the constant from (39.22). Consequently,

$$(41.13) \qquad a_w(u,u) \ge S_1(u,u) - |S_2(u,u)| \ge (c_0 - c_1 \max_{|\delta| \le k} c_\delta) \, \|u;H\|^2$$

and the bilinear form a_w is H-elliptic if the constants c_δ from (39.22) are sufficiently small,

$$(41.14) \qquad c_\delta < \frac{c_0}{c_1} \, , \qquad |\delta| \le k \, .$$

Thus, we can use the Lax-Milgram Theorem 39.5 and obtain assertions about the existence and uniqueness of w-weak solutions of boundary value problems for elliptic operators $\mathscr{L}$ of higher orders provided the weight w has property (P_1) with sufficiently small constants c_α in (39.22) [e.g., as small as formula (41.14) indicates; but this estimate can be improved since the calculations made above are very rough].

(ii) Property (P_2) from Subsection 40.2 can be modified, too, in a way which again allows to derive the boundedness and H-ellipticity of the bilinear form $a_w(u,v)$. Therefore, let us say that *the weight* w *has property* $(P_2)_k$ ($k \ge 1$) if there exist weight functions $w_1,w_2,\ldots,w_k$ and positive constants $c_{1\delta}$, $c_{2\delta}$ ($1 \le |\delta| \le k$) such that

$$(41.15) \qquad |D^\delta w(x)|^2 \le c_{1\delta}^2 \, w(x) \, w_{|\delta|}(x) \qquad \text{for a.e.} \quad x \in \Omega$$

and

$$(41.16) \qquad \|D^\delta u; L^2(\Omega; w_{|\delta|})\| \le c_{2\delta} \|u; W^{|\alpha|,2}(\Omega;w)\|$$

for every $u \in V$ with $\gamma + \delta = \alpha$, $1 \le |\alpha| \le k$, $1 \le |\delta| \le k$; [*)] V is a subspace of $H = W^{k,2}(\Omega;w)$ and depends on the *type* of the boundary value problem considered (e.g. V = H for the Neumann problem and $V = W_0^{k,2}(\Omega;w)$ for the Dirichlet problem).

Now the reader can easily see that the bilinear form $a_w(u,v)$ is continuous on V × V and V-elliptic provided the constants $c_{1\delta}$, $c_{2\delta}$ in (41.15),

---

[*)] Obviously property $(P_2)_1$ is exactly property (P_2^*) from Remark 40.7, with w_0 replaced by w_1 .

(41.16) are sufficiently small. The proof of these two assertions follows the ideas of the proof made for $k = 1$ in Subsections 40.6 - 40.9 and is left to the reader; we only show roughly how the V-ellipticity can be derived :

Using the notation from (41.12) we obtain in view of (41.15), the Hölder inequality and (41.16) the estimate

$$
\begin{aligned}
|S_2(u,u)| &= \left| \sum_{\substack{|\alpha|,|\beta|\leq k \\ |\alpha|\geq 1}} \sum_{\substack{\gamma+\delta=\alpha \\ |\delta|\geq 1}} c_{\gamma\delta} \int_\Omega a_{\alpha\beta}\, D^\beta u\, D^\gamma u\, D^\delta w\, dx \right| \\[2ex]
&\leq \sum_{\alpha,\beta} \sum_{\gamma+\delta=\alpha} c_{\gamma\delta}\, \|a_{\alpha\beta};\, L^\infty(\Omega)\|\, c_{1\delta} \int_\Omega |D^\beta u|\, w^{1/2}\, |D^\gamma u|\, w^{1/2}_{|\delta|}\, dx \\[2ex]
&\leq \sum_{\alpha,\beta} \sum_{\gamma+\delta=\alpha} c_{\gamma\delta}\, \tilde{c}_{\alpha\beta}\, c_{1\delta}\, \|D^\beta u;\, L^2(\Omega;w)\| \cdot \|D^\gamma u;\, L^2(\Omega;w_{|\delta|})\| \\[2ex]
&\leq \sum_{\alpha,\beta} \sum_{\gamma+\delta=\alpha} c_{\gamma\delta}\, \tilde{c}_{\alpha\beta}\, c_{1\delta}\, c_{2\delta}\, \|D^\beta u;\, L^2(\Omega;w)\| \cdot \|u;\, W^{|\alpha|,2}(\Omega;w)\| \\[2ex]
&\leq c_1 \max_{|\delta|\leq k} (c_{1\delta}\, c_{2\delta})\, \|u;\, W^{k,2}(\Omega;w)\|^2
\end{aligned}
$$

and it follows $-$ as in (41.13) $-$ that $a_w(u,u) \geq \tilde{c}\|u;H\|^2$ for $u \in V$ if $c_{1\delta} c_{2\delta} < c_0/c_1$.

(iii) *The power type weight* $w(x) = d_M^\epsilon(x)$ *has property* $(P_2)_k$: It follows from (37.17) that we can take $w_{|\delta|}(x) = d_M^{\epsilon-2|\delta|}(x)$, so that inequality (41.15) holds, and inequality (41.16) expresses the imbedding theorems mentioned in Subsection 37.4 and also, several times, in Subsection 37.7 [see e.g. (37.10)]. Therefore, we can assert that *there is an open interval* I *containing the origin and such that for* $\epsilon \in I$ *there exists one and only one* d_M^ϵ-*-weak solution* $u \in W^{k,2}(\Omega;d_M,\epsilon)$ *of the Dirichlet problem for the operator* $\mathscr{L}$ *of order* 2k provided $u_0 \in W^{k,2}(\Omega;d_M,\epsilon)$ and $F \in \left[W_0^{k,2}(\Omega;d_M,\epsilon)\right]^*$. As concerns other boundary value problems, one again has to check the intersection of the interval I mentioned with the set of those ϵ's for which the imbeddings of the type $V \cap W^{k,2}(\Omega;d_M,\epsilon) \subsetneq L^2(\Omega;d_M,\epsilon-2k)$ hold $-$ similarly as for the case $k = 1$.

Chapter V

ELLIPTIC PROBLEMS WITH "BAD" COEFFICIENTS

In the literature, many papers appear which deal with equations of the
type

$$(*) \qquad \sum_{|\alpha| \leq 2k} a_\alpha(x)\, x_N^{\varepsilon(\alpha)}\, D^\alpha u = f(x)$$

defined, say, on the half-space $\mathbf{R}_N^+ = \{x = (x_1,\ldots,x_N)\,,\ x_N > 0\}$, or with
equations which can be transformed to the form $(*)$. Usually, the coefficients a_α are assumed to have some "good" behaviour (e.g., there exist constants c_1 , c_2 such that $0 < c_1 \leq a_\alpha(x) \leq c_2 < \infty$), and the "bad" behaviour is
expressed by the term $x_N^{\varepsilon(\alpha)}$, $\varepsilon(\alpha) \in \mathbf{R}$. The "complete" coefficient
$a_\alpha(x) x_N^{\varepsilon(\alpha)}$ then becomes *singular*, if $\varepsilon(\alpha) < 0$, or *degenerates*, if $\varepsilon(\alpha) > 0$.
Here, the singularity or degeneration is concentrated on the *boundary* $\partial \mathbf{R}_N^+ =$
$= \{x = (x_1,\ldots,x_N),\ x_N = 0\}$.

In this chapter, we will deal with equations for which the singular or
degenerating behaviour of the coefficients is *general*, not only of the type of
a power, and can appear elsewhere in $\overline{\Omega}$, not only on the boundary.

Section 14 . *S i n g u l a r   a n d   d e g e n e r a t e*
e q u a t i o n s   -   a   s i m p l e   c a s e

§ 42 . A n   e x a m p l e . F o r m u l a t i o n   o f   t h e
p r o b l e m

<u>42.1 INTRODUCTION.</u> In the foregoing chapter we investigated *elliptic* partial
differential equations and the main reason for seeking a solution in weighted
spaces was to eliminate the "bad" behaviour of the right hand side in the equation and/or in the boundary conditions. Now we shall deal with equations *whose*
ellipticity is violated in some sense.

Let us again consider the (formal) differential operator of order $2k$

$$(42.1) \qquad (\mathcal{L}u)(x) = \sum_{|\alpha|,\,|\beta| \leq k} (-1)^{|\alpha|}\, D^\alpha\big(a_{\alpha\beta}(x)\, D^\beta u(x)\big)$$

defined on a domain $\Omega \subset \mathbf{R}^N$. Weak solutions of boundary value problems for
such an operator are sought in the classical Sobolev space

$$W^{k,2}(\Omega)$$

provided the coefficients $a_{\alpha\beta}$ are *bounded*, i.e.,

$$(42.2) \qquad a_{\alpha\beta} \in L^\infty(\Omega)\,, \qquad |\alpha|,\ |\beta| \leq k\,,$$

and the operator $\mathcal{L}$ is *elliptic*, e.g. in the following strong (algebraic) sense :

$$(42.3) \qquad \sum_{|\alpha|,|\beta| \leq k} a_{\alpha\beta}(x)\, \xi_\alpha\, \xi_\beta \geq c_0 \sum_{|\gamma| \leq k} |\xi_\gamma|^2$$

for all real vectors $\xi = \{\xi_\gamma,\ |\gamma| \leq k\}$ and for a.e. $x \in \Omega$ with an ellipticity constant $c_0 > 0$.

If some (or both) of the conditions (42.2), (42.3) are violated, then the classical theory of weak solutions in Sobolev spaces cannot be used in general. In this case it is (sometimes!) possible to save the situation by *introducing an appropriate w e i g h t e d Sobolev space*. Let us clarify this approach on a simple example.

42.2. <u>EXAMPLE</u>. On the domain $\Omega \subset \mathbf{R}^N$ we consider the *second order* differential operator (k = 1) of the following special form

$$(42.4) \qquad (\mathcal{L}u)(x) = -\sum_{i=1}^{N} \frac{\partial}{\partial x_i}\left(a_i(x)\, \frac{\partial u(x)}{\partial x_i}\right) + a_0(x)\, u(x) \ ,$$

which is associated with the bilinear form

$$(42.5) \qquad a(u,v) = \sum_{i=1}^{N} \int_\Omega a_i(x)\, \frac{\partial u(x)}{\partial x_i}\, \frac{\partial v(x)}{\partial x_i}\, dx + \int_\Omega a_0(x)\, u(x)\, v(x)\, dx \ .$$

The expression

$$(42.6) \qquad a(u,u) = \sum_{i=1}^{N} \int_\Omega \left|\frac{\partial u(x)}{\partial x_i}\right|^2 a_i(x)\, dx + \int_\Omega |u(x)|^2\, a_0(x)\, dx$$

defines the *square of a norm in* $W^{1,2}(\Omega)$, i.e., the equivalence relation

$$\sqrt{a(u,u)} \simeq \|u;\ W^{1,2}(\Omega)\|$$

holds provided the following conditions are fulfilled :

$$(42.7) \qquad a_i(x) \geq c_0 > 0 \quad \text{and} \quad a_i(x) \leq c_1 < \infty$$

for $i = 0,1,\ldots,N$ and for a.e. $x \in \Omega$.

Now, let us suppose that some of the conditions (42.7) are *violated*, i.e. let us admit coefficients which can grow to infinity or tend to zero somewhere in $\overline{\Omega}$. This means that *singular coefficients* are admissible, i.e., that for some index i ,

$$|a_i(x)| < \infty \quad \text{a.e. in}\ \ \Omega \ , \quad \text{but}\quad |a_i(x)| \to \infty \quad \text{for}\ \ x \to x_0^* \in \overline{\Omega} \ ,$$

and also such coefficients are allowed which violate the ellipticity condition to the *degenerate-elliptic case*, i.e., such coefficients that for some index j ,

$$a_j(x) > 0 \quad \text{a.e. in} \quad \Omega \text{ , but } \quad a_j(x) \to 0 \quad \text{for} \quad x \to x^{**} \in \overline{\Omega} \text{ .}$$

Typical examples of such "bad behaving" coefficients are coefficients of the form

$$a_i(x) = |x - x^*|^\varepsilon$$

with $x^* \in \overline{\Omega}$, or more generally,

$$a_i(x) = d_M^\varepsilon(x) = \left[\text{dist } (x,M)\right]^\varepsilon$$

with $M \subset \overline{\Omega}$: For $\varepsilon > 0$ the degenerating case appears while for $\varepsilon < 0$ we obtain singular coefficients.

In the just mentioned cases, nevertheless, the expression $\sqrt{a(u,u)}$ defines a norm in the *weighted Sobolev space*

$$(42.8) \qquad W^{1,2}(\Omega;S)$$

(see Subsection 0.4 with $p = 2$) where the collection S of weight functions is determined directly by the coefficients of the operator $\mathcal{L}$ from (42.4) :

$$(42.9) \qquad S = \{a_0, a_1, \ldots, a_N\} \text{ .}$$

We only assume that all the coefficients a_i are *weight functions*, i.e., measurable and a.e. in Ω positive functions [see Subsection 0.4 where the set of such weight functions was denoted by $W(\Omega)$ – see (0.3) – so that

$$(42.10) \qquad a_i \in W(\Omega) \text{ , } \quad i = 0,1,\ldots,N \text{]}$$

and, moreover,

$$(42.11) \qquad a_i \in L^1_{loc}(\Omega) \quad \text{and} \quad \frac{1}{a_i} \in L^1_{loc}(\Omega)$$

which ensures that the space $W^{1,2}(\Omega;S)$ from (42.8) is a Banach space and that the space $W_0^{1,2}(\Omega;S)$ [= the closure of the set $C_0^\infty(\Omega)$ with respect to the norm $\sqrt{a(u,u)}$] is meaningful – see A. KUFNER, B. OPIC [4], [6]; compare with formula (40.1) and with Remark 40.3 (i).

So we have

$$(42.12) \qquad a(u,u) = \|u; \ W^{1,2}(\Omega;S)\|^2 \text{ .}$$

Further, the Hölder inequality implies that

$$\left| \int_\Omega a_i \frac{\partial u}{\partial x_i} \frac{\partial v}{\partial x_i} \, dx \right| \leq \int_\Omega \sqrt{a_i(x)} \left|\frac{\partial u}{\partial x_i}\right| \sqrt{a_i(x)} \left|\frac{\partial v}{\partial x_i}\right| \, dx$$

$$\leq \left(\int_\Omega \left|\frac{\partial u}{\partial x_i}\right|^2 a_i \, dx\right)^{1/2} \left(\int_\Omega \left|\frac{\partial v}{\partial x_i}\right|^2 a_i \, dx\right)^{1/2} \leq \|u; \ W^{1,2}(\Omega;S)\| \cdot \|v; \ W^{1,2}(\Omega;S)\|$$

and since the other terms in (42.5) can be estimated analogously, we have

$$(42.13) \qquad |a(u,v)| \leq (N + 1) \ \|u; \ W^{1,2}(\Omega;S)\| \cdot \|v; \ W^{1,2}(\Omega;S)\| \text{ .}$$

Inequality (42.13) and identity (42.12) show that the bilinear form
$a(u,v)$ from (42.5) is *bounded (continuous)* on $W^{1,2}(\Omega;S) \times W^{1,2}(\Omega;S)$ and
$W^{1,2}(\Omega;S)$-*elliptic*. Consequently, all the conditions of the Lax–Milgram Theorem
39.5 are fulfilled and we can easily derive the *existence and uniqueness of a
weak solution of a boundary value problem for the operator* $\mathscr{L}$ *from (42.4)
in the weighted space* $W^{1,2}(\Omega;S)$ with S given by (42.9).

We point out the fact that the weight S appearing in the space
$W^{1,2}(\Omega;S)$ was determined *directly by the coefficients of the operator* $\mathscr{L}$
from (42.4). Now, we extend the considerations made for the special second
order differential operator $\mathscr{L}$ from (42.4) to the *general* differential operator $\mathscr{L}$ of order $2k$ from (42.1).

<u>42.3.</u> <u>CONDITIONS ON THE COEFFICIENTS</u> $a_{\alpha\beta}$. Let us consider the differential
operator $\mathscr{L}$ from (42.1) and assume that its coefficients $a_{\alpha\beta}$ satisfy the
following conditions :

A.1 $\qquad a_{\alpha\alpha} \in W(\Omega)$ for $|\alpha| \leq k$.

A.2 $\qquad a_{\alpha\alpha} \in L^1_{loc}(\Omega)$, $\quad \dfrac{1}{a_{\alpha\alpha}} \in L^1_{loc}(\Omega)$ for $|\alpha| \leq k$.

A.3 $\qquad$ There is a constant $c_1 > 0$ such that for all $|\alpha|, |\beta| \leq k$, $\alpha \neq \beta$,

(42.14) $\qquad |a_{\alpha\beta}(x)| \leq c_1 \sqrt{a_{\alpha\alpha}(x)a_{\beta\beta}(x)}$ for a.e. $x \in \Omega$.

A.4 $\qquad$ There is a constant $c_2 > 0$ such that for all real vectors
$\quad \xi = \{\xi_\gamma, |\gamma| \leq k\}$ and for a.e. $x \in \Omega$,

(42.15) $$\sum_{|\alpha|,|\beta|\leq k} a_{\alpha\beta}(x)\xi_\alpha\xi_\beta \geq c_2 \sum_{|\gamma|\leq k} a_{\gamma\gamma}(x)\xi_\gamma^2 .$$

<u>42.4.</u> <u>REMARKS</u>. (i) The first condition in **A**.2 is an improvement of condi-
tion (42.2), so that also *singular coefficients* are admitted (with, of course,
l i m i t e d growth).

(ii) Condition **A**.4 – i.e., inequality (42.15) – together with condition **A**.1 improve the ellipticity condition (42.3). The coefficients $a_{\gamma\gamma}$
appearing in the right hand side of (42.15) express the possible *degeneration*.

(iii) In Example 42.2 we have had $a_{\alpha\beta}(x) \equiv 0$ for $\alpha \neq \beta$ so that all
conditions **A**.1 – **A**.4 are fulfilled.

Conditions **A**.1 and **A**.2 allow to introduce an appropriate weighted space.

42.5. THE WEIGHTED SOBOLEV SPACE. We will consider the space $W^{k,2}(\Omega;S)$ with

$$(42.16) \qquad S = \{a_{\alpha\alpha}(x);\ |\alpha| \le k\}\ ,$$

i.e.

$$W^{k,2}(\Omega;S) = \{u = u(x),\ x \in \Omega;\ \int_{\Omega} |D^{\alpha}u(x)|^2\, a_{\alpha\alpha}(x)\, dx < \infty\ \text{ for }\ |\alpha| \le k\}\ .$$

Since the functions $a_{\alpha\alpha}$ are weight functions (condition **A**.1), the definition is meaningful. Due to the second condition in **A**.2, $W^{k,2}(\Omega;S)$ is a complete normed linear space, i.e., a *Banach (Hilbert) space* under the norm

$$(42.17) \qquad \|u;\ W^{k,2}(\Omega;S)\| = \left(\sum_{|\alpha| \le k} \int_{\Omega} |D^{\alpha}u(x)|^2\, a_{\alpha\alpha}(x)\, dx \right)^{1/2}$$

$$= \left(\sum_{|\alpha| \le k} \|D^{\alpha}u;\ L^2(\Omega;a_{\alpha\alpha})\|^2 \right)^{1/2}\ .$$

Simultaneously we introduce the space $W_0^{k,2}(\Omega;S)$ as the closure of the set $C_0^{\infty}(\Omega)$ with respect to the norm (42.17). The space $W_0^{k,2}(\Omega;S)$ is normed again by the expression (42.17); its definition is meaningful in view of the first condition in **A**.2.

42.6. THE DIRICHLET PROBLEM. Let $\mathcal{L}$ be the differential operator from (42.1) and $a(u,v)$ the corresponding bilinear form given by the formula

$$(42.18) \qquad a(u,v) = \sum_{|\alpha|,\,|\beta| \le k} \int_{\Omega} a_{\alpha\beta}(x)\, D^{\beta}u(x)\, D^{\alpha}v(x)\, dx\ .$$

Let the coefficients $a_{\alpha\beta}$ satisfy conditions **A**.1 – **A**.4 and let $W^{k,2}(\Omega;S)$ and $W_0^{k,2}(\Omega;S)$ be the weighted spaces from Subsection 42.5. Further, let u_0 be a *given* function from $W^{k,2}(\Omega;S)$ and F a *given* continuous linear functional from the dual space $\left[W_0^{k,2}(\Omega;S)\right]^*$.

A function $u \in W^{k,2}(\Omega;S)$ is called a *weak solution of the Dirichlet problem* (for the differential operator $\mathcal{L}$) *in the weighted space* $W^{k,2}(\Omega;S)$ if it satisfies

$$(42.19) \qquad u - u_0 \in W_0^{k,2}(\Omega;S)\ ,$$

$$(42.20) \qquad a(u,v) = \langle F,v \rangle\ \text{ for every }\ v \in W_0^{k,2}(\Omega;S)\ .$$

§ 43 . E x i s t e n c e   t h e o r e m

43.1. THEOREM. *Let the coefficients* $a_{\alpha\beta}$ *of the operator* $\mathcal{L}$ *from* (42.1) *be defined on a domain* $\Omega \subset \mathbb{R}^N$ *and satisfy conditions* **A**.1 - **A**.4 *from Subsection 42.3. Let the function* $u_0 \in W^{k,2}(\Omega;S)$ *and the functional* $F \in [W_0^{k,2}(\Omega;S)]^*$ *be given.*

Then there exists one and only one weak solution u *of the Dirichlet problem for the operator* $\mathcal{L}$ *in the space* $W^{k,2}(\Omega;S)$ *(in the sense of Definition 42.6). Moreover, there is a positive constant* c *independent of* u_0 *and* F *such that*

$$(43.1) \qquad \|u;\ W^{k,2}(\Omega;S)\| \leq c\big(\|u_0;\ W^{k,2}(\Omega;S)\| + \|F;\ [W_0^{k,2}(\Omega;S)]^*\|\big) \ .$$

P r o o f : (i) Using property **A**.3, the Hölder inequality and the definition of the norm in the space $W^{k,2}(\Omega;S)$ — see (42.17) — we obtain the estimate

$$
\begin{aligned}
|a(u,v)| &\leq \sum_{|\alpha|,|\beta|\leq k} \int_\Omega |a_{\alpha\beta}|\ |D^\beta u|\ |D^\alpha v|\ dx \\[4pt]
&\leq \max\,(c_1,1) \sum_{|\alpha|,|\beta|\leq k} \int_\Omega |D^\beta u|\ \sqrt{a_{\beta\beta}}\ |D^\alpha v|\ \sqrt{a_{\alpha\alpha}}\ dx \\[4pt]
(43.2) \qquad &\leq \max\,(c_1,1) \sum_{|\alpha|,|\beta|\leq k} \|D^\beta u;\ L^2(\Omega;a_{\beta\beta})\| \cdot \|D^\alpha v;\ L^2(\Omega;a_{\alpha\alpha})\| \\[4pt]
&\leq c_3 \|u;\ W^{k,2}(\Omega;S)\| \cdot \|v;\ W^{k,2}(\Omega;S)\|
\end{aligned}
$$

which holds for arbitrary functions $u,\ v \in W^{k,2}(\Omega;S)$ [c_1 is the constant from (42.14)].

(ii) Inequalities (43.2) imply at the same time that the expression $a(u_0,v)$ with a *fixed* $u_0 \in W^{k,2}(\Omega;S)$ defines a continuous linear functional on the space $W^{k,2}(\Omega;S)$ as well as on its subspaces, in particular on the subspace $H = W_0^{k,2}(\Omega;S)$.

(iii) Taking $\xi_\alpha = D^\alpha u(x)$ in (42.15) and integrating the resulting inequality over Ω , we immediately obtain the estimate

$$(43.3) \qquad a(u,u) \geq c_2 \|u;\ W^{k,2}(\Omega;S)\|^2$$

which holds for every function $u \in W^{k,2}(\Omega;S)$ and therefore also for every $u \in W_0^{k,2}(\Omega;S)$.

(iv) In view of the inequalities (43.2) and (43.3) we can apply the Lax-Milgram Theorem 39.5 with the bilinear form $b(u,v) = a(u,v)$, with the Hilbert space $H = W_0^{k,2}(\Omega;S)$ and with the functional $h \in H^*$ chosen in the form

$$(43.4) \qquad \langle h,v \rangle = \langle F,v \rangle - a(u_0,v) \ .$$

Here $F \in H^*$ and $u_0 \in W^{k,2}(\Omega;S)$ are respectively the functional and the function from the definition of the weak solution of the Dirichlet problem (Subsection 42.6), and by point (ii) of this proof, h indeed belongs to H^*. Consequently, by Theorem 39.5 there is a uniquely determined function $\tilde{u} \in H$

such that

(43.5) $\qquad a(\tilde{u},v) = \langle h,v \rangle \qquad$ for every $\ v \in H$.

Now, let us put

(43.6) $\qquad u = \tilde{u} + u_0$.

Then $\ u - u_0 = \tilde{u} \in H = W_0^{k,2}(\Omega;S)$, i.e., condition (42.19) from Subsection
42.6 is fulfilled. Moreover, condition (42.20) from this Subsection is fulfilled, too, since (43.5) together with (43.4) yields

$$a(u,v) = a(\tilde{u} + u_0,\ v) = a(\tilde{u},v) + a(u_0,v) = \langle h,v \rangle + a(u_0,v) = \langle F,v \rangle .$$

Consequently, the function u from (43.6) is the required, uniquely determined
weak solution of the Dirichlet problem in $W^{k,2}(\Omega;S)$.

$\qquad$ Moreover, Theorem 39.5 yields the estimate

$$\|\tilde{u};H\| \le \frac{1}{c_2} \|h;H^*\|$$

- see (39.19), and since $\ \|u;H\| \le \|\tilde{u};H\| + \|u_0;H\|\ $ in virtue of (43.6) and
$\|h;H^*\| \le \|F;H^*\| + c_3\|u_0;H\|\ $ in virtue of (43.4) and (43.2), we finally have
the estimate (43.1) with $\ c = \max\ (1/c_2,\ 1 + c_3/c_2)$.

43.2. <u>OTHER BOUNDARY VALUE PROBLEMS</u>. We have formulated the theorem about
the existence and uniqueness of a weak solution for the *Dirichlet problem* where
the role of the space H from Theorem 39.5 is played by the space $W_0^{k,2}(\Omega;S)$.
Nonetheless, since all the assumptions of Theorem 39.5 $-$ the continuity on
$H \times H$ and the H-ellipticity $-$ are fulfilled for $H = W^{k,2}(\Omega;S)$ [see points
(i) and (iii) of the proof of Theorem 43.1] , it is evident that other boundary
value problems can be treated in a similar way and the corresponding assertions
about the existence and uniqueness of a weak solution can be obtained, too.

$\qquad$ Let us point out that in some generalizations which will be mentioned later it is more important that the Dirichlet problem is dealt with : In these
generalizations, some *imbedding theorems* for the weighted spaces will be needed,
and as we have seen in the foregoing Chapter IV, such imbeddings hold sometimes
under rather restrictive conditions if spaces $W^{k,2}(\Omega;S)$ are considered (i.e.,
spaces which represent the space H in the case of the *Neumann problem*) while
the imbeddings for the space $W_0^{k,2}(\Omega;S)$ (corresponding to the D i r i c h -
l e t problem) are easier to handle.

43.3. <u>REMOVING CONDITION **A**.4</u>. *Condition* **A**.4 guaranteeing the H-ellipticity
of the bilinear form $a(u,v)$ $-$ see (43.3) $-$ already *follows from condition*
A.3 *provided the constant* c_1 *in* (42.14) *is sufficiently small.*

$\qquad$ Indeed, we have

$$\sum_{|\alpha|,|\beta|\le k} a_{\alpha\beta}(x)\xi_\alpha\xi_\beta = \sum_{\alpha=\beta} + \sum_{\alpha\ne\beta} \ge \sum_{\alpha=\beta} - \left|\sum_{\alpha\ne\beta}\right|$$

$$\ge \sum_{|\alpha|\le k} a_{\alpha\alpha}(x)\xi_\alpha^2 - \sum_{\alpha\ne\beta} |a_{\alpha\beta}(x)\xi_\alpha\xi_\beta| \ .$$

The last sum can be estimated by means of (42.14) and of the inequality $ab \le \frac{1}{2}(a^2 + b^2)$ as follows :

$$\sum_{\alpha\ne\beta} |a_{\alpha\beta}(x)\xi_\alpha\xi_\beta| \le c_1 \sum_{\alpha\ne\beta} \sqrt{a_{\alpha\alpha}(x)}\ |\xi_\alpha|\ \sqrt{a_{\beta\beta}(x)}\ |\xi_\beta|$$

$$\le \frac{1}{2}\, c_1 \sum_{\alpha\ne\beta} \left(a_{\alpha\alpha}(x)\xi_\alpha^2 + a_{\beta\beta}(x)\xi_\beta^2\right)$$

$$\le \frac{1}{2}\, c_1 2(\kappa - 1) \sum_{|\alpha|\le k} a_{\alpha\alpha}(x)\xi_\alpha^2$$

where κ is the number of multiindices of length at most k, $\kappa = (N+k)!/(N!k!)$. Hence

$$\sum_{|\alpha|,|\beta|\le k} a_{\alpha\beta}(x)\xi_\alpha\xi_\beta \ge \left(1 - c_1(\kappa - 1)\right) \sum_{|\alpha|\le k} a_{\alpha\alpha}(x)\xi_\alpha^2$$

which means that *for*

$$(43.7) \qquad\qquad c_1 < \frac{1}{\kappa - 1}$$

condition **A**.4 *is fulfilled, i.e., inequality* (42.15) *holds with the positive constant* $c_2 = 1 - c_1(\kappa - 1)$.

43.4. <u>REMARK</u>. Conditions **A**.1 - **A**.4 represent the *simplest* assumptions that immediately yield the existence theorem. Since these conditions are simple, they are very rough, too, and we will therefore present some (again rather obvious) generalizations of these conditions.

Let us point out one a d v a n t a g e of this simple case : *No assumptions about the set* $\Omega \subset \mathbf{R}^N$ *have been needed.*

§ 44 . W e a k e n i n g   c o n d i t i o n s **A**.1 - **A**.4

44.1. <u>CONDITION **A**.1.</u> In our foregoing consideration, we have already assumed that *all* "diagonal" coefficients $a_{\alpha\alpha}$ of the differential operator $\mathscr{L}$ from (42.1) are positive almost everywhere. However, it is evident that all the arguments used in the proof of Theorem 43.1 remain valid if *some* of the coefficients $a_{\alpha\alpha}$ *vanish*, provided the corresponding weighted space is meaningful. Thus condition **A**.1 can be replaced by the following one :

A.1* Denote by $\mathcal{Ж}$ the set of those multiindices α , $|\alpha| \le k$, for which

$a_{\alpha\alpha} \in W(\Omega)$, and let $a_{\beta\beta}(x) \equiv 0$ for $\beta \notin \mathbb{K}$ ($|\beta| \leq k$). Let the set $\mathbb{K}$ contain at least one multiindex of the length k and let the expression

$$(44.1) \qquad \|u\|_{\mathbb{K}} = \left(\sum_{\alpha \in \mathbb{K}} \|D^{\alpha}u;\ L^2(\Omega;a_{\alpha\alpha})\|^2 \right)^{1/2}$$

be a *norm* on the linear space $W^{k,2}(\Omega;\tilde{S})$ defined as the set of all $u = u(x)$ such that $\|u\|_{\mathbb{K}} < \infty$.

Conditions **A**.2 - **A**.4 do not change, except that condition **A**.2 is relevant only for $\alpha \in \mathbb{K}$ and that it suffices to perform the summation on the right hand side of (42.15) only over $\alpha \in \mathbb{K}$.

All what was said in Subsection 43.3 remains valid also in this case, with the only change which consists in replacing the number κ in (43.7) by the number of multiindices contained in the set $\mathbb{K}$.

44.2. <u>REMARK</u>. The generalization proposed in the foregoing Subsection 44.1 is rather formal. The difficult point is the assumption that the expression (44.1) is a norm (in any case it is at least a *seminorm*). The difference between $\|u\|_{\mathbb{K}}$ and the norm defined in (42.17) is in the set of the summation indices : In (42.17) we sum up over all $|\alpha| \leq k$, in (44.1) only over $\alpha \in \mathbb{K}$. The notation $W^{k,2}(\Omega;\tilde{S})$, which is rather inconsistent, should emphasize the fact that we make use of a r e s t r i c t e d collection $\tilde{S}$ of weight functions : $\tilde{S} = \{a_{\alpha\alpha}, \alpha \in \mathbb{K}\}$, while S in $W^{k,2}(\Omega;S)$ means $S = \{a_{\alpha\alpha}, |\alpha| \leq k\}$.

If $\|u\|_{\mathbb{K}}$ from (44.1) is not a norm, we cannot use the approach described. But sometimes we can overcome this gap, adding - roughly speaking - to the expression (44.1) some further terms which turn it into a norm. See Example 44.8 below.

44.3. <u>CONDITION</u> **A**.4. Let us consider the weighted space $W^{k,2}(\Omega;S)$ introduced in Subsection 42.5 and assume that there exists a *certain subset* $\mathbb{K}_1$ of the set of all multiindices of length k such that the expression

$$(44.2) \qquad \|u\|_{\mathbb{K}_1} = \left(\sum_{\alpha \in \mathbb{K}_1} \|D^{\alpha}u;\ L^2(\Omega;a_{\alpha\alpha})\|^2 \right)^{1/2}$$

is again *a norm on the space* $W_0^{k,2}(\Omega;S)$, *equivalent to the original norm* $\|u;W^{k,2}(\Omega;S)\|$ introduced in (42.17). Then condition **A**.4 can be evidently replaced by the following weaker condition :

A.4* There is a constant $c_2^* > 0$ such that for all real vectors $\xi = \{\xi_{\gamma}, |\gamma| \leq k\}$ and for a.e. $x \in \Omega$

$$(44.3) \qquad \sum_{|\alpha|,|\beta|\leq k} a_{\alpha\beta}(x)\, \xi_\alpha\, \xi_\beta \geq c_2^* \sum_{\gamma \in \mathbb{K}_1} a_{\gamma\gamma}(x)\, \xi_\gamma^2 \, .$$

__44.4. REPLACING THE SPACE__ $W^{k,2}(\Omega;S)$. The collection S of the weight functions from (42.16) is determined d i r e c t l y by the "diagonal" coefficients $a_{\alpha\alpha}$ of the differential operator $\mathcal{L}$. Consider now *another collection*

$$(44.4) \qquad S^* = \{w_\alpha^*(x) \in W(\Omega) \, , \ |\alpha| \leq k\}$$

which has the following property : *The spaces*

$$W^{k,2}(\Omega;S) \quad \text{and} \quad W^{k,2}(\Omega;S^*)$$

are d i f f e r e n t (in the sense that the corresponding norms
$\| u; \ W^{k,2}(\Omega;S) \|$ and $\| u; \ W^{k,2}(\Omega;S^*) \|$ are *not equivalent*) *while*

$$(44.5) \qquad W_0^{k,2}(\Omega;S) = W_0^{k,2}(\Omega;S^*)$$

(in the sense that the two norms just mentioned *are equivalent on the set*
$C_0^\infty(\Omega)$).

Since the space H , to which we apply Theorem 39.5, is the space
$W_0^{k,2}(\Omega;S)$, we can, in view of (44.5), use the collection S^* as well.

The a d v a n t a g e of this approach, which at first sight seems to be only formal, consists in the fact that it allows for greater variability in the choice of the function u_0 which represents the right hand side in the boundary conditions. In fact, it may happen that the given function u_0 does not belong to the space $W^{k,2}(\Omega;S)$ (defined d i r e c t l y by the differential operator $\mathcal{L}$) but does belong to the space $W^{k,2}(\Omega;S^*)$ for which (44.5) holds. Naturally, in this case we have to solve the Dirichlet problem in that space to which u_0 belongs, i.e., in the space $W^{k,2}(\Omega;S^*)$.

This problem will be illustrated in Example 44.6 below. Let us notice that, if the function u_0 belongs simultaneously to *two different* spaces $W^{k,2}(\Omega;S)$ and $W^{k,2}(\Omega;S^*)$ [the corresponding norms being equivalent on $C_0^\infty(\Omega)$, so that the spaces $W_0^{k,2}(\Omega;S)$ and $W_0^{k,2}(\Omega;S^*)$ *coincide*] we can solve the Dirichlet problem in the former as well as in the latter space. In both cases we obtain the *same solution*, for we construct them by means of a uniquely determined function $\tilde{u} \in W_0^{k,2}(\Omega;S) = W_0^{k,2}(\Omega;S^*)$ – cf. identity (44.5); $\tilde{u}$ is the function from (43.5), (43.6).

Obviously, in virtue of (44.5) it is important that we consider the *Dirichlet problem*.

__44.5. CONDITIONS A.3, A.4.__ The choice of two different collections S and S^* of weight functions that give the same "nulled" space (44.5), described in

the foregoing Subsection 44.4, allows for greater variability also in the coefficients of the operator $\mathcal{L}$: in virtue of the equivalence of the norms on
the "nulled" spaces we can replace some of the inequalities (42.14) and (42.15)
by the corresponding inequalities of the type

$$|a_{\alpha\beta}(x)| \leq c_1 \sqrt{w_\alpha^*(x)\, w_\beta^*(x)} \quad , \qquad \alpha \neq \beta \; ,$$

and

$$\sum_{|\alpha|, |\beta| \leq k} a_{\alpha\beta}(x)\, \xi_\alpha\, \xi_\beta \geq c_2 \sum_{|\gamma| \leq k} w_\gamma^*(x)\, \xi_\gamma^2$$

[see (44.4) for w_γ^*].

<u>44.6.</u> <u>EXAMPLE.</u> Consider a plane domain ($N = 2$) and take the halfplane
$\{(x_1,x_2),\ x_2 > 1\}$ for Ω . Further, consider the second order differential
operator

$$(44.6) \qquad \mathcal{L}\,u = - \frac{\partial}{\partial x_1}\Big(x_2^{-2}\,\frac{\partial u}{\partial x_1}\Big) - \frac{\partial}{\partial x_2}\Big(x_2^2\,\frac{\partial u}{\partial x_2}\Big) + u \; .$$

This is an operator of the type considered in Example 42.2, and it follows from
this Example that the weighted space corresponding to our operator $\mathcal{L}$ is the
space $W^{1,2}(\Omega;S)$ with the collection

$$(44.7) \qquad S = \{1, x_2^{-2}, x_2^2\} \; ,$$

normed by the square root of

$$(44.8) \qquad \|u;\ W^{1,2}(\Omega;S)\|^2 = \int_\Omega |u|^2\, dx + \int_\Omega \Big|\frac{\partial u}{\partial x_1}\Big|^2\, x_2^{-2}\, dx + \int_\Omega \Big|\frac{\partial u}{\partial x_2}\Big|^2\, x_2^2\, dx \; .$$

Let us now solve the Dirichlet problem for the operator $\mathcal{L}$ with the
boundary condition given by the function $u_0(x_1,x_2) = (x_1^2 + 1)^{-1}$. In this
case we have

$$u_0 \notin W^{1,2}(\Omega;S) \; .$$

Therefore, let us consider still another collection of weight functions

$$(44.9) \qquad S^* = \{x_2^{-3/2}, x_2^{-2}, x_2^2\}$$

normed by the square root of

$$(44.10) \qquad \|u;\ W^{1,2}(\Omega;S^*)\|^2 = \int_\Omega |u|^2\, x_2^{-3/2}\, dx + \int_\Omega \Big|\frac{\partial u}{\partial x_1}\Big|^2\, x_2^{-2}\, dx + \int_\Omega \Big|\frac{\partial u}{\partial x_2}\Big|^2\, x_2^2\, dx \; .$$

Since the function $u_0(x_1,x_2) = (x_1^2 + 1)^{-1}$ belongs to $W^{1,2}(\Omega;S^*)$, we have
$W^{1,2}(\Omega;S) \neq W^{1,2}(\Omega;S^*)$. On the other hand, the spaces $W_0^{1,2}(\Omega;S)$ and
$W_0^{1,2}(\Omega;S^*)$ *coincide*, since the respective norms are equivalent on $C_0^\infty(\Omega)$:
Indeed, for every $u \in C_0^\infty(\Omega)$ we have

(44.11)
$$\int_\Omega |u|^2 \, x_2^{-3/2} \, dx \le \int_\Omega |u|^2 \, dx \le 4 \int_\Omega \left|\frac{\partial u}{\partial x_2}\right|^2 x_2^2 \, dx \;,$$

where the first inequality follows from the fact that $x_2 > 1$ for $x = (x_1,x_2)$ $\in \Omega$, and the second follows by integration with respect to x_1 over $\mathbb{R}$ from the one-dimensional inequality

$$\int_1^\infty |u(x_1,x_2)|^2 \, dx_2 \le 4 \int_1^\infty \left|\frac{\partial u}{\partial x_2}(x_1,x_2)\right|^2 x_2^2 \, dx_2$$

which is the Hardy inequality (0.32) for $p = \varepsilon = 2$ [note that $u(x_1,x_2) \in$ $\in C_0^\infty(1,\infty)$ as a function of x_2 for every $x_1 \in \mathbb{R}$]. Now, from the first inequality in (44.11) we have

(44.12)
$$\|u; \; W^{1,2}(\Omega;S^*)\|^2 \le \|u; \; W^{1,2}(\Omega;S)\|^2$$

while the second inequality in (44.11) implies

(44.13)
$$\|u; \; W^{1,2}(\Omega;S)\|^2 \le \int_\Omega \left|\frac{\partial u}{\partial x_1}\right|^2 x_2^{-2} \, dx + 5 \int_\Omega \left|\frac{\partial u}{\partial x_2}\right|^2 x_2^2 \, dx \le 5\|u; \; W^{1,2}(\Omega;S^*)\|^2 .$$

Inequalities (44.12) and (44.13) express the desired equivalence of the norms.

Consequently, according to Subsection 44.4, we can solve the Dirichlet problem for $\mathscr{L}$ from (44.6) with the boundary condition $u = u_0$, $u_0(x_1,x_2) =$ $= (x_1^2 + 1)^{-1}$, in the space $W^{1,2}(\Omega;S^*)$.

<u>44.7. REMARKS.</u> (i) The foregoing example is rather unconvincing. Indeed, it concerns the solution of the (formal) Dirichlet problem

$$\mathscr{L} u = f \quad \text{on} \quad \Omega = \left\{(x_1,x_2) \;,\; x_2 < 1 \;,\; x_1 \in \mathbb{R}\right\},$$
$$u(x) = (x_1^2 + 1)^{-1} \quad \text{on} \quad \partial\Omega = \left\{(x_1,x_2) \;,\; x_2 = 1 \;,\; x_1 \in \mathbb{R}\right\},$$

in the *weak* sense. This means that we have to extend the right hand side in the boundary condition, namely, the function $g(x_1) = (x_1^2 + 1)^{-1}$, from $\partial\Omega$ to the whole domain Ω in such a way that the *extended function* u_0 [which then appears in condition (42.19) of the definition of the weak solution] belongs to the weighted Sobolev space under consideration. There are, of course, many possibilities how to extend the function g from $\partial\Omega$ to Ω . In Example 44.6 we extended the function g to the half-plane Ω putting $u_0(x_1,x_2) =$ $= (x_1^2 + 1)^{-1}$; but this extension fails to be an element of $W^{1,2}(\Omega;S)$, with S from (44.7), and therefore we solved the problem (w e a k l y) in the more convenient space $W^{1,2}(\Omega;S^*)$ with S^* given by (44.9). However, it is evident that *another extension* of g - e.g. to the function $u_0^*(x_1,x_2) =$ $= h(x_2)(x_1^2 + 1)^{-1}$ with a suitable smooth function $h(x_2)$ such that $h(1) = 1$ and $h(x_2) = 0$ for $x_2 \ge 2$ - leads to a function $u_0^* \in W^{1,2}(\Omega;S)$ and makes it possible to solve the problem (weakly) in the initial space $W^{1,2}(\Omega;S)$.

(ii) The operator $\mathcal{L}$ in Example 44.6 serves also as a specimen of an operator in which degenerating and singular coefficients appear *simultaneously* : the coefficient x_2^{-2} vanishes at infinity (= d e g e n e r a t i o n) while the coefficient x_2^2 is unbounded (= s i n g u l a r i t y).

44.8. **EXAMPLE.** Consider again a plane domain (N = 2) and choose the square $(0,1) \times (0,1)$ for the domain Ω . Further, consider the *fourth order* differential operator (k = 2)

$$(44.14) \qquad \mathcal{L} u = \frac{\partial^2}{\partial x_1 \partial x_2} \left[\frac{\partial^2 u}{\partial x_1 \partial x_2} \right] .$$

In this case, the corresponding bilinear form is

$$(44.15) \qquad a(u,v) = \int_\Omega \frac{\partial^2 u}{\partial x_1 \partial x_2} \frac{\partial^2 v}{\partial x_1 \partial x_2} \, dx \ ,$$

and from the "diagonal" coefficients $a_{\alpha\alpha}$, $|\alpha| \leq 2$, only the coefficient $a_{(1,1)(1,1)}(x) \equiv 1$ is non-zero. If we proceeded as in Subsection 44.1, we should have to take the multiindex $(1,1)$ as the (only) element of the set $\mathbb{K}$ from condition **A.1***. However, then the other requirement of condition **A.1*** is not fulfilled : The expression $\|u\|_{\mathbb{K}}$ from (44.1), which is in our case the expression

$$\sqrt{a(u,u)} \ = \ \left[\int_\Omega \left| \frac{\partial^2 u}{\partial x_1 \partial x_2} \right|^2 dx \right]^{1/2} \ ,$$

is not a norm since $\sqrt{a(u,u)}$ vanishes e.g. for any constant or linear function u .

Consequently, the approach from Subsection 44.1 cannot be used and we have to modify our method in a certain way. For this reason, let us consider a *new weighted space* $W^{2,2}(\Omega;\tilde{S})$ where the collection $\tilde{S}$ consists of *two* non-zero weight functions $w_{(1,1)}(x) = a_{(1,1)(1,1)}(x) \equiv 1$ and $w_{00}(x) \equiv 1$ [i.e., we extend the set $\mathbb{K}$ of multiindices mentioned above to the set $\tilde{\mathbb{K}} = \{(0,0),(1,1)\}$]. This new space $W^{2,2}(\Omega;\tilde{S})$ is normed by the square root of

$$(44.16) \qquad \|u\|_{\tilde{\mathbb{K}}}^2 = \int_\Omega |u|^2 \, dx + \int_\Omega \left| \frac{\partial^2 u}{\partial x_1 \partial x_2} \right|^2 dx .$$

Obviously, we have the estimate

$$(44.17) \qquad a(u,u) \leq \|u\|_{\tilde{\mathbb{K}}}^2 \quad (= \|u;\ W^{2,2}(\Omega;\tilde{S})\|^2).$$

Let us show that *the expressions* $\sqrt{a(u,u)}$ *and* $\|u\|_{\tilde{\mathbb{K}}}$ are e q u i v a l e n t n o r m s on $W_0^{2,2}(\Omega;\tilde{S})$.

216

Indeed, if we choose $u \in C_0^\infty(\Omega)$ and use the fact that, in virtue of this inclusion, $u(0,x_2) = 0$ and $\frac{\partial u}{\partial x_1}(x_1,0) = 0$, then from the obvious formulas

$$|u(x_1,x_2)|^2 = \left| \int_0^{x_1} \frac{\partial u}{\partial x_1}(t,x_2) \, dt \right|^2 \le \left(\int_0^1 \left| \frac{\partial u}{\partial x_1}(t,x_2) \right| dt \right)^2 ,$$

$$\left| \frac{\partial u}{\partial x_1}(x_1,x_2) \right|^2 = \left| \int_0^{x_2} \frac{\partial^2 u}{\partial x_1 \partial x_2}(x_1,s) \, ds \right|^2 \le \left(\int_0^1 \left| \frac{\partial^2 u}{\partial x_1 \partial x_2}(x_1,s) \right| ds \right)^2$$

— using for the last integral in each of the two above inequalities the Hölder inequality and then integrating the resulting inequalities over Ω — we obtain the estimate

$$(44.18) \qquad \int_\Omega |u|^2 \, dx \le \int_\Omega \left| \frac{\partial^2 u}{\partial x_1 \partial x_2} \right|^2 dx = a(u,u) .$$

This estimate together with (44.16) implies that

$$(44.19) \qquad \|u\|_{\widetilde{\mathfrak{M}}}^2 \le 2a(u,u) \quad \text{for every } u \in C_0^\infty(\Omega) ,$$

and this together with (44.17) guarantees the equivalence of the expressions $\sqrt{a(u,u)}$ and $\|u\|_{\widetilde{\mathfrak{M}}}$ on $W_0^{2,2}(\Omega;\tilde{S})$.

At the same time, inequality (44.19) says that the bilinear form $a(u,v)$ from (44.15) is $W_0^{2,2}(\Omega;\tilde{S})$-elliptic. Since the continuity of $a(u,v)$ on $W_0^{2,2}(\Omega;\tilde{S}) \times W_0^{2,2}(\Omega;\tilde{S})$ follows immediately from (44.15) and (44.16) by the Hölder inequality, we have shown that the conditions of Theorem 39.5 are fulfilled, and consequently, *there exists a uniquely determined weak solution of the Dirichlet problem for the operator* $\mathcal{L}$ *from* (44.14) *in the m o d i f i - e d space* $W^{2,2}(\Omega;\tilde{S})$.

44.9. <u>REMARKS.</u> (i) The weighted space $W^{2,2}(\Omega;\tilde{S})$ from the previous example is actually not a *weighted* space [we used weights of the type $w_\alpha(x) \equiv 1$] but rather an *anisotropic* one (with the *dominating mixed derivative* $\partial^2 u / \partial x_1 \partial x_2$). We have chosen this example because of its s i m p l i c i t y and we shall present another similar example involving a more complicated weight (see Example 44.10). On the other hand, Example 44.8 shows that the approach mentioned in this chapter can be useful also in cases where weight functions do *not necessarily* appear. See also Example 44.11 below.

(ii) The method used in Example 44.8 improved the approach from Subsection 44.1, *extending the set of multiindices* $\mathfrak{M}$ to the set $\tilde{\mathfrak{M}}$.

The success of this modification is guaranteed only when we are able to estimate the "newly added" terms of $\|u\|_{\widetilde{\mathfrak{M}}}$ by the original expression $\|u\|_{\mathfrak{M}}$.

In Example 44.8, the estimate (44.18) was this important tool, and in the following Examples 44.10 and 44.11, we will again meet this situation - see formula (44.25).

Estimates of such a type will also be the basic tool for the method described in the next Section 15.

(iii) The notation $W^{2,2}(\Omega;\tilde{S})$ used in Example 44.8 was rather inconsistent and did not comply with the notation introduced in Subsection 42.5 [see also Subsection 0.4, formula (0.4)] where we assumed that *all* weight functions w_α (i.e., $a_{\alpha\alpha}$) from the collection S belong to $W(\Omega)$, and so, are p o - s i t i v e a.e. in Ω . The notation $W^{2,2}(\Omega;\tilde{S})$ [or $W^{k,2}(\Omega;\tilde{S})$ from Subsection 44.1] can be regarded as a slight generalization of the usual concept, which consists in the fact that also vanishing weights w_α are admitted : Instead of viewing $\tilde{S}$ as the set $\{w_{(0,0)},w_{(1,1)}\}$ - or $\{w_\alpha \, , \, \alpha \in \mathbb{K}\}$ in Subsection 44.1 - we understand it as follows : $\tilde{S} = \{w_\alpha \, , \, |\alpha| \leq 2\}$, where $w_\alpha(x) \equiv 0$ for $\alpha \neq (0,0),\ (1,1)$ - or $\tilde{S} = \{w_\alpha \, , \, |\alpha| \leq k\}$ with $w_\alpha(x) \equiv 0$ for $\alpha \notin \mathbb{K}$ in Subsection 44.1.

This inconsistent notation will be used also later in some places.

<u>44.10.</u> <u>EXAMPLE.</u> Consider again a plane domain (N = 2) choosing the first quadrant $(0,\infty) \times (0,\infty)$ for the domain Ω . Further, consider the fourth order operator (k = 2)

$$(44.20) \qquad \mathcal{L}u = \frac{\partial^2}{\partial x_1 \partial x_2} \left(x_1^{\delta_1} x_2^{\delta_2} \frac{\partial^2 u}{\partial x_1 \partial x_2} \right) - \frac{\partial}{\partial x_1} \left(x_1^{\gamma_1} x_2^{\gamma_2} \frac{\partial u}{\partial x_1} \right)$$

$$- \frac{\partial}{\partial x_2} \left(x_1^{\beta_1} x_2^{\beta_2} \frac{\partial u}{\partial x_2} \right) \, .$$

The corresponding bilinear form is

$$(44.21) \qquad a(u,v) = \int_\Omega \frac{\partial^2 u}{\partial x_1 \partial x_2} \frac{\partial^2 v}{\partial x_1 \partial x_2} x_1^{\delta_1} x_2^{\delta_2} \, dx + \int_\Omega \frac{\partial u}{\partial x_1} \frac{\partial v}{\partial x_1} x_1^{\gamma_1} x_2^{\gamma_2} \, dx$$

$$+ \int_\Omega \frac{\partial u}{\partial x_2} \frac{\partial v}{\partial x_2} x_1^{\beta_1} x_2^{\beta_2} \, dx \, .$$

Here, from the "diagonal" coefficients $a_{\alpha\alpha}$, $|\alpha| \leq 2$, *only three* coefficients are non-zero : $a_{\alpha\alpha}(x) = x_1^{\delta_1} x_2^{\delta_2}$ for $\alpha = (1,1)$, $a_{\alpha\alpha}(x) = x_1^{\gamma_1} x_2^{\gamma_2}$ for $\alpha =$ = (1,0) and $a_{\alpha\alpha}(x) = x_1^{\beta_1} x_2^{\beta_2}$ for $\alpha = (0,1)$. Consequently, the set $\mathbb{K}$ from condition **A.**1 consists of three elements : $\mathbb{K} = \{(1,0),(0,1),(1,1)\}$. But, similarly as in Example 44.8, the expression $\|u\|_{\mathbb{K}} = \sqrt{a(u,u)}$ from (44.1) is not a norm, and therefore, we cannot use the approach of Subsection 44.1 directly and have to add some new non-zero weight function.

(i) We add the weight function

$$(44.22) \qquad w_\alpha(x) = x_1^{\delta_1 - 2} \, x_2^{\delta_2 - 2} \qquad \text{for} \quad |\alpha| = 0$$

and consider the weighted space $W^{2,2}(\Omega;\tilde{S})$ where the collection $\tilde{S}$ is chosen — in accordance with the slight generalization mentioned in Remark 44.9 (iii) — as follows :

$$(44.23) \qquad \tilde{S} = \{w_\alpha \, , \, |\alpha| \le 2\} = \{w_{(0,0)}, w_{(1,0)}, w_{(0,1)}, w_{(2,0)}, w_{(1,1)}, w_{(0,2)}\}$$

with $w_{(0,0)}$ from (44.22), $w_\alpha = a_{\alpha\alpha}$ for $\alpha = (1,0)$, $(0,1)$ and $(1,1)$ and $w_\alpha(x) \equiv 0$ for $\alpha = (2,0)$ and $(0,2)$.

The choice of the added weight function $w_{(0,0)}$ was made to purpose : If we suppose that $\delta_1 \ne 1$, $\delta_2 \ne 1$ and use twice the Hardy inequality (0.32) [with $p = 2$ and $\varepsilon = \delta_1$ for the function $u(\cdot,x_2)$ and then with $p = 2$ and $\varepsilon = \delta_2$ for the function $\frac{\partial u}{\partial x_1}(x_1,\cdot)$], we obtain for every $u \in C_0^\infty(\Omega)$ the relations

$$(44.24) \qquad \int_\Omega |u(x_1,x_2)|^2 \, x_1^{\delta_1 - 2} \, x_2^{\delta_2 - 2} \, dx_1 \, dx_2$$

$$= \int_0^\infty \int_0^\infty \left(\int |u(x_1,x_2)|^2 \, x_1^{\delta_1 - 2} \, dx_1 \right) x_2^{\delta_2 - 2} \, dx_2$$

$$\le \int_0^\infty \left(\frac{4}{|\delta_1 - 1|^2} \int_0^\infty \left| \frac{\partial u}{\partial x_1}(x_1,x_2) \right|^2 x_1^{\delta_1} \, dx_1 \right) x_2^{\delta_2 - 2} \, dx_2$$

$$= \frac{4}{|\delta_1 - 1|^2} \int_0^\infty \left(\int_0^\infty \left| \frac{\partial u}{\partial x_1}(x_1,x_2) \right|^2 x_2^{\delta_2 - 2} \, dx_2 \right) x_1^{\delta_1} \, dx_1$$

$$\le \frac{4}{|\delta_1 - 1|^2} \int_0^\infty \left(\frac{4}{|\delta_2 - 1|^2} \int_0^\infty \left| \frac{\partial^2 u}{\partial x_1 \partial x_2}(x_1,x_2) \right|^2 x_2^{\delta_2} \, dx_2 \right) x_1^{\delta_1} \, dx_1$$

$$= \frac{16}{|\delta_1 - 1|^2 |\delta_2 - 1|^2} \int_\Omega \left| \frac{\partial^2 u}{\partial x_1 \partial x_2} \right|^2 x_1^{\delta_1} \, x_2^{\delta_2} \, dx_1 \, dx_2 \ .$$

This inequality is the important estimate mentioned in Remark 44.9 (ii) : we can rewrite (44.24) in the form

$$(44.25) \qquad \int_\Omega |u(x_1,x_2)|^2 \, x_1^{\delta_1 - 2} \, x_2^{\delta_2 - 2} \, dx \le \tilde{c} \, \|u\|_{\mathbb{X}}^2 = \tilde{c} \, a(u,u) \ ,$$

i.e., we have estimated the term added in the norm of $W^{2,2}(\Omega;\tilde{S})$ by $\|u\|_{\mathbb{X}}$, and since

$$\|u; \ W^{2,2}(\Omega;\tilde{S})\|^2 = a(u,u) + \int_{\Omega} |u|^2 \ x_1^{\delta_1-2} \ x_2^{\delta_2-2} \ dx_1 \ dx_2 \ ,$$

by virtue of (44.25) we obtain the estimate

(44.26) $\quad \|u; \ W^{2,2}(\Omega;\tilde{S})\|^2 \leq (\tilde{c} + 1)a(u,u)$

which hold for every $u \in C_0^{\infty}(\Omega)$, i.e., for every $u \in W_0^{2,2}(\Omega;S)$ with $\tilde{c} + 1 = $ $= 1 + 16|\delta_1 - 1|^{-2} \ |\delta_2 - 1|^{-2}$. However, this means that the bilinear form $a(u,v)$ from (44.21) is $W_0^{2,2}(\Omega;\tilde{S})$-elliptic. Since it can be easily shown by means of the Hölder inequality that the form $a(u,v)$ is continuous on $W_0^{2,2}(\Omega;\tilde{S}) \times W_0^{2,2}(\Omega;\tilde{S})$, too, we actually arrived - via Theorem 39.5 - at the following assertion :

If $\delta_1 \neq 1$, $\delta_2 \neq 1$ *and the collection* $\tilde{S}$ *is chosen according to* (44.23), *then there is a uniquely determined weak solution of the Dirichlet problem for the operator* $\mathcal{L}$ *from* (44.20) *in the weighted space* $W^{2,2}(\Omega;\tilde{S})$.

At the same time, inequality (44.26) together with the obvious inequality $a(u,u) \leq \|u; \ W^{2,2}(\Omega;\tilde{S})\|^2$ asserts that for $\delta_1 \neq 1$, $\delta_2 \neq 1$ the expressions $\|u; \ W^{2,2}(\Omega;\tilde{S})\|$ and $\sqrt{a(u,u)}$ are equivalent norms on $W_0^{2,2}(\Omega;\tilde{S})$ [but not on $W^{2,2}(\Omega;\tilde{S})$!].

(ii) In the considerations made in point (i) we have not at all used those parts of the norm that involve the first derivatives $\partial u/\partial x_1$, $\partial u/\partial x_2$. Therefore, let us ask if the Dirichlet problem for the operator $\mathcal{L}$ from (44.20) can be handled if we consider the space $W^{2,2}(\Omega;\tilde{S})$ where we put

(44.27) $\quad \tilde{S} = \{w_{(0,0)}(x) = x_1^{\delta_1-2} \ x_2^{\delta_2-2}, \ w_{(1,1)}(x) = x_1^{\delta_1} \ x_2^{\delta_2}, \ w_{\alpha}(x) \equiv 0$
$\qquad\qquad$ for all other $|\alpha| \leq 2\}$;

i.e., the norm is expressed by

$$\|u; \ W^{2,2}(\Omega;\tilde{S})\|^2 = \int_{\Omega} \left(|u|^2 \ x_1^{\delta_1-2} \ x_2^{\delta_2-2} + \left|\frac{\partial^2 u}{\partial x_1 \partial x_2}\right|^2 \ x_1^{\delta_1} \ x_2^{\delta_2} \right) dx_1 \ dx_2 \ .$$

In the same way as in point (i) it can be shown that the bilinear form $a(u,v)$ from (44.21) is $W_0^{2,2}(\Omega;\tilde{S})$-elliptic. Problems can arise when we want to verify whether $a(u,v)$ is continuous on $W_0^{2,2}(\Omega;\tilde{S}) \times W_0^{2,2}(\Omega;\tilde{S})$.

Indeed : If the inequality

(44.28) $\quad |a(u,v)| \leq c_1 \|u; \ W^{2,2}(\Omega;\tilde{S})\| \cdot \|v; \ W^{2,2}(\Omega;\tilde{S})\|$

were satisfied for every $u, v \in W_0^{2,2}(\Omega;\tilde{S})$, then

$$|a(u,u)| \leq c_1 \|u; \ W^{2,2}(\Omega;\tilde{S})\|^2$$

would hold as well. This in particular would mean that for $u \in C_0^{\infty}(\Omega)$ we should have the estimate

$$\int_\Omega \left|\frac{\partial u}{\partial x_1}\right|^2 x_1^{\gamma_1} x_2^{\gamma_2} \, dx_1 \, dx_2 \leq c_1 \left[\int_\Omega |u|^2 x_1^{\delta_1 - 2} x_2^{\delta_2 - 2} \, dx_1 \, dx_2 \right.$$

$$\left. + \int_\Omega \left|\frac{\partial^2 u}{\partial x_1 \partial x_2}\right|^2 x_1^{\delta_1} x_2^{\delta_1} \, dx_1 \, dx_2 \right]$$

as well as an analogous estimate with

$$\int_\Omega \left|\frac{\partial u}{\partial x_2}\right|^2 x_1^{\beta_1} x_2^{\beta_2} \, dx_1 \, dx_2$$

on the left hand side. However, there are counterexamples available demonstrating that these estimates can hold only if

$$(44.29) \qquad \gamma_1 = \delta_1 \ , \quad \gamma_2 = \delta_2 - 2 \ , \quad \beta_2 = \delta_2 \ , \quad \beta_1 = \delta_1 - 2 \ .$$

If (44.29) fails to hold, (44.28) is not fulfilled, either. Conversely, using the Hardy inequality we can show that the continuity condition (44.28) really holds provided conditions (44.29) (and the conditions $\delta_1 \neq 1$, $\delta_2 \neq 1$) are fulfilled.

Consequently, we can assert that *a weak solution of the Dirichlet problem for the operator $\mathcal{L}$ from (44.20) in the space $W^{2,2}(\Omega;\tilde{S})$ with the collection $\tilde{S}$ from (44.27) exists if $\delta_1 \neq 1$, $\delta_2 \neq 1$ and (44.29) hold.*

44.11. <u>EXAMPLE.</u> Consider the Dirichlet problem for the Laplace operator $-\Delta$. This is the operator from Example 42.2 with $a_0(x) \equiv 0$, $a_1(x) = a_2(x) = \cdots = a_N(x) \equiv 1$. Apparently, we should work with the c l a s s i c a l Sobolev spaces $W^{1,2}(\Omega)$ and $W_0^{1,2}(\Omega)$. However, if the domain Ω is u n b o u n - d e d then the corresponding bilinear form

$$a(u,v) = \sum_{i=1}^{N} \int_\Omega \frac{\partial u}{\partial x_i} \frac{\partial v}{\partial x_i} \, dx$$

in general fails to be $W_0^{1,2}(\Omega)$-elliptic, i.e., there need not exist a constant $c > 0$ such that

$$a(u,u) \geq c \|u; \, W^{1,2}(\Omega)\|^2$$

holds for all functions $u \in C_0^\infty(\Omega)$. This follows from the fact that for certain types of unbounded domains, the expression $\sqrt{a(u,u)}$ is not an equivalent norm on the space $W_0^{1,2}(\Omega)$.

Hence, in such a case the Lax-Milgram Theorem 39.5 cannot be used in connection with the classical Sobolev spaces. Nevertheless, it can be shown that for a suitably chosen weight function b_0 - e.g. $b_0(x) = (1 + |x|)^\epsilon$ with $\epsilon \leq -2$ - the expression $\sqrt{a(u,u)}$ will be an equivalent norm on the space

$W_0^{1,2}(\Omega;S)$ provided we choose $S = \{w_\alpha , |\alpha| \leq 1\}$ with $w_\alpha = b_0$ for $|\alpha| = 0$ and $w_\alpha(x) \equiv 1$ for $|\alpha| = 1$. We have

$$\|u; W^{1,2}(\Omega;S)\|^2 = \int_\Omega |u|^2 b_0 \, dx + a(u,u)$$

and the equivalence of this expression with $a(u,u)$ for $u \in C_0^\infty(\Omega)$ shows that the form $a(u,y)$ is $W_0^{1,2}(\Omega;S)$-elliptic. Consequently, we can prove the existence of a weak solution of the Dirichlet problem for the Laplace operator in the space $W^{1,2}(\Omega;S)$.

Section 15 . *S i n g u l a r   a n d   d e g e n e r a t e   e q u a t i o n s*
- a   m o r e   c o m p l i c a t e d   c a s e

§ 45 . C o n d i t i o n s   o n   t h e   c o e f f i c i e n t s

45.1. <u>INTRODUCTION.</u> In Section 14 we worked mainly with collections S of weight functions whose elements were determined by the "diagonal" coefficients $a_{\alpha\alpha}$ of the differential operator $\mathscr{L}$ from (42.1). At first, we assumed that *all* the coefficients $a_{\alpha\alpha}$ belong to $W(\Omega)$ (Subsection 42.3, condition **A.**1), and then we weakened this condition by admitting that *some of the coefficients* $a_{\alpha\alpha}$ *vanish* (Subsection 44.1, condition **A.**1*). In this second case, it was sometimes necessary to enrich the set of non-vanishing coefficients $a_{\alpha\alpha}$ by some further appropriate weight functions w_α (Examples 44.8, 44.10 and 44.11).

Nevertheless, coefficients $a_{\alpha\alpha}$ which are negative or change their signs are admissible, too, although they neither belong to $W(\Omega)$ nor vanish identically. Of course, some additional requirements are necessary in these cases.

Let us start with a simple example.

45.2. <u>EXAMPLE.</u> Consider the differential operator $\mathscr{L}$ from (42.4) - see Example 42.2 - changing now the assumption (42.10) : We assume that (42.10) holds only for $i = 1,\ldots,N$, i.e.,

(45.1) $a_i \in W(\Omega)$, $i = 1,\ldots,N$,

while, concerning the coefficient a_0 , we assume that

(45.2) $a_0 = - \lambda b_0$ with $b_0 \in W(\Omega)$, $\lambda \geq 0$

(λ being a constant); conditions (42.11) remain valid.

In this case, the expression $\sqrt{a(u,u)}$ with $a(u,u)$ given by (42.6) fails to have all the properties of the norm and consequently, the weighted space

$W^{1,2}(\Omega;S)$ can no more be introduced in a natural way. Therefore we choose, instead of the collection S from (42.9), a new collection

(45.3) $$S = \{b_0, a_1, \ldots, a_N\}$$

and consider the spaces $W^{1,2}(\Omega;S)$ and $W_0^{1,2}(\Omega;S)$ with this new collection, introducing the norm by the formula

(45.4) $$\|u; W^{1,2}(\Omega;S)\|^2 = \int_\Omega |u|^2 \, b_0 \, dx + \sum_{i=1}^N \int_\Omega \left|\frac{\partial u}{\partial x_i}\right|^2 a_i \, dx \ .$$

In the same way as in Example 42.2 we prove that

$$|a(u,v)| \leq (N + \lambda)\|u; W^{1,2}(\Omega;S)\| \cdot \|v; W^{1,2}(\Omega;S)\|$$

for every $u, v \in W^{1,2}(\Omega;S)$ and, a fortiori, for $u, v \in W_0^{1,2}(\Omega;S)$. Consequently, the first condition of the Lax-Milgram Theorem 39.5 is proved. It remains to find out when the second condition of this theorem, namely the $W_0^{1,2}(\Omega;S)$-ellipticity, is fulfilled.

To this end, let us assume i n a d d i t i o n that the following estimate holds for all functions $u \in W_0^{1,2}(\Omega;S)$ with a constant $c_0 > 0$ independent of u :

(45.5) $$\|u; L^2(\Omega;b_0)\|^2 \leq c_0 \sum_{i=1}^N \left\|\frac{\partial u}{\partial x_i} ; L^2(\Omega;a_i)\right\|^2 \ .$$

Then we have from (45.4) that

(45.6) $$\|u; W^{1,2}(\Omega;S)\|^2 \leq (c_0 + 1) \sum_{i=1}^N \left\|\frac{\partial u}{\partial x_i} ; L^2(\Omega;a_i)\right\|^2 \ ,$$

i.e., the expression $\|u; W^{1,2}(\Omega;S)\|$ from (45.4) and the expression

$$|||u; W^{1,2}(\Omega;S)||| = \left(\sum_{i=1}^N \left\|\frac{\partial u}{\partial x_i} ; L^2(\Omega;a_i)\right\|^2\right)^{1/2}$$

are equivalent norms on $W_0^{1,2}(\Omega;S)$.

If we use the inequalities (45.5) and (45.6), we obtain

$$a(u,u) = \sum_{i=1}^N \left\|\frac{\partial u}{\partial x_i} ; L^2(\Omega;a_i)\right\|^2 - \lambda\|u; L^2(\Omega;b_0)\|^2$$

$$\geq (1 - \lambda c_0) \sum_{i=1}^N \left\|\frac{\partial u}{\partial x_i} ; L^2(\Omega;a_i)\right\|^2 \geq \frac{1 - \lambda c_0}{c_0 + 1} \|u; W^{1,2}(\Omega;S)\|^2$$

and consequently, the bilinear form $a(u,v)$ is $W_0^{1,2}(\Omega;S)$-elliptic provided the constant $(1 - \lambda c_0)/(c_0 + 1)$ is positive,

(45.7) $$\lambda c_0 < 1 \ , \text{ i.e., } \ 0 \leq \lambda < \frac{1}{c_0} \ .$$

Consequently, we can assert that also for an operator $\mathcal{L}$ of the form (42.4) with a "non-positive" coefficient a_0 of the form (45.2), *there is one and only one weak solution of the Dirichlet problem in the space* $W^{1,2}(\Omega;S)$ *with* S *given by* (45.3) *provided the additional condition* (45.5) *is fulfilled and the constants* c_0 *in* (45.5) *and* λ *in* (45.2) *satisfy* (45.7).

45.3. REMARKS. (i) Condition (45.5) reminds property (P_2) from Subsection 40.2 - see inequality (40.6). Again, we can roughly say that condition (45.5) expresses an imbedding of the type

$$(45.8) \qquad \overset{\circ}{W}_0^{1,2}(\Omega;S) \subsetneqq L^2(\Omega;b_0)$$

with the *imbedding constant* c_0 . If this constant is sufficiently small, the coefficient a_0 can be - according to (45.2) and (45.7) - "negative enough".

(ii) Condition (45.5) indicates that if some coefficients $a_{\alpha\alpha}$ appear in the differential operator $\mathcal{L}$ which *are not* from $W(\Omega)$, then we should be able to estimate the terms [of the norm in the weighted space] that correspond to these "bad" coefficients by the other terms [which correspond to coefficients from $W(\Omega)$]. In other words, among all "diagonal" coefficients $a_{\alpha\alpha}$ we have to determine certain coefficients which are *decisive*. The situation is similar to that occuring in the case of c l a s s i c a l elliptic equations: in this latter case we use the classical Sobolev spaces and the decisive coefficients are the coefficients $a_{\alpha\alpha}$ with $|\alpha| = k$, i.e., the coefficients appearing in $\mathcal{L}u$ at the derivatives of the highest order.

In our more general case it is not so easy to determine the decisive coefficients. There has to be at least one coefficient $a_{\alpha\alpha}$ with $|\alpha| = k$ among then, but as some of the foregoing examples indicate, in some cases we need also terms corresponding to derivatives of lower orders. The next Subsection represents an attempt at describing how to choose the set of important coefficients $a_{\alpha\alpha}$, i.e., the decisive set of multiindices.

45.4. COEFFICIENTS OF THE DIFFERENTIAL OPERATOR AND THE WEIGHT FUNCTIONS.
Let us again consider the (formal) differential operator $\mathcal{L}$ of order $2k$ from (42.1), i.e.,

$$(45.9) \qquad (\mathcal{L}u)(x) = \sum_{|\alpha|,|\beta|\leq k} (-1)^{|\alpha|} D^\alpha\big(a_{\alpha\beta}(x)\, D^\beta u(x)\big) \ .$$

Further, let us denote by $\mathbb{K}_0$ the set of all such multiindices α , $|\alpha| \leq k$, for which $a_{\alpha\alpha}$ are weight functions :

$$(45.10) \qquad a_{\alpha\alpha} \in W(\Omega) \quad \text{for} \quad \alpha \in \mathbb{K}_0 \ .$$

Let the following condition be fulfilled.

B.1   The set $\mathbb{K}_0$ contains at least one multiindex of length k ; moreover, $a_{\alpha\alpha} \in L^1_{loc}(\Omega)$ for $\alpha \in \mathbb{K}_0$.

Now, let us choose a set $\mathbb{K}_1 \subset \mathbb{K}_0$ so that the following condition is fulfilled :

B.2   For $\alpha \in \mathbb{K}_0$, there is a constant $c_\alpha > 0$ such that

$$(45.11) \qquad \|D^\alpha u;\ L^2(\Omega;a_{\alpha\alpha})\|^2 \leq c_\alpha \sum_{\gamma \in \mathbb{K}_1} \|D^\gamma u;\ L^2(\Omega;a_{\gamma\gamma})\|^2$$

holds for every $u \in C_0^\infty(\Omega)$.

Denote the sum on the right hand side of (45.11) by $\|u\|^2_{\mathbb{K}_1}$, i.e.

$$(45.12) \qquad \|u\|_{\mathbb{K}_1} = \left(\sum_{\gamma \in \mathbb{K}_1} \|D^\gamma u;\ L^2(\Omega;a_{\gamma\gamma})\|^2 \right)^{1/2} ;$$

the expression $\|\cdot\|_{\mathbb{K}_1}$ possesses all the properties of a *seminorm*.

The set $\mathbb{K}_1$ can be chosen in various ways; condition (45.11) will certainly be fulfilled if we set $\mathbb{K}_1 = \mathbb{K}_0$, but this trivial choice is unsatisfactory. However, we naturally aim at choosing the set $\mathbb{K}_1$ *as small as possible;* practically, this means that we try to estimate the *greatest possible* number of terms of the form $\|D^\alpha u;\ L^2(\Omega;a_{\alpha\alpha})\|^2$ by combinations of the *smallest possible* number of analogous terms with other multiindices. When doing this we will assume that the set $\mathbb{K}_1$ chosen satisfies the following conditions :

B.3 $\dfrac{1}{a_{\alpha\alpha}} \in L^1_{loc}(\Omega)$ for $\alpha \in \mathbb{K}_1$.

B.4   There exists a constant $c_1 > 0$ such that for $\alpha,\ \beta \in \mathbb{K}_1$

$$(45.13) \qquad |a_{\alpha\beta}(x)| \leq c_1 \sqrt{a_{\alpha\alpha}(x)a_{\beta\beta}(x)} \qquad \text{for a.e. } x \in \Omega .$$

B.5   There exists a constant $c_2 > 0$ such that

$$(45.14) \qquad a(u,u) \geq c_2 \|u\|^2_{\mathbb{K}_1}$$

holds for all functions $u \in C_0^\infty(\Omega)$; here $a(u,v)$ is the bilinear form corresponding to the operator $\mathcal{L}$ — see (42.18).

Now, let us choose a *set* $\mathbb{K}_2 \subset \{\alpha ,\ |\alpha| \leq k\} \setminus \mathbb{K}_1$ and *weight functions* $w_\beta \in W(\Omega)$ for $\beta \in \mathbb{K}_2$ so that the following conditions are satisfied :

B.6 $w_\beta \in L^1_{loc}(\Omega)$ and $\dfrac{1}{w_\beta} \in L^1_{loc}(\Omega)$ for $\beta \in \mathbb{K}_2$;

further, there exist constants $\tilde{c}_\beta > 0$ such that for $\beta \in \mathbb{K}_2$ and for all $u \in C_0^\infty(\Omega)$ the inequalities

$$(45.15) \qquad \left\| D^\beta u; \ L^2(\Omega; w_\beta) \right\|^2 \leq \tilde{c}_\beta \| u \|_{\mathbb{K}_1}^2$$

hold.

B.7   For every pair $\alpha, \beta \in \mathbb{K}_1 \cup \mathbb{K}_2$ there exists a constant $c_{\alpha\beta} > 0$ such that

$$(45.16) \qquad \left| a_{\alpha\beta}(x) \right| \leq c_{\alpha\beta} \sqrt{w_\alpha(x) w_\beta(x)} \quad \text{for a.e. } x \in \Omega \ ;$$

here we put

$$(45.17) \qquad w_\gamma(x) = a_{\gamma\gamma}(x) \quad \text{for } \gamma \in \mathbb{K}_1 \ .$$

If $\alpha \notin \mathbb{K}_1 \cup \mathbb{K}_2$ or $\beta \notin \mathbb{K}_1 \cup \mathbb{K}_2$, then $a_{\alpha\beta}(x) \equiv 0$.

<u>45.5.</u> <u>REMARKS.</u>   (i) The foregoing subsection may impress the reader as rather too intricate and cumbersome, but it actually provides i n s t r u c t i o n s how to proceed in a particular case when constructing the sets $\mathbb{K}_1$ and $\mathbb{K}_2$ and choosing the functions w_β for $\beta \in \mathbb{K}_2$. This also explains why we explicitly formulated condition **B.**4, although it is included in the first part of condition **B.**7 : Usually, we have to follow the steps described in Subsections 45.4, and if we do not succeed in fulfilling the (simpler) condition **B.**4, our approach is unpracticable and it is of no use constructing the set $\mathbb{K}_2$ and verifying conditions **B.**5, **B.**6 and **B.**7.

(ii) Conditions **B.**1 - **B.**7 have a little reminded conditions **A.**1 - **A.**4 from Subsection 42.3, they are only a little more involved. However, two points are to be accentuated :

(ii-1) The estimates (45.11) and (45.15) are new and have no counterpart among the **A**-conditions; they express the requirement of distinguishing certain decisive multiindices with help of which the remaining terms can be estimated - see Remark 45.3 (ii).

(ii-2) Instead of condition **A.**4 which was an algebraic condition describing the (degenerating) ellipticity of the differential operator $\mathscr{L}$ - see (42.15) - we now have a certain "integral" ellipticity condition **B.**5 - see (45.14). We will mention some algebraic ellipticity condition later, see Subsection 46.3.

<u>45.6.</u> <u>THE WEIGHTED SOBOLEV SPACE.</u> Having chosen the sets $\mathbb{K}_1$ and $\mathbb{K}_2$ from Subsection 45.4, let us choose a set $\mathbb{K}$ so that

$$(45.18) \qquad \mathbb{K}_1 \subset \mathbb{K} \subset \mathbb{K}_1 \cup \mathbb{K}_2$$

holds; further, we denote

(45.19) $$S = \{w_\alpha;\ \alpha \in \mathbb{K}\}$$

[for $\alpha \in \mathbb{K}_1$, we choose w_α according to (45.17), i.e., $w_\alpha = a_{\alpha\alpha}$] and set

(45.20) $$\|u;\ W^{k,2}(\Omega;S)\| = \left(\sum_{\alpha \in \mathbb{K}} \|D^\alpha u;\ L^2(\Omega;w_\alpha)\|^2\right)^{1/2}$$

[as concerns the *formal* introduction of the space $W^{k,2}(\Omega;S)$ as the set of $u = u(x)$, $x \in \Omega$, for which $\|u;\ W^{k,2}(\Omega;S)\| < \infty$, we use the convention mentioned in Remark 44.9 (iii), setting $w_\alpha(x) \equiv 0$ if $\alpha \notin \mathbb{K}$].

Now, we assume that the set $\mathbb{K}$ is chosen in such a way that the following condition is fulfilled :

B.8   The linear space $W^{k,2}(\Omega;S)$ just introduced is a (complete) Hilbert space under the norm (45.20).

If condition **B**.8 is fulfilled, then we can introduce the corresponding "nulled" space $W_0^{k,2}(\Omega;S)$ in the usual way, i.e., as the closure of the set $C_0^\infty(\Omega)$ with respect to the norm (45.20); it follows from **B**.1 and **B**.6 that $w_\alpha \in L_{loc}^1(\Omega)$ for $\alpha \in \mathbb{K}$, so that $C_0^\infty(\Omega)$ is a subset of $W^{k,2}(\Omega;S)$ and the closure makes sense.

45.7.   <u>REMARKS.</u>   (i) As $\mathbb{K} \subset \mathbb{K}_1 \cup \mathbb{K}_2$, it follows from (45.15) that for all $u \in C_0^\infty(\Omega)$ the inequality

(45.21) $$\|u;\ W^{k,2}(\Omega;S)\|^2 \le c_3 \|u\|_{\mathbb{K}_1}^2$$

is satisfied with the constant $c_3 = 1 + \sum_{\beta \in \mathbb{K}_2} \tilde{c}_\beta > 0$ independent of u [see (45.12) for $\|u\|_{\mathbb{K}_1}$]. As $\mathbb{K}_1 \subset \mathbb{K}$, the converse inequality $\|u\|_{\mathbb{K}_1} \le \|u;\ W^{k,2}(\Omega;S)\|$ holds as well. So we have shown that *the seminorm* $\|\cdot\|_{\mathbb{K}_1}$ *is a norm on the space* $W_0^{k,2}(\Omega;S)$, *equivalent to the norm* (45.20).

(ii)   It immediately follows from (45.14) and (45.21) that *the bilinear form* $a(u,v)$ *is* $W_0^{k,2}(\Omega;S)$-*elliptic* :   Indeed, we have

(45.22) $$a(u,u) \ge \frac{c_2}{c_3} \|u;\ W^{k,2}(\Omega;S)\|^2$$

for every $u \in C_0^\infty(\Omega)$. Condition **B**.6 and, above all, inequality (45.15), thus play a crucial role in our considerations.

§ 46. Existence theorem. Some generalizations. Examples

Now, we are able to state the main result of Section 15 :

__46.1.__ __THEOREM.__ *Let* $\mathcal{L}$ *be the differential operator from* (42.1) *and* $W^{k,2}(\Omega;S)$ *the weighted Sobolev space from Subsection* 45.6. *Let the coeffi-cients* $a_{\alpha\beta}$ *of the operator* $\mathcal{L}$ *and the weight functions* w_γ *satisfy con-ditions* __B__.1 – __B__.8. *Let the function* $u_0 \in W^{k,2}(\Omega;S)$ *and the functional* $F \in \left[W_0^{k,2}(\Omega;S)\right]^*$ *be given.*

Then there exists one and only one weak solution u *of the Dirichlet problem for the operator* $\mathcal{L}$ *in the space* $W^{k,2}(\Omega;S)$ (in the sense of De-finition 42.6) *and, moreover, the estimate* (43.1) *holds.*

The p r o o f is completely analogous to that of Theorem 43.1. The $W_0^{k,2}(\Omega;S)$-ellipticity of the bilinear form $a(u,v)$ from (42.18) follows from (45.22), and conditions __B__.7, __B__.6 and __B__.2 guarantee the continuity of $a(u,v)$ on $W_0^{k,2}(\Omega;S) \times W_0^{k,2}(\Omega;S)$. Consequently, the assertions follow by the Lax--Milgram Theorem 39.5.

__46.2.__ __REMARK.__ When choosing the set $\mathbb{K}$ we have, with regard to (45.18), a certain liberty, so that generally we can construct *various* spaces $W^{k,2}(\Omega;S)$ (more precisely, various collections S of weight functions depending on the choice of the set $\mathbb{K}$ of multiindices). Naturally, Remark 45.7 (i) implies that the corresponding "nulled" spaces $W_0^{k,2}(\Omega;S)$ *coincide* (in the sense that the corresponding norms are equivalent) for all admissible choices of the set $\mathbb{K}$. This fact can be used when choosing the function u_0 that represents the boundary conditions; the situation is the same as in Subsection 44.4.

__46.3.__ __THE ALGEBRAIC CONDITION OF ELLIPTICITY.__ *Algebraic* conditions of the type of inequality (42.15) *imply* conditions in an "integral form" of the type of inequality (45.14) and their verification is sometimes easier. Therefore, let us introduce a certain analogue of condition __A__.4 which can be used instead of condition __B__.5 :

__B__.5* There exists a constant $c_2 > 0$ such that for all real vectors $\xi = = \{\xi_\gamma, \ |\gamma| \le k\}$ and for a.e. $x \in \Omega$,

$$(46.1) \qquad \sum_{|\alpha|,|\beta|\le k} a_{\alpha\beta}(x)\, \xi_\alpha\, \xi_\beta \ge c_2 \sum_{\gamma \in \mathbb{K}_1} a_{\gamma\gamma}(x)\, \xi_\gamma^2 \ .$$

First of all, it is evident that (46.1) immediately implies inequality (45.14) : it suffices to choose $\xi_\alpha = D^\alpha u(x)$ in (46.1) and to integrate the resulting inequality over the set Ω . Consequently, condition **B**.5 is fulfilled provided condition **B**.5* is.

Of course, the algebraic condition **B**.5* is *substantially more restrictive* : it immediately follows from (46.1) that

$$a_{\alpha\alpha}(x) \geq 0 \quad \text{a.e. in} \quad \Omega$$

for *all* α such that $|\alpha| \leq k$ [not only for $\alpha \in \mathbb{K}_0$, where the property $a_{\alpha\alpha} \in W(\Omega)$ was a condition for constructing the set $\mathbb{K}_0$ - see (45.10)]. Indeed, if for some $\beta \notin \mathbb{K}_0$ the coefficient $a_{\beta\beta}$ were negative on a set of positive measure, then inequality (46.1) with the choice $\xi = \{\xi_\alpha = 0$ for $\alpha \neq \beta$, $\xi_\beta = 1\}$ would yield a contradiction.

On the other hand, however, condition **B**.5 admits also "diagonal" coefficients $a_{\alpha\alpha}$ which are negative or change signs - see **Example** 45.2 or Example 46.4 below.

<u>46.4.</u> <u>EXAMPLE.</u> Let Ω be a plane domain ($N = 2$) and let us choose

$$(46.2) \qquad \mathcal{L}u = \frac{\partial^4 u}{\partial x_1^4} - 2\lambda \frac{\partial^4 u}{\partial x_1^2 \partial x_2^2} + \frac{\partial^4 u}{\partial x_2^4} , \quad \lambda > 0 .$$

Thus, we now have $a_{(2,0)(2,0)}(x) = a_{(0,2)(0,2)}(x) \equiv 1$, $a_{(1,1)(1,1)}(x) \equiv -2\lambda$, $a_{\alpha\beta}(x) \equiv 0$ for all other α , β with $|\alpha| \leq 2$, $|\beta| \leq 2$. The "diagonal" coefficient $a_{(1,1)(1,1)}$ is negative and hence it does not belong to $W(\Omega)$; this is why $\mathbb{K}_0 = \{(2,0),(0,2)\}$. Since condition **B**.2 should hold, we cannot reduce the set $\mathbb{K}_0$, hence $\mathbb{K}_1 = \mathbb{K}_0$. Condition **B**.5* naturally fails to hold since inequality (46.1) has the form

$$\xi_{(2,0)}^2 - 2\lambda\xi_{(1,1)}^2 + \xi_{(0,2)}^2 \geq c_2(\xi_{(2,0)}^2 + \xi_{(0,2)}^2)$$

and this condition evidently is violated for $\xi_{(2,0)} = \xi_{(0,2)} = 0$, $\xi_{(1,1)} = 1$. Nonetheless, condition **B**.5 - i.e., inequality (45.14) - is satisfied for small λ's . Indeed,

$$\|u\|_{\mathbb{K}_1}^2 = \int\limits_{\Omega} \left(\left| \frac{\partial^2 u}{\partial x_1^2} \right|^2 + \left| \frac{\partial^2 u}{\partial x_2^2} \right|^2 \right) dx$$

and

$$(46.3) \qquad a(u,u) = \|u\|_{\mathbb{K}_1}^2 - 2\lambda \int\limits_{\Omega} \left| \frac{\partial^2 u}{\partial x_1 \partial x_2} \right|^2 dx .$$

Since we consider the Dirichlet problem for the operator $\mathcal{L}$ from (46.2), we are working with functions $u \in C_0^\infty(\Omega)$. Therefore, we can extend these functions by zero onto the whole plane $\mathbb{R}^2$ and integrate always over $\mathbb{R}^2$. Passing to the Fourier transform we easily prove

(46.4)
$$\int_\Omega \left|\frac{\partial^2 u}{\partial x_1 \partial x_2}\right|^2 dx \leq \frac{1}{2} \int_\Omega \left(\left|\frac{\partial^2 u}{\partial x_1^2}\right|^2 + \left|\frac{\partial^2 u}{\partial x_2^2}\right|^2\right) dx = \frac{1}{2} \|u\|_{\mathbb{K}_1}^2 .$$

Hence and from (46.3) we obtain

$$a(u,u) \geq (1 - \lambda) \|u\|_{\mathbb{K}_1}^2 \quad \text{for} \quad u \in C_0^\infty(\Omega) ;$$

thus for $\lambda < 1$ condition **B**.5 - i.e., inequality (45.14) - is fulfilled.

Now, let us see how we can choose, in our case, the set $\mathbb{K}_2$ and the weight functions w_β for $\beta \in \mathbb{K}_2$. Since $a_{(1,1)(1,1)}(x) = -2\lambda \neq 0$, the multiindex $(1,1)$ does not belong to $\mathbb{K}_0$ ($= \mathbb{K}_1$). Condition **B**.7 implies that necessarily $(1,1) \in \mathbb{K}_2$, and the weight function $w_{(1,1)}$ must satisfy (45.16), i.e.,

(46.5)
$$|-2\lambda| \leq c^* \sqrt{w_{(1,1)}(x) w_{(1,1)}(x)} = c^* w_{(1,1)}(x) \quad \text{for a.e.} \quad x \in \Omega .$$

Simultaneously, (45.15) must hold, that is,

$$\int_\Omega \left|\frac{\partial^2 u}{\partial x_1 \partial x_2}\right|^2 w_{(1,1)}(x) \, dx \leq \tilde{c} \, \|u\|_{\mathbb{K}_1}^2 .$$

Taking into account (46.4) we see that this inequality is fulfilled by the weight function $w_{(1,1)}(x) \equiv 1$, which at the same time satisfies also condition (46.5). Thus, if we choose

$$\mathbb{K}_2 = \{(0,0),(1,1)\} , \quad w_{(1,1)}(x) \equiv 1 , \quad w_{(0,0)}(x) = (1 + |x|)^\varepsilon$$

($\varepsilon \leq -2$; $\varepsilon = 0$ if the domain Ω is bounded), condition **B**.8 will be fulfilled as well. In view of Remark 46.2, the existence of a weak solution of the Dirichlet problem for the operator $\mathcal{L}$ from (46.2) is guaranteed in *two* spaces $W^{2,2}(\Omega;S)$: either with the choice

$$S = \{w_\alpha, \ |\alpha| \leq 2\} \quad \text{where} \quad w_\alpha(x) = (1 + |x|)^\varepsilon \quad \text{for} \quad |\alpha| = 0 ,$$

$$w_\alpha(x) \equiv 0 \quad \text{for} \quad |\alpha| = 1 , \quad w_\alpha(x) \equiv 1 \quad \text{for} \quad |\alpha| = 2 ,$$

which corresponds to the set $\mathbb{K} = \mathbb{K}_1 \cup \mathbb{K}_2$, or with

$$S = \{w_\alpha, \ |\alpha| \leq 2\} \quad \text{where} \quad w_\alpha(x) = (1 + |x|)^\varepsilon \quad \text{for} \quad |\alpha| = 0 ,$$

$$w_{(2,0)}(x) = w_{(0,2)}(x) \equiv 1 , \quad w_\alpha(x) \equiv 0 \quad \text{otherwise}$$

which corresponds to the set $\mathbb{K} = \mathbb{K}_1 \cup \{0,0\}$. The choice $\mathbb{K} = \mathbb{K}_1$ is *inadmissible* since in general condition **B**.8 would be violated.

If the domain Ω is bounded, then $\mathbb{K}_2$ may be chosen so that $\mathbb{K}_1 \cup \mathbb{K}_2 = \{(0,0),(1,0),(0,1),(2,0),(1,1),(0,2)\} = \{\alpha; \ |\alpha| \leq 2\}$, here we take $w_\alpha(x) \equiv 1$ for $\alpha \in \mathbb{K}_2$. In this case we in fact solve the Dirichlet problem for the operator $\mathcal{L}$ from (46.2) in the c l a s s i c a l Sobolev space $W^{2,2}(\Omega)$.

Here it has always been essential that $0 < \lambda < 1$. Let us note that for these values the operator $\mathscr{L}$ from (46.2) is not elliptic in the classical sense, since there is a non-zero vector ξ such that

$$\sum_{|\alpha|,|\beta|\leq 2} a_{\alpha\beta}(x)\, \xi_\alpha\, \xi_\beta = \xi^2_{(2,0)} - 2\lambda\, \xi^2_{(1,1)} + \xi^2_{(0,2)} = 0 \ .$$

<u>46.5.</u> <u>EXAMPLE.</u> Let us consider the domain Ω as well as the operator $\mathscr{L}$ from Example 44.8. Here $\mathbb{K}_0 = \{(1,1)\}$ since $a_{(1,1)(1,1)}(x) \equiv 1$ is the only non-zero coefficient. We again have to choose $\mathbb{K}_1 = \mathbb{K}_0$. It can be easily verified (using the arguments of Example 44.8) that the choice

$$\mathbb{K}_2 = \{(0,0),(1,0),(0,1)\} \ , \quad w_\alpha(x) \equiv 1 \ \text{ for } \ \alpha \in \mathbb{K}_2$$

leads to the desired result, i.e., to the solvability of the Dirichlet problem in the corresponding space $W^{2,2}(\Omega;S)$ with $S = \{w_\alpha,\ \alpha \in \mathbb{K}\}$ and $\mathbb{K}$ determined by the inclusions

$$\mathbb{K}_1 \cup \{(0,0)\} \subset \mathbb{K} \subset \mathbb{K}_1 \cup \mathbb{K}_2 \ .$$

Let us recall that the space $W^{2,2}(\Omega;\tilde{S})$ from Example 44.8 corresponds to the choice $\mathbb{K} = \mathbb{K}_1 \cup \{(0,0)\}$.

<u>46.6.</u> <u>EXAMPLE.</u> Let us consider the domain Ω as well as the operator $\mathscr{L}$ from Example 44.10. There we had $a_{(1,1)(1,1)}(x) = x_1^{\delta_1} x_2^{\delta_2}$, $a_{(1,0)(1,0)}(x) = x_1^{\gamma_1} x_2^{\gamma_2}$, $a_{(0,1)(0,1)}(x) = x_1^{\beta_1} x_2^{\beta_2}$, and the other coefficients $a_{\alpha\beta}$ vanished identically. Hence $\mathbb{K}_0 = \{(1,0)(0,1),(1,1)\}$. We have shown in point (i) of Example 44.10 that the choice $\mathbb{K}_1 = \mathbb{K}_0$ and $\mathbb{K}_2 = \{(0,0)\}$ with $w_{(0,0)}(x) = x_1^{\delta_1-2} x_2^{\delta_2-2}$ leads to the desired existence result with $\mathbb{K} = \mathbb{K}_1 \cup \mathbb{K}_2$ (for $\delta_1 \neq 1$, $\delta_2 \neq 1$).

So far as conditions (44.29) of point (ii) of Example 44.10 are fulfilled it suffices to choose $\mathbb{K}_1 = \{(1,1)\}$ (i.e., we have $\mathbb{K}_1 \subset \mathbb{K}_0$, $\mathbb{K}_1 \neq \mathbb{K}_0$) $\mathbb{K}_2 = \{(0,0)\}$ with the weight function $w_{(0,0)}$ as above, and $\mathbb{K} = \mathbb{K}_1 \cup \mathbb{K}_2$.

If only the first or the second pair of conditions (44.29) is fulfilled, we arrive at the result even with the choice $\mathbb{K}_1 = \{(0,1)(1,1)\}$ or $\mathbb{K}_1 = \{(1,0),(1,1)\}$, respectively, provided we choose $\mathbb{K}_2$ and $w_{(0,0)}$ as above and put $\mathbb{K} = \mathbb{K}_1 \cup \mathbb{K}_2$ again.

<u>46.7.</u> <u>ONE MORE EXAMPLE.</u> On the half plane $\Omega = \{(x_1,x_2),\ x_2 > 0\}$ consider the second order differential operator

$$(46.6) \qquad \mathscr{L}u = -\,\Delta u - \frac{\lambda}{x_2}\,\frac{\partial u}{\partial x_2} \ , \quad \lambda \in \mathbf{R} \ .$$

Here $a_{(1,0)(1,0)}(x) = a_{(0,1)(0,1)}(x) \equiv 1$, $a_{(0,0)(0,1)}(x) = -\lambda/x_2$. Hence $\mathbb{K}_0 = \mathbb{K}_1 = \{(1,0),(0,1)\}$. Further

$$(46.7) \qquad a(u,u) = \int_\Omega \left(\left| \frac{\partial u}{\partial x_1} \right|^2 + \left| \frac{\partial u}{\partial x_2} \right|^2 \right) dx - \lambda \int_\Omega \frac{\partial u}{\partial x_2} \, u \, \frac{1}{x_2} \, dx \ .$$

The first integral on the right hand side is (the square of) the seminorm $\|u\|^2_{\mathbb{K}_1}$; the second integral can be estimated by the Hölder inequality and by the Hardy inequality (0.32) with respect to x_2 for $p = 2$ and $\varepsilon = 0$:

$$\left| \int_\Omega \frac{\partial u}{\partial x_2} \, u \, \frac{1}{x_2} \, dx \right| \leq \left(\int_\Omega \left| \frac{\partial u}{\partial x_2} \right|^2 dx \right)^{1/2} \left(\int_\Omega u^2 \, \frac{1}{x_2^2} \, dx \right)^{1/2}$$

$$\leq 2 \int_\Omega \left| \frac{\partial u}{\partial x_2} \right|^2 dx \leq 2 \, \|u\|^2_{\mathbb{K}_1} \ ,$$

and we have

$$a(u,u) \geq (1 - 2|\lambda|) \, \|u\|^2_{\mathbb{K}_1} \quad \text{for} \quad u \in C_0^\infty(\Omega) \ .$$

In this way, we are able to prove that for $|\lambda| < \frac{1}{2}$ condition **B**.5 is fulfilled. Let us show that *condition* **B**.5 *is fulfilled for e v e r y* $\lambda \leq 0$.

Indeed, if $u \in C_0^\infty(\Omega)$, then

$$\frac{\partial}{\partial x_2}(u^2 \, \frac{1}{x_2} \,) = 2u \, \frac{\partial u}{\partial x_2} \, \frac{1}{x_2} - u^2 \, \frac{1}{x_2^2}$$

and consequently

$$0 = \int_\Omega \frac{\partial}{\partial x_2}(u^2 \, \frac{1}{x_2}) \, dx = 2 \int_\Omega u \, \frac{\partial u}{\partial x_2} \, \frac{1}{x_2} \, dx - \int_\Omega u^2 \, \frac{1}{x_2^2} \, dx \ .$$

This implies

$$\int_\Omega \frac{\partial u}{\partial x_2} \, u \, \frac{1}{x_2} \, dx \geq 0 \quad \text{for every} \quad u \in C_0^\infty(\Omega)$$

which in view of (46.7) yields

$$a(u,u) \geq \|u\|^2_{\mathbb{K}_1}$$

for $\lambda \leq 0$. Further, it can be seen that there is a number $\tilde{\lambda} \geq \frac{1}{2}$ such that condition **B**.5 is *not* valid for all $\lambda \geq \tilde{\lambda}$. Hence, our approach cannot be applied to the case $\lambda \geq \tilde{\lambda}$.

However, this difficulty can be avoided if we consider, instead of the operator $\mathscr{L}$ from (46.6), the operator $\mathscr{L}_1$ given by the formula

$$\mathscr{L}_1 u = - \frac{\partial}{\partial x_1}(x_2^\lambda \, \frac{\partial u}{\partial x_1}) - \frac{\partial}{\partial x_2}(x_2^\lambda \, \frac{\partial u}{\partial x_2}) \ .$$

This operator is very simply connected with the operator $\mathscr{L}$:

$$\mathscr{L}_1 u = x_2^\lambda \, \mathscr{L} u \ .$$

Moreover, here $\mathbb{K}_0 = \mathbb{K}_1 = \{(1,0),(0,1)\}$, $a_{\alpha\alpha}(x) = x_2^\lambda$ for $|\alpha| = 1$, and the bilinear form a_1 corresponding to the operator $\mathscr{L}_1$ satisfies

$$a_1(u,u) = \|u\|_{\mathbb{K}_1}^2 = \int_\Omega \left(\left|\frac{\partial u}{\partial x_1}\right|^2 + \left|\frac{\partial u}{\partial x_2}\right|^2 \right) x_2^\lambda \, dx$$

so that condition **B.5** presents no difficulties. If we choose, in addition, $\mathbb{K}_2 = \{(0,0)\}$ with $w_{(0,0)}(x) = x_2^{\lambda-2}$, then the (weak) solvability of the Dirichlet problem for the operator $\mathscr{L}_1$ is guaranteed in the space $W^{1,2}(\Omega;S)$ with $\mathbb{K} = \mathbb{K}_1 \cup \mathbb{K}_2$ and $S = \{x_2^{\lambda-2}, x_2^\lambda, x_2^\lambda\}$ provided $\lambda \neq 1$: Indeed, for such λ also condition **B.6** is fulfilled, since as a consequence of the Hardy inequality (0.32) (with respect to x_2 and for $p = 2$, $\varepsilon = \lambda$) we obtain

$$\|u; L^2(\Omega;w_{(0,0)})\|^2 = \int_\Omega |u|^2 \, x_2^{\lambda-2} \, dx = \int_{-\infty}^\infty \left(\int_0^\infty |u(x_1,x_2)|^2 \, x_2^{\lambda-2} \, dx_2 \right) dx_1$$

$$\leq \int_{-\infty}^\infty \left(\frac{4}{(\lambda-1)^2} \int_0^\infty \left|\frac{\partial u}{\partial x_2}(x_1,x_2)\right|^2 x_2^\lambda \, dx_2 \right) dx_1 = \frac{4}{(\lambda-1)^2} \int_\Omega \left|\frac{\partial u}{\partial x_2}\right|^2 x_2^\lambda \, dx$$

$$\leq \frac{4}{(\lambda-1)^2} \|u\|_{\mathbb{K}_1}^2 \; ;$$

and this is in fact inequality (45.15).

46.8. <u>**WEAKENING THE CONDITIONS ON THE SET**</u> $\mathbb{K}_0$. Let us again consider the differential operator $\mathscr{L}$ from (45.9) and assume that one of the "diagonal" coefficients, say $a_{\delta\delta}$, exhibits the following behaviour :

$$(46.8) \qquad \begin{aligned} a_{\delta\delta}(x) &> 0 \quad \text{for} \quad x \in \Omega_0 \, , \quad \Omega_0 \subset \Omega \, , \\ a_{\delta\delta}(x) &\equiv 0 \quad \text{for} \quad x \in \Omega \setminus \Omega_0 \; ; \end{aligned}$$

we assume moreover that both sets Ω_0 , $\Omega \setminus \Omega_0$ have a *positive measure*. Then $a_{\delta\delta}$ *does not belong to* $W(\Omega)$ and therefore $\delta \notin \mathbb{K}_0$. However, since $a_{\delta\delta}(x) \equiv 0$ does not hold, either, we necessarily have $\delta \in \mathbb{K}_2$ and hence

$$(46.9) \qquad \int_{\Omega_0} |D^\delta u(x)|^2 \, a_{\delta\delta}(x) \, dx \leq c_\delta^* \|u\|_{\mathbb{K}_1}^2 \; .$$

This inequality is a consequence of conditions **B.7** and **B.6** : according to **B.7** there is a weight function $w_\delta \in W(\Omega)$ such that $|a_{\delta\delta}(x)| \leq c_{\delta\delta} w_\delta(x)$, and according to **B.6** we have $\|D^\delta u; L^2(\Omega;w_\delta)\|^2 \leq \tilde{c}_\delta \|u\|_{\mathbb{K}_1}^2$.

If condition (46.9) is not satisfied, our approach is inapplicable; nevertheless, it can be modified in the following way : We include the multi-index δ in the set $\mathbb{K}_0$ and form the set $\mathbb{K}_1$ again with help of condition **B.2** [for $\alpha = \delta$ formula (46.8) implies that on the left hand side of (45.11) we actually have only the integral over Ω_0]. The further steps are the same as above, with our modification of the set $\mathbb{K}_0$ manifesting itself in the other conditions **B.3** — **B.8** : for instance, condition **B.7** implies that the coef-

233

ficients $a_{\alpha\delta}$ and $a_{\delta\alpha}$ vanish for $x \in \Omega \smallsetminus \Omega_0$ [cf. (45.16) with $w_\delta = a_{\delta\delta}$].

<u>46.9.</u> <u>EXAMPLE.</u> As an example of an operator $\mathscr{L}$ for which condition (46.9) is violated, let us consider the following fourth order differential operator on a plane domain :

$$u = \frac{\partial^4 u}{\partial x_1^4} + \frac{\partial^2}{\partial x_2^2} \left(a(x) \frac{\partial^2 u}{\partial x_2^2} \right) ,$$

where $a(x)$ is a function of the type (46.8) with the following property : There is a point $x_0 \in \Omega_0$, a neighbourhood $U(x_0) \subset \Omega_0$ and a positive constant c_0 such that $a(x) \geq c_0$ for $x \in U(x_0)$. According to the modification of our approach introduced in Subsection 46.8, we then have $\mathbb{K}_1 = \mathbb{K}_0 =$ $= \{(2,0),(0,2)\}$ and

$$\|u\|_{\mathbb{K}_1}^2 = \int_\Omega \left| \frac{\partial^2 u}{\partial x_1^2} \right|^2 dx + \int_{\Omega_0} \left| \frac{\partial^2 u}{\partial x_2^2} \right|^2 a(x) \ dx .$$

Section 16 . *Strong singularities and strong degeneration*

§ 47 . M o d i f i e d   s p a c e s .   E x i s t e n c e   t h e o r e m

<u>47.1.</u> <u>MOTIVATION.</u> In our foregoing considerations, we tried to weaken the conditions imposed on the coefficients $a_{\alpha\beta}$ of our differential operator of order $2k$,

$$(47.1) \qquad (\mathscr{L}u)(x) = \sum_{|\alpha|,|\beta|\leq k} (-1)^{|\alpha|} D^\alpha \big(a_{\alpha\beta}(x) \, D^\beta u(x) \big) .$$

However, we have not been able to weaken the requirement that the weights w_α (which are equal to $a_{\alpha\alpha}$ in some cases) fulfil the condition

$$(47.2) \qquad \frac{1}{w_\alpha} \in L_{loc}^1 (\Omega)$$

which ensures that the weighted spaces $W^{k,2}(\Omega;S)$ considered are really *Banach* spaces, i.e., *complete* linear normed spaces, and the condition

$$(47.3) \qquad w_\alpha \in L_{loc}^1 (\Omega)$$

which makes it possible to introduce the space $W_0^{k,2}(\Omega;S)$ — important for the Dirichlet problem — as the closure of $C_0^\infty(\Omega)$ with respect to the norm of $W^{k,2}(\Omega;S)$. These restrictions for the weight functions appeared in condition **A.**2 and, in the more complicated case, in conditions **B.**1, **B.**3 and **B.**6.

On the other hand, in a number of the foregoing examples we have met weight functions and coefficients of the type

(47.4) $$w(x) = \left[\text{dist}(x,M)\right]^{\varepsilon}$$

M being a subset of $\bar{\Omega}$ and ε a real number. Equations with such coefficients and spaces with such weights occur most frequently in applications. For
$\varepsilon < 0$, such weights grow to infinity in a neighbourhood of M and therefore,
an equation with coefficients of this type is said to have a *singularity* on
M . On the contrary, for $\varepsilon > 0$ the function w converges to zero for
$x \to x_0 \in M$ and an equation with such coefficients is said to *degenerate* on M .

If M is a part of the boundary of Ω ,
$$M \subset \partial\Omega \ ,$$
then coefficients (weights) of the type (47.4) "get spoiled" solely in the
neighbourhood of $\partial\Omega$ and therefore, there are no difficulties with conditions
of the type (47.2) and (47.3). Quite another situation occurs if
$$M \subset \Omega \ .$$
Under the condition meas $M = 0$ the requirement (47.3) is fulfilled for
$\varepsilon > 0$, but for negative ε's it holds only provided $\varepsilon > -\varepsilon_0$ where $\varepsilon_0 > 0$
depends on the dimension of the set M ; similarly the requirement (47.2) is
not fulfilled for $\varepsilon > \varepsilon_0$. Hence we can say that for certain sufficiently
large $|\varepsilon|$, weight functions of the type (47.4) do not fulfil conditions of
the type **A**.2 if $M \subset \Omega$ (or, at least, $M \cap \Omega$ is non-empty). In such cases
we speak about coefficients (weights) with s t r o n g *singularity* or
s t r o n g *degeneration* on M (i.e., *inside* Ω) and the approach described
in the foregoing sections is inapplicable since the sets $W^{k,2}(\Omega;S)$ and
$W_0^{k,2}(\Omega;S)$, important for our consideration, do not have the properties we
need or even do not make sense at all.

Nevertheless, our approach can be modified to suit even such cases. Roughly
speaking, it is necessary to consider the set $\Omega_0 = \Omega \setminus M$ instead of Ω , so
that M then becomes part of the boundary $\partial\Omega_0$.

47.2. __A MODIFIED DEFINITION OF THE SPACE__ $W^{k,2}(\Omega;S)$. Let us assume that the
condition

(47.5) $$\frac{1}{w} \in L^1(\Omega)$$

is violated for a weight function $w \in W(\Omega)$, and denote

(47.6) $$P_2(w) = \left\{ x \in \Omega; \int_{\Omega \cap U(x)} w^{-1}(y)\, dy = \infty \text{ for every neighbourhood } U(x) \text{ of } x \right\} .$$

Let $\mathbb{K}$ be a subset of the set of all multiindices α of length at most
k and suppose that $\mathbb{K}$ contains the multiindex $\theta = (0,\ldots,0)$ and at least
one multiindex of length k . Further, let S be a collection of weight functions associated with the set $\mathbb{K}$,

(47.7)
$$S = \left\{ w_\alpha \in W(\Omega), \quad \alpha \in \mathbb{K} \right\}$$

and denote by $W^{k,2}(\Omega;S)$ the linear set of all functions $u = u(x)$, $x \in \Omega$, for which the expression

$$\|u; W^{k,2}(\Omega;S)\| = \left(\sum_{\alpha \in \mathbb{K}} \|D^\alpha u; L^2(\Omega;w_\alpha)\|^2 \right)^{1/2}$$

(47.8)
$$= \left(\sum_{\alpha \in \mathbb{K}} \int_\Omega |D^\alpha u(x)|^2 \, w_\alpha(x) \, dx \right)^{1/2}$$

is finite.

As follows from examples in A. KUFNER, B. OPIC [4], [6], it is the set

(47.9)
$$\Omega_0 = \bigcup_{\alpha \in \mathbb{K}} P_2(w_\alpha)$$

with $P_2(w_\alpha)$ from (47.6) being the "bad" set which causes the *noncompleteness* of the corresponding weighted space $W^{k,2}(\Omega;S)$.

Let us denote

(47.10)
$$\Omega^* = \Omega \setminus \Omega_0 \, .$$

Since Ω_0 is closed (see A. KUFNER, B. OPIC [4], Lemma 3.2), Ω^* is an open set in $\mathbf{R}^N$ and it follows from the definition that

$$\frac{1}{w_\alpha} \in L^1_{loc}(\Omega^*) \, .$$

Therefore, the space $W^{k,2}(\Omega^*;S)$ is meaningful and, moreover, it is a Banach space.

Using these results, *we d e f i n e the weighted Sobolev space*
$$W^{k,2}(\Omega;S)$$
as the space $W^{k,2}(\Omega^*;S)$ *with* Ω^* *from* (47.10) *provided the condition* $1/w_\alpha \in L^1_{loc}(\Omega)$ *is* not *satisfied for some* $\alpha \in \mathbb{K}$.

47.3. REMARK. Obviously, $P_2(w) = \emptyset$ if w satisfies condition (47.5). Consequently, the set Ω_0 from (47.9) is empty if $1/w_\alpha \in L^1_{loc}(\Omega)$ for *all* $\alpha \in \mathbb{K}$, and the "new" space $W^{k,2}(\Omega^*;S)$ coincides with the "old" one. Therefore, it is reasonable to use for the "new" space the same notation $W^{k,2}(\Omega;S)$ as for the "old" one; hopefully, this licence will cause no confusion.

47.4. THE SPACE $W_0^{k,2}(\Omega;S)$ AND ITS MODIFICATION. Usually, the space $W_0^{k,2}(\Omega;S)$ is introduced as the closure of the set $C_0^\infty(\Omega)$ with respect to the norm (47.8) assuming that, in addition to the condition

$$\frac{1}{w_\alpha} \in L^1_{loc}(\Omega) \quad \text{for all} \quad \alpha \in \mathbb{K} \, ,$$

the following condition is fulfilled :

$$(47.11) \qquad w_\alpha \in L^1_{loc}(\Omega) \quad \text{for all } \alpha \in \mathfrak{K} .$$

This last condition guarantees that

$$(47.12) \qquad C^\infty_0(\Omega) \subset W^{k,2}(\Omega;S) .$$

Obviously $W^{k,2}_0(\Omega;S)$ is again a Banach space under the norm (47.8).

If (47.11) is violated, then inclusion (47.12) is meaningless [it can be shown that (47.11) is necessary and sufficient for (47.12) — see A. KUFNER, B. OPIC [4], Lemma 4.4] and therefore, the space $W^{k,2}_0(\Omega;S)$ cannot be introduced. Then we proceed as follows.

We denote, for $w \in W(\Omega)$,

$$(47.13) \qquad P_0(w) = \left\{ x \in \Omega; \quad \int\limits_{\Omega \cap U(x)} w(y)\, dy = \infty \quad \text{for every} \right.$$
$$\left. \text{neighbourhood } U(x) \text{ of } x \right\} .$$

Obviously, $P_0(w) = \emptyset$ if $w \in L^1_{loc}(\Omega)$. Further, we introduce the set

$$(47.14) \qquad \Omega_1 = \bigcup_{\alpha \in \mathfrak{K}} P_0(w_\alpha) ;$$

then Ω_1 is closed in Ω and $w_\alpha \in L^1_{loc}(\Omega \setminus \Omega_1)$ for every $\alpha \in \mathfrak{K}$.

If Ω^* is the set from (47.10) — i.e., $\Omega^* = \Omega \setminus \Omega_0$ with Ω_0 from (47.9) — we denote

$$(47.15) \qquad \Omega^{**} = \Omega \setminus \Omega_1$$

and *d e f i n e the space*

$$W^{k,2}_0(\Omega;S)$$

as the closure of the set

$$(47.16) \qquad V = \left\{ f; \ f = g_{|\Omega^*} , \ g \in C^\infty_0(\Omega^{**}) \right\}$$

with respect to the norm (47.8) [considered as the norm of the "new" space $W^{k,2}(\Omega^*,S)$!].

Again, $W^{k,2}_0(\Omega;S)$ is a Banach space : the assumption $f = g_{|\Omega^*}$ with $g \in C^\infty_0(\Omega^{**})$ guarantees that $V \subset W^{k,2}(\Omega;S)$, so that the closure is meaningful, and since $W^{k,2}(\Omega;S)$ is defined to be the space $W^{k,2}(\Omega^*,S)$ — see Subsection 47.2 — the completeness of $W^{k,2}_0(\Omega;S)$ as a closed set in a Banach space is guaranteed as well.

47.5. THE DIRICHLET PROBLEM : FORMULATION, EXISTENCE AND UNIQUENESS THEOREM.

In the foregoing Subsections 47.2 and 47.4 we have introduced the spaces $W^{k,2}(\Omega;S)$ and $W^{k,2}_0(\Omega;S)$ without any further assumptions on the weight function w_α appearing in the collection S — the only requirement now is

$$w_\alpha \in W(\Omega) \ .$$

This enables us to proceed in complete analogy with the foregoing Sections 14
and 15. We can

- *introduce the concept of a weak solution* of the Dirichlet problem as in Subsection 42.6,
- *formulate conditions on the coefficients* $a_{\alpha\beta}$ of the differential operator
$\mathscr{L}$ as in Subsection 42.3 in the simple case (of course, w i t h o u t
condition **A**.2) or as in Subsection 45.4 in the more complicated case
(again w i t h o u t the requirement $a_{\alpha\alpha} \in L^1_{loc}(\Omega)$ for $\alpha \in \mathbb{K}_0$ in condition **B**.1 , without condition **B**.3 and w i t h o u t the first part of
condition **B**.6),
- *introduce the appropriate weighted Sobolev spaces as* in Subsections 42.5
and/or 45.6 and finally
- *prove* - via the Lax-Milgram Theorem 39.5 - *the corresponding theorem about
the existence and uniqueness of a weak solution* in the relevant weighted
space as in Subsections 43.1 and/or 46.1.

The only thing which we have to keep in mind is that we are working with the
m o d i f i e d weighted spaces $W^{k,2}(\Omega;S)$ and $W_0^{k,2}(\Omega;S)$.

§ 48 .   E x a m p l e s .   R e m a r k s

<u>48.1. INTRODUCTION.</u> The foregoing considerations show that in fact we are
considering - in a new setting - a boundary value problem not on Ω but
on $\Omega^* = \Omega \setminus \Omega_0$ · see (47.10) and (47.9). If the corresponding weight functions are *continuous* in Ω then it can be shown (see A. KUFNER, B. OPIC [4],
Theorem 3.3 and Lemma 4.6) that the set Ω_0 from (47.9) as well as the set
Ω_1 from (47.14) *are of measure zero;* in this case, we can consider Ω_0 and
Ω_1 as parts of the boundary of the domain of definition.

All will be seen more clearly from the following examples, in which we
shall work with the plane domain $\Omega = (-1,1) \times (-1,1)$ (i.e., $N = 2$) and with
the differential operator

$$(48.1) \qquad \mathscr{L}\,u = - \sum_{i=1}^{2} \frac{\partial}{\partial x_i} \left(a(x) \frac{\partial u}{\partial x_i} \right) + a(x)u$$

(i.e., $k = 1$). In this case, the natural space in which we shall look for a
weak solution is the space $W^{1,2}(\Omega;S)$ with the collection

$$S = \{a,a,a\}$$

and with the norm

$$(48.2) \qquad \|u;\ W^{1,2}(\Omega;S)\|^2 = \int_\Omega \left(\left|\frac{\partial u}{\partial x_1}\right|^2 + \left|\frac{\partial u}{\partial x_2}\right|^2 + |u|^2 \right) a(x)\ dx \ .$$

We denote

$$\Omega_+ = \{x \in \Omega;\ x_1 > 0\}\ ,\quad \Omega_- = \{x \in \Omega;\ x_1 < 0\}\ ,$$
(48.3)
$$\Gamma = \{(x_1,0);\ 0 < x_1 < 1\}\ .$$

<u>48.2.</u> <u>EXAMPLE</u> (s t r o n g   s i n g u l a r i t y). Let us take

(48.4)
$$a(x) = a(x_1,x_2) = \begin{cases} x_2^{-2} & \text{for } x \in \Omega_+\ , \\ |x_2|^{-\lambda} & \text{for } x \in \Omega_- \end{cases}$$

with $0 < \lambda < 1$ for the coefficient a in (48.1). Since $\frac{1}{a}$ belongs to $L_{loc}^1(\Omega)$, the set Ω_0 from (47.9) is empty (i.e., $\Omega^* = \Omega$) and *the space* $W^{1,2}(\Omega;S)$ *is well defined and complete*. On the other hand, $a \notin L_{loc}^1(\Omega)$ and the set Ω_1 from (47.14) is the segment Γ from (48.3). We say that on Γ a *strong singularity* of the coefficient a occurs. In accordance with Sub-section 47.4, we define $W_0^{1,2}(\Omega;S)$ as the closure of $C_0^\infty(\Omega \setminus \overline{\Gamma})$.

The weak solution of the Dirichlet problem for the operator $\mathcal{L}$ from (48.1) is a function $u \in W^{1,2}(\Omega;S)$ for which

(48.5)
$$\int\limits_\Omega a(x)\left(\frac{\partial u}{\partial x_1}\frac{\partial v}{\partial x_1} + \frac{\partial u}{\partial x_2}\frac{\partial v}{\partial x_2} + uv\right)\,dx = <F,v>$$

for all $v \in C_0^\infty(\Omega \setminus \overline{\Gamma})$. Since v vanishes in a neighbourhood of $\overline{\Gamma}$, we can consider the identity (48.5) on $\Omega^{**} = \Omega \setminus \overline{\Gamma}$ instead on Ω . Further, the boun-dary condition is expressed by the requirement

$$u - u_0 \in W_0^{1,2}(\Omega;S)$$

with a prescribed function $u_0 \in W^{1,2}(\Omega;S)$ $[= W^{1,2}(\Omega^{**},S)$ since meas $\overline{\Gamma} =$ = meas $(\Omega \setminus \Omega^{**}) = 0$]. These facts suggest the idea that we have to prescribe a boundary condition not only on $\partial\Omega$, but also on $\overline{\Gamma}$ since $\partial\Omega^{**} = \partial\Omega \cup \overline{\Gamma}$. But in fact we automatically have $u_{|\Gamma} = 0$ (in the sense of traces) since also u_0 necessarily has a zero trace on Γ as a consequence of the fact that the weight function $a(x)$ is of the form $[\text{dist}(x,\Gamma)]^\varepsilon$ with $\varepsilon = -2 < -1$ (see Subsec-tion 48.7 below).

Let us mention that a singularity of $a(x)$ appears on the *whole* segment
(48.6)
$$I = \{(x_1,0);\ -1 < x_1 < 1\}$$
but on $I \setminus \overline{\Gamma}$ the singularity is *weak* thanks to the condition $0 < \lambda < 1$.

<u>48.3.</u> <u>EXAMPLE</u> (s t r o n g   d e g e n e r a t i o n). Let us take

(48.7)
$$a(x) = a(x_1,x_2) = \begin{cases} x_2^2 & \text{for } x \in \Omega_+\ , \\ |x_2|^\lambda & \text{for } x \in \Omega_- \end{cases}$$

with $0 < \lambda < 1$ for the coefficient a in (48.1) [this function is the reciprocal of the function from (48.4)]. Here we have a strong degeneration on the segment Γ (and a weak one on $I \setminus \overline{\Gamma}$) since the condition $\frac{1}{a} \in L^1_{loc}(\Omega)$ is *not* fulfilled. Therefore, the space $W^{1,2}(\Omega;S)$ is in fact the space $W^{1,2}(\Omega \setminus \overline{\Gamma} ;S)$, and $W_0^{1,2}(\Omega;S)$ is the closure of the restriction of functions from $C_0^\infty(\Omega)$ to $\Omega^* = \Omega \setminus \overline{\Gamma}$.

48.4. REMARKS. (i) A comparison of Examples 48.2 and 48.3 shows that the behaviour of the solutions $u \in W^{1,2}(\Omega;S)$ differs on Γ : In Example 48.2 we necessarily have $u_{|\Gamma} = 0$; in Example 48.3 we have no information and no requirement — the "trace from above" (for $x_2 \to 0+$) can be completely different from the "trace from below" (for $x_2 \to 0-$). See again Subsection 48.7 below.

(ii) Combining the considerations from Examples 48.2 and 48.3, we can construct examples in which strong singularities appear on one part of Ω [i.e., on the set Ω_0 from (47.9)] while strong degeneration appears on another part of Ω [i.e., on the set Ω_1 from (47.14)]. Moreover, both phenomena can take place *on the same set* ($\Omega_1 = \Omega_0$) as the following example shows.

48.5. EXAMPLE (s t r o n g s i n g u l a r i t y t o g e t h e r w i t h s t r o n g d e g e n e r a t i o n). Let us take

$$(48.8) \qquad a(x) = a(x_1,x_2) = \begin{cases} e^{-1/x_2} & \text{for } x \in \Omega_+ , \\ 1 & \text{for } x \in \Omega_- \end{cases}$$

for the coefficient a in (48.1). In this case we have $\frac{1}{a} \notin L^1_{loc}(\Omega)$ as well as $a \notin L^1_{loc}(\Omega)$. The sets Ω_0 and Ω_1 coincide with the segment $\overline{\Gamma}$ and therefore, we define $W^{1,2}(\Omega;S)$ as the space $W^{1,2}(\Omega \setminus \overline{\Gamma}; S)$ and $W_0^{1,2}(\Omega;S)$ as the closure of the set $C_0^\infty(\Omega \setminus \overline{\Gamma})$.

In this case, we have a strong singularity on Γ from below (i.e., for $x_2 \to 0-$) and a strong degeneration from above (i.e., for $x_2 \to 0+$). In view of the definition of the space $W^{1,2}(\Omega;S)$, we should consider Γ as part of the boundary of the domain of definition $\Omega^{**} = \Omega \setminus \Omega_1 = \Omega \setminus \overline{\Gamma}$. But in this case, for $u \in W^{1,2}(\Omega;S)$ we automatically have a zero "trace on Γ from below" and no condition for a "trace on Γ from above". The arguments are analogous to those in Example 48.2 and Remark 48.4 (i).

48.6. BOUNDARY CONDITIONS. In the weak formulation of the Dirichlet problem the boundary condition(s) are replaced by condition

$$(48.9) \qquad u - u_0 \in W_0^{k,2}(\Omega;S)$$

 — see Subsection 42.6, formula (42.19).

This condition "imitates" the condition which occurs in the definition of
a weak solution of the Dirichlet problem in c l a s s i c a l (= non-weighted) Sobolev spaces :

$$(48.10) \qquad u - u_0 \in W_0^{k,2}(\Omega)$$

(see, e.g. J. NEČAS [1], Chap. 1, formula (2.16a), and Chap. 3, formula (2.5a),
or K. REKTORYS [1], Definition 32.2). Condition (48.10) has a natural interpretation in terms of traces of a function from $W^{k,2}(\Omega)$ on the boundary $\partial\Omega$:
it means that

$$(48.11) \qquad D^\alpha(u - u_0)\big|_{\partial\Omega} = 0 \quad \text{for} \quad |\alpha| \le k - 1 \ ,$$

i.e., that " $D^\alpha u = D^\alpha u_0$ on $\partial\Omega$ " for $|\alpha| \le k - 1$ in the sense of traces.
(Of course, we have to assume that the boundary $\partial\Omega$ is "sufficiently smooth".)

Unfortunately, the knowledge of the properties of traces for w e i g h -
t e d Sobolev spaces is rather incomplete as yet. Therefore, an interpretation
of condition (48.9) similar to (48.11) can be transferred to weighted spaces
only to a very limited extent and for special weights. Let us illustrate this
fact by a simple example.

<u>48.7.   EXAMPLE.</u> On the plane domain $\Omega = (0,1) \times (0,1)$, let us consider the
operator

$$\mathscr{L} u = - \frac{\partial}{\partial x_1}\left(x_1^\varepsilon \frac{\partial u}{\partial x_1}\right) - \frac{\partial}{\partial x_2}\left(x_1^\varepsilon \frac{\partial u}{\partial x_2}\right) + x_1^\varepsilon u \ ,$$

i.e., the operator (48.1) with the special choice $a(x) = a(x_1,x_2) = x_1^\varepsilon$, $\varepsilon \in \mathbb{R}$.
The weighted space corresponding to this operator is the space $W^{1,2}(\Omega;S)$ with
the collection $S = \left\{x_1^\varepsilon,\ x_1^\varepsilon,\ x_1^\varepsilon\right\}$, and since the "bad behaviour" of the coefficient x_1^ε is concentrated on the set

$$M = \left\{(0,x_2),\ 0 < x_2 < 1\right\} \ ,$$

which is a part of $\partial\Omega$, we have no problems with conditions of the type **A**.2.
Consequently, we can proceed as in Example 42.2, and the existence (and uniqueness) of a weak solution of the Dirichlet problem for $\mathscr{L}$ in the space
$W^{1,2}(\Omega;S)$ [$= W^{1,2}(\Omega;d_M,\varepsilon)$ in the notation of, e.g., Chapter IV] is guaranteed for *every* $\varepsilon \in \mathbb{R}$.

Concerning the trace on the boundary $\partial\Omega$ of a function $v \in W^{1,2}(\Omega;S)$,
i.e., $v \in W^{1,2}(\Omega;d_M,\varepsilon)$, we can assert the following facts (see [I], Sections
9.13 – 9.18) :

(i) For $\varepsilon < 1$, the trace $v\big|_{\partial\Omega}$ exists and belongs to the space
$L^2(\partial\Omega)$; moreover, for $\varepsilon \le - 1$ this trace *vanishes on* M , i.e.,

$$v\big|_M = 0 \quad \text{for} \quad \varepsilon \le -1 \ .$$

(ii) For $\varepsilon \geq 1$, the trace exists on $\partial\Omega \smallsetminus M$ and belongs to $L^2(\partial\Omega \smallsetminus M)$, while on M *its existence is not guaranteed.*

Consequently, the "boundary condition" (48.9), i.e., in our case, the condition

(48.12) $\qquad u - u_0 \in W_0^{1,2}(\Omega;d_M,\varepsilon)$,

means that

(i) on $\partial\Omega \smallsetminus M$, we have $u = u_0$ for every $\varepsilon \in \mathbb{R}^{\cdot}$;

(ii) on M , we have

(ii-1) $u = u_0$ for $-1 < \varepsilon < 1$,

(ii-2) $u = 0$ for $\varepsilon \leq -1$ [since necessarily $u_{0|M} = 0$ for this ε],

(ii-3) *no condition* imposed on u for $\varepsilon \geq 1$.

So, we see that even in this simple case the problem of interpretation of
the boundary condition (48.12) is difficult. Evidently, it will be still more
complicated for higher order equations and for more general weights.

48.8. **REMARK.** In the foregoing example, we stated that for $\varepsilon \geq 1$, the existence of a trace (on M) of a function $v \in W^{1,2}(\Omega;d_M,\varepsilon)$ is not guaranteed.
Indeed, we can construct such functions $v(x_1,x_2) \in W^{1,2}(\Omega;d_M,\varepsilon)$ that v are
not bounded in a neighbourhood of the set M , i.e.

$$\lim_{x_1 \to 0+} |v(x_1,x_2)| = \infty   \text{ for a.e. } x_2 \in (0,1)$$

(see, e.g., [I], Example 9.17). Nonetheless, the behaviour of the function v
in the neighbourhood of M can be described more precisely : namely, for
every $v \in W^{1,2}(\Omega;d_M,\varepsilon)$ we have

(48.13) $$\lim_{x_1 \to 0+} x_1^\lambda |v(x_1,x_2)| = 0   \text{ for a.e. } x_2 \in (0,1)$$

where

$$\lambda > \frac{\varepsilon - 1}{2} .$$

This concerns the case $\varepsilon \geq 1$; however, relation (48.13) holds even for
$\varepsilon \leq -1$ with λ satisfying the inequality

$$\frac{\varepsilon - 1}{2} < \lambda < 0 .$$

48.9. **OTHER BOUNDARY VALUE PROBLEMS.** In Sections 14 – 16 , we dealt mainly
with the Dirichlet problem; in this case, it was the space $W_0^{k,2}(\Omega;S)$ which
played the role of the space H from the Lax–Milgram Theorem 39.5. Obviously,
other boundary value problems can be treated in the same way by using – in-
stead of the space $W_0^{k,2}(\Omega;S)$ – a certain wider class $V \subset W^{k,2}(\Omega;S)$ [with

$V = W^{k,2}(\Omega;S)$ for the Neumann problem]. The fact that is rather restrictive as concerns the practicability of our approach in these cases is that a number of conditions [like **B**.2, **B**.5, **B**.6 – particularly inequality (45.15), etc.] can be no more required for functions from $C_0^\infty(\Omega)$ but for some wider classes of functions. This leads, in the end, to the requirement that certain imbedding theorems [of the type (45.8)] should take place not for the "nulled" spaces $W_0^{k,2}(\Omega;S)$, but for, say, $W^{k,2}(\Omega;S)$. Since these latter imbeddings need more restrictive assumptions about the weight functions considered, the impact of our approach is restricted accordingly. The situation is similar to that of the Neumann problem in Section 12 where the restrictive condition $N - m \geq$ $\geq 2k + 1$ appeared due to the imbedding theorems used.

48.10. <u>REMARK.</u> In this chapter, we followed closely the ideas developed in the papers A. KUFNER, B. OPIC [1], [5]. The same remark concerns also the first part of the next Chapter VI.

Chapter VI

NONLINEAR DIFFERENTIAL EQUATIONS

Section 17 . *P r o b l e m s   w i t h   " b a d   c o e f f i c i e n t s "*

§ 49 . F o r m u l a t i o n   o f   t h e   p r o b l e m .   S o m e
       a u x i l i a r y   r e s u l t s

<u>49.1.</u> INTRODUCTION. In the foregoing Chapter V, we considered a l i n e a r partial differential operator $\mathscr{L}$ of order $2k$ and constructed a suitable weighted Sobolev space $W^{k,2}(\Omega;S)$ in which some boundary value problem (mainly, the D i r i c h l e t problem) for $\mathscr{L}$ was uniquely weakly solvable. The collection S of weight functions $w_\alpha \in W(\Omega)$ was determined by the operator $\mathscr{L}$ or, more precisely, by its coefficients $a_{\alpha\beta}$.

Now, we shall consider n o n l i n e a r differential operators of the form

$$(49.1) \qquad (\mathscr{N}u)(x) = \sum_{|\alpha|\leq k} (-1)^{|\alpha|} D^\alpha a_\alpha\big(x;\, \delta_k u(x)\big)$$

with

$$(49.2) \qquad \delta_k u(x) = \{D^\beta u(x);\ |\beta| \leq k\} ,$$

where $a_\alpha = a_\alpha(x;\xi)$ are given functions (= the "coefficients" of the operator

$\mathscr{N}$) defined on $\Omega \times \mathbf{R}^{\kappa}$. [Here κ denotes the number of components of the vector function $\delta_k u$; we have $\kappa = (N + k)!/(N!k!) - $ see Subsection 43.3.]

While in the *linear* case the operator $\mathscr{L}$ was *given* and we tried to find out *which weighted spaces* (i.e., *which weight functions*) are suitable for, say, the Dirichlet problem to be solved in them, in the *nonlinear* case we shall proceed in the r e v e r s e d w a y : We shall assume that *the weighted spaces* $W^{k,p}(\Omega;S)$ and $W_0^{k,p}(\Omega;S)$ with $p > 1$ *are given* (i.e., that the collection S of weight functions w_α is prescribed – for the corresponding definitions see Subsection 0.4) and our aim is to show *what operators* $\mathscr{N}$ [i.e., *what functions* $a_\alpha = a_\alpha(x;\xi)$] are suitable for the Dirichlet problem to be solved (in a weak sense) in these spaces.

<u>49.2.</u> <u>SOME NOTATIONS.</u> (i) Let $\mathbb{K}$ be a subset of the set $\{\alpha; |\alpha| \le k\}$ of all N-dimensional multiindices of length at most k and let $\mathbb{K}$ have the following property : $\mathbb{K}$ contains the zero multiindex $\theta = (0,\ldots,0)$ and at least one multiindex of the length k .

(ii) Let S be a collection of weight functions associated with the set $\mathbb{K}$,

$$(49.3) \qquad S = \left\{ w_\alpha = w_\alpha(x); \ w_\alpha \in W(\Omega), \ \alpha \in \mathbb{K} \right\} \ .$$

We assume that for a given number p ,

$$p > 1 \ ,$$

we have

$$(49.4) \qquad w_\alpha^{-1/(p-1)} \in L^1_{loc}(\Omega) \qquad \text{for all} \ \alpha \in \mathbb{K}$$

and

$$(49.5) \qquad w_\alpha \in L^1_{loc}(\Omega) \qquad \text{for all} \ \alpha \in \mathbb{K} \ .$$

(iii) We define the weighted Sobolev space

$$W^{k,p}(\Omega;S)$$

as the set of all functions $u = u(x)$, $x \in \Omega$, such that

$$(49.6) \qquad \begin{aligned} \|u; \ W^{k,p}(\Omega;S)\|^p &= \sum_{\alpha \in \mathbb{K}} \int_\Omega |D^\alpha u(x)|^p \, w_\alpha(x) \ dx \\ &= \sum_{\alpha \in \mathbb{K}} \|D^\alpha u; \ L^p(\Omega;w_\alpha)\|^p < \infty \ ; \end{aligned}$$

in view of condition (49.4), it is a *reflexive Banach space*. Further, we introduce a subspace of $W^{k,p}(\Omega;S)$, the "nulled" space

$$W_0^{k,p}(\Omega;S) \ ,$$

as the closure of $C_0^\infty(\Omega)$ with respect to the norm $\|\cdot; \ W^{k,p}(\Omega;S)\|$ defined by (49.6); in view of condition (49.5), this space is well defined.

49.3. __REMARKS.__ (i) Analogously as in Section 16, conditions (49.4) and (49.5) can be avoided by an appropriate modification of the definition of the weighted spaces. Namely, if we proceed as in Subsection 47.2, introduce for $w \in W(\Omega)$ the set

$$P_p(w) = \left\{ x \in \Omega; \quad \int_{\Omega \cap U(x)} w^{-1/(p-1)}(y) \, dy = \infty \quad \text{for every} \right.$$

$$\text{neighbourhood} \quad U(x) \quad \text{of} \quad x \Big\}$$

and define

$$\Omega_0 = \bigcup_{\alpha \in \mathbb{K}} P_p(w_\alpha) \, ,$$

we can define $W^{k,p}(\Omega;S)$ as the space $W^{k,p}(\Omega \smallsetminus \Omega_0; S)$. Since $w_\alpha^{-1/(p-1)} \in$ $\in L^1_{loc}(\Omega \smallsetminus \Omega_0)$, we avoided condition (49.4). Further, completely analogously as in Subsection 47.4, we can avoid condition (49.5) by defining the set Ω_1 by formula (47.14) and introducing the space $W_0^{k,p}(\Omega;S)$ as the closure of the set V from (47.16) with respect to the norm of the space $W^{k,p}(\Omega;S)$ [already understood in the above mentioned modified sense, i.e., as $W^{k,p}(\Omega \smallsetminus \Omega_0; S)$].

The reader can assume that all the weighted spaces which will be used in this section are of this modified type.

(ii) A comparison with the definition of the space $W^{k,p}(\Omega;S)$ in Subsection 0.4 shows that in view of the form of the set $\mathbb{K}$, we have in mind a certain a n i s o t r o p i c version of the spaces considered. This change is more or less formal; we will assume some _anisotropy_ also for the differential operator, changing slightly the definition from (49.1).

49.4. __THE DIFFERENTIAL OPERATOR.__ Let $\mathbb{K}$ be the set of multiindices from Subsection 49.2 (i). We shall consider nonlinear partial differential operators of order $2k$ of the form

(49.7) $$(\mathcal{N}u)(x) = \sum_{\alpha \in \mathbb{K}} (-1)^{|\alpha|} D^\alpha a_\alpha(x; \delta_{\mathbb{K}} u(x))$$

where

(49.8) $$\delta_{\mathbb{K}} u(x) = \{D^\beta u(x); \beta \in \mathbb{K}\} \, .$$

We denote by m the number of elements of the set $\mathbb{K}$, i.e., the number of components of the vector function $\delta_{\mathbb{K}} u$ from (49.8), and assume that

$$a_\alpha = a_\alpha(x;\xi) \, , \quad \alpha \in \mathbb{K}$$

(the "coefficients" of the operator $\mathcal{N}$) are real functions defined on $\Omega \times \mathbb{R}^m$ - i.e., defined for a.e. $x \in \Omega$ and for all $\xi = \{\xi_\beta; \beta \in \mathbb{K}\} \in \mathbb{R}^m$ - which have the following properties :

(i) They satisfy the _Carathéodory condition_, i.e., they are m e a s u - r a b l e in x for every $\xi \in \mathbb{R}^m$ and c o n t i n u o u s in ξ for a.e.

$x \in \Omega$.

(ii) They satisfy the (*weighted*) *growth condition*

$$(49.9) \qquad |a_\alpha(x;\xi)| \leq w_\alpha^{1/p}(x)\left[g_\alpha(x) + c_\alpha \sum_{\beta \in \text{Ж}} |\xi_\beta|^{p-1} w_\beta^{1/q}(x)\right]$$

where $g_\alpha \in L^q(\Omega)$ with $q = p/(p-1)$ and $c_\alpha \geq 0$ are certain (given) functions and constants while w_α are elements of the (prescribed) collection S from (49.3).

The class of all such functions $a_\alpha = a_\alpha(x;\xi)$ will be denoted by

$$(49.10) \qquad CAR(p,S) .$$

Further, we introduce the form

$$(49.11) \qquad a(u,v) = \sum_{\alpha \in \text{Ж}} \int_\Omega a_\alpha\bigl(x;\ \delta_{\text{Ж}}\, u(x)\bigr) D^\alpha v(x)\ dx$$

associated with the differential operator $\mathcal{N}$; the form $a(u,v)$ is, in general, linear only with respect to the variable v .

49.5. THE DIRICHLET PROBLEM. Let Ж and S be respectively the set of multiindices and the corresponding collection of weight functions from Subsection 49.2. Let $\mathcal{N}$ be the (nonlinear) differential operator from (49.7) with coefficients $a_\alpha = a_\alpha(x;\xi) \in CAR(p,S)$. Finally, let a function $u_0 \in$ $\in W^{k,p}(\Omega;S)$ and a functional $F \in \left[W_0^{k,p}(\Omega;S)\right]^*$ be given.

We shall say that a function $\hat{u} = u + u_0 \in W^{k,p}(\Omega;S)$ is a *weak solution of the Dirichlet problem* for the operator $\mathcal{N}$ (with the "right hand side" F and "boundary data" u_0) if

$$(49.12) \qquad u = \hat{u} - u_0 \in W_0^{k,p}(\Omega;S)$$

and

$$(49.13) \qquad a(\hat{u},v) = <F,v> \quad \text{for every}\ v \in W_0^{k,p}(\Omega;S) .$$

49.6. REMARK. Using the definition of the form $a(u,v)$ — see (49.11) — and the fact that $\hat{u} = u + u_0$, we can rewrite (49.13) in the following manner

$$(49.14) \qquad \sum_{\alpha \in \text{Ж}} \int_\Omega a_\alpha\bigl(x;\ \delta_{\text{Ж}}\, u(x) + \delta_{\text{Ж}}\, u_0(x)\bigr) D^\alpha v(x)\ dx = <F,v> .$$

We will see later that — due to the fact that $a_\alpha \in CAR(p,S)$ — the expression $a(u,v)$ is meaningful for $u,\ v \in W^{k,p}(\Omega;S)$ (cf. § 50 below).

49.7. EXAMPLE. The operator

$$(\mathscr{N}u)(x) = \sum_{\alpha \in \mathbb{K}} (-1)^{|\alpha|} D^{\alpha} \Big[|D^{\alpha}u(x)|^{P-1} \text{ sgn } D^{\alpha}u(x) \, w_{\alpha}(x) \Big]$$

with $w_{\alpha} \in W(\Omega)$ is a typical representative of operators of the type (49.7).
Its "coefficients"

$$a_{\alpha}(x;\xi) = |\xi_{\alpha}|^{P-1} \text{ sgn } \xi_{\alpha} \, w_{\alpha}(x)$$

obviously belong to the class $CAR(p,S)$ from (49.10); in particular, the
growth condition (49.9) is satisfied with $g_{\alpha}(x) \equiv 0$, $c_{\alpha} = 1$. Analogously as
in the linear case, the coefficients $w_{\alpha}(x)$ express a certain degeneration or
singularity on those parts of $\overline{\Omega}$ on which w_{α} tend to zero or to infinity,
respectively.

49.8. REMARK. It is easily seen that the mapping $\Phi : L^P(\Omega) \to L^P(\Omega;w)$ defined by

$$(49.15) \qquad \Phi(u) = uw^{-1/p} , \quad w \in W(\Omega) ,$$

is an *isometric isomorphism* between the spaces considered and connects weighted
and nonweighted spaces. Using this mapping one can prove the following two
assertions about continuous linear functionals on $W^{k,p}(\Omega;S)$ and about Nemyckij operators on weighted spaces $L^P(\Omega;w)$, modifying in an obvious way the
proofs of the corresponding assertions for the nonweighted case (see, e.g.,
A. KUFNER, O. JOHN, S. FUČÍK [1], Sec. 5.9 for Lemma 49.9 below and S. FUČÍK,
A. KUFNER [1], for Lemma 49.10 below). The notation in these Lemmas (the set $\mathbb{K}$,
its "dimension" m , the collection S etc.) are taken from the foregoing Subsections.

49.9. LEMMA (continuous linear functionals). *Let F be a functional from the
dual space $[W^{k,p}(\Omega;S)]^*$. Then there exists an m-tuple*

$$(49.16) \qquad Q = \{f_{\alpha} \in L^q(\Omega;w_{\alpha}^{-1}); \, \alpha \in \mathbb{K}\} , \quad q = \frac{p}{p-1} ,$$

such that

$$(49.17) \qquad <F,v> = \sum_{\alpha \in \mathbb{K}} \int_{\Omega} f_{\alpha} \, D^{\alpha}v \, w_{\alpha}^{1/p - 1/q} \, dx$$

and

$$(49.18) \qquad \| F; [W^{k,p}(\Omega;S)]^* \| = \inf \Big\{ \sum_{\alpha \in \mathbb{K}} \| f_{\alpha}; L^q(\Omega;w_{\alpha}^{-1}) \|^q \Big\}^{1/q}$$

*where the infimum is taken over all m-tuples Q of the form (49.16) such that
the representation (49.17) takes place.*

49.10. LEMMA (Nemyckij operators). *Let $\Omega \subset \mathbb{R}^N$, $m \in \mathbb{N}$, $p > 1$. Let
$h(x;\xi)$ be a function defined for a.e. $x \in \Omega$ and all $\xi \in \mathbb{R}^m$ which satisfies*

the Carathéodory condition [see Subsection 49.4 (i)]. *Let* $H(u_1,\ldots,u_m)$ *be the so-called* N e m y c k i j   o p e r a t o r *generated by the function* h , *i.e.*

$$H(u_1,\ldots,u_m)(x) = h\bigl(x;\, u_1(x),\ldots u_m(x)\bigr) \ , \quad x \in \Omega \ .$$

Let w, $w_j \in W(\Omega)$, $j = 1,\ldots,m$.

If $(u_1,\ldots,u_m) \in \prod_{j=1}^{m} L^p(\Omega;w_j)$ *then*

$$H(u_1,\ldots,u_m) \in L^q(\Omega;w^{-1}) \ , \quad q = \frac{p}{p-1} \ ,$$

if and only if the following condition is fulfilled : There exist a function $g \in L^q(\Omega)$ *and a constant* $c \geq 0$ *such that for a.e.* $x \in \Omega$ *and all* $\xi \in \mathbf{R}^m$

$$(49.19) \qquad \bigl|h(x;\, \xi_1,\ldots,\xi_m)\bigr| \leq w^{1/q}(x)\Bigl[g(x) + c \sum_{j=1}^{m} |\xi_j|^{p-1} \, w_j^{1/q}(x)\Bigr] \ .$$

If condition (49.19) is fulfilled, then the Nemyckij operator H *is a continuous mapping from* $\prod_{j=1}^{m} L^p(\Omega;w_j)$ *into* $L^q(\Omega;w^{-1})$.

§ 50 .   T h e   m a i n   e x i s t e n c e   t h e o r e m

The main tool for deriving an assertion about the existence and uniqueness of a weak solution of the Dirichlet problem for the nonlinear operator $\mathcal{N}$ from (49.7) will be the *theory of monotone operators,* namely, the following Browder's theorem whose proof can be found, e.g., in J. L. LIONS [1] (Chap. 2, Theorem 2.1); cf. also S. FUČÍK, A. KUFNER [1], Theorem 29.5.

<u>50.1.</u> <u>THEOREM</u> (F. E. BROWDER). *Let* X *be a reflexive Banach space. Let* T *be an operator defined on* X *with values in the dual space* X* , *and let the following conditions be satisfied :*

(a)   T *is a   b o u n d e d   operator,* i.e., the image of any bounded subset of the space X is a bounded subset of the space X* ;

(b)   *the operator* T *is   d e m i c o n t i n u o u s ,* i.e., for arbitrary $u_0 \in X$ and any sequence $\{u_n\}_{n=1}^{\infty}$ of elements of the space X such that

$$u_n \to u_0 \quad in \ X$$

we have

$$Tu_n \rightharpoonup Tu_0 \quad in \ X^* \ (weakly);$$

(c)   *the operator* T *is   c o e r c i v e ,* i.e.,

$$(50.1) \qquad \lim_{\|u;X\| \to \infty} \frac{\langle Tu,u\rangle}{\|u;X\|} = \infty \ ;$$

(d) *the operator* T *is m o n o t o n e on the space* X , i.e., *for*
all u, v $\in$ X we have

(50.2)   < Tu - Tv, u - v > $\geq$ 0 .

Then the equation
(50.3)   Tu = f

has at least one solution u $\in$ X *for every* f $\in$ X* . *If, moreover, inequality*
(50.2) *is strict for all* u, v $\in$ X , u $\neq$ v , *then equation* (50.3) *has preci-*
sely one solution u $\in$ X *for every* f $\in$ X* .

The main theorem of Section 17 now reads as follows.

<u>50.2.</u> <u>THEOREM</u> (existence and uniqueness of a weak solution). *Let* Ω *be an*
open set in $\mathbb{R}^N$, p > 1 , $\mathbb{K}$ *and* S *respectively the set of multiindices*
and the collection of weight functions w_α *from Subsection* 49.2 (i), (ii),
$W^{k,p}(\Omega;S)$ *and* $W_0^{k,p}(\Omega;S)$ *the corresponding weighted Sobolev spaces from Sub-*
section 49.2 (iii). *Let* $\mathcal{N}$ *be the nonlinear differential operator of order*
2k *from* (49.7) *and let its coefficients* $a_\alpha = a_\alpha(x;\xi)$ *satisfy the following*
conditions :

(50.4) $a_\alpha \in CAR(p,S)$ *for* $\alpha \in \mathbb{K}$;

for a.e. x $\in \Omega$ *and all* $\xi, \eta \in \mathbb{R}^N$ *the inequalities*

(50.5) $\sum_{\alpha \in \mathbb{K}} \left[a_\alpha(x;\xi) - a_\alpha(x;\eta) \right] (\xi_\alpha - \eta_\alpha) \geq 0$,

(50.6) $\sum_{\alpha \in \mathbb{K}} a_\alpha(x;\xi) \xi_\alpha \geq c_1 \sum_{\alpha \in \mathbb{K}} |\xi_\alpha|^p w_\alpha(x)$

hold with a constant $c_1 > 0$.

Then there exists at least one weak solution $\hat{u} \in W^{k,p}(\Omega;S)$ *of the Di-*
richlet problem for the operator $\mathcal{N}$ (in the sense of Subsection 49.5).

If the inequality in (50.5) *is strict for* $\xi \neq \eta$, *then the weak solution*
$\hat{u}$ *is uniquely determined.*

<u>50.3.</u> <u>REMARK.</u> If we suppose that $\mathbb{K} = \{\alpha; |\alpha| \leq k\}$ and $w_\alpha(x) \equiv 1$ for all
$\alpha \in \mathbb{K}$, i.e., for all $|\alpha| \leq k$, then we obtain a special case of Theorem 50.2.
This special case represents the usual application of the theory of monotone
operators to the (weak) solution of boundary value problems in the c l a s -
s i c a l Sobolev spaces $W^{k,p}(\Omega)$. Our theorem is a slight extension of these
classical results, which can be found, e.g., in the above mentioned books J. L.
LIONS [1] and S. FUČÍK, A. KUFNER [1]. We want to show here what properties of
the coefficients $a_\alpha(x;\xi)$ - expressed here mainly by the inequalities (49.9),
(50.5)   and   (50.6)   -   allow to extend the existence results mentioned

to weighted spaces. Accordingly, we will call condition (50.5) the *monotonicity condition* and (50.6) the *(weighted) coercivity condition*.

50.4. EXAMPLE. The operator $\mathcal{N}$ from Example 49.7 with coefficients $a_\alpha(x;\xi) = |\xi_\alpha|^{p-1} \operatorname{sgn} \xi_\alpha \, w_\alpha(x)$ obviously fulfils conditions (50.5) and (50.6) (the latter with the equality sign and with $c_1 = 1$). Therefore, according to Theorem 50.2, there exists a uniquely determined weak solution of the Dirichlet problem for this model operator.

50.5. PROOF OF THEOREM 50.2. (i) Let us consider the form $a(u,v)$ from (49.11), associated with the differential operator $\mathcal{N}$ from (49.7), i.e.

$$(50.7) \qquad a(u,v) = \sum_{\alpha \in \mathbb{M}} \int_\Omega a_\alpha\big(x; \delta_{\mathbb{M}} u(x)\big) D^\alpha v(x) \, dx \ ,$$

and define functions h_α by the formula

$$(50.8) \qquad h_\alpha(x;\xi) = a_\alpha(x;\xi)\big[w_\alpha(x)\big]^{1/q-1/p} \ , \quad \alpha \in \mathbb{M} .$$

Since $a_\alpha \in \mathrm{CAR}(p,S)$ by (50.4), we conclude that h_α satisfies the Carathéodory condition [see Subsection 49.4 (i)] and, in view of the growth condition (49.9),

$$(50.9) \qquad |h_\alpha(x;\xi)| \leq w_\alpha^{1/q}(x)\big[g_\alpha(x) + c_\alpha \sum_{\beta \in \mathbb{M}} |\xi_\beta|^{p-1} w_\beta^{1/q}(x)\big] \ .$$

Lemma 49.10 implies − see (49.19) with $w = w_\alpha$ − that the operator $H_\alpha(u)(x) = h_\alpha\big(x; \{u_\beta(x); \beta \in \mathbb{M}\}\big)$ is a continuous Nemyckij operator from the product $\prod_{\beta \in \mathbb{M}} L^p(\Omega;w_\beta)$ into $L^q(\Omega;w_\alpha^{-1})$. Particularly, the function $f_\alpha(x) = h_\alpha\big(x; \delta_{\mathbb{M}} u(x)\big)$ belongs to $L^q(\Omega;w_\alpha^{-1})$ for $u \in W^{k,p}(\Omega;S)$.

Since

$$a(u,v) = \sum_{\alpha \in \mathbb{M}} \int_\Omega h_\alpha\big(x; \delta_{\mathbb{M}} u(x)\big) D^\alpha v(x) \, w_\alpha^{1/p-1/q}(x) \, dx$$

$$= \sum_{\alpha \in \mathbb{M}} \int_\Omega f_\alpha(x) \, D^\alpha v(x) \, w_\alpha^{1/p-1/q}(x) \, dx \ ,$$

we obtain from Lemma 49.9 that $a(u,v)$ is (for u f i x e d) the value of a continuous linear functional on $W^{k,p}(\Omega;S)$. We denote this functional by $\tilde{T}u$, since it depends on u , and so we have

$$a(u,v) = \langle\tilde{T}u,v\rangle \quad \text{for} \quad u, v \in W^{k,p}(\Omega;S) \ .$$

Since u was fixed but a r b i t r a r y , we have constructed an operator

$$(50.10) \qquad \tilde{T} : W^{k,p}(\Omega;S) \to \big[W^{k,p}(\Omega;S)\big]^* \ .$$

From (49.18) we have

$$
(50.11) \qquad \left\| \tilde{T}u; \, [W^{k,P}(\Omega;S)]^* \right\| \leq \left(\sum_{\alpha \in \mathbb{M}} \left\| f_\alpha; \, L^q(\Omega;w_\alpha^{-1}) \right\|^q \right)^{1/q}
$$

$$
\leq c_3 \left(1 + \left\| u; \, W^{k,P}(\Omega;S) \right\|^P \right)^{1/q}
$$

since inequality (50.9) implies — together with the inequality $(a_1 + \ldots + a_r)^q \leq r^{q-1}(a_1^q + \ldots + a_r^q)$ and with the fact that $q(p-1) = p$ — that

$$
\left\| f_\alpha; \, L^q(\Omega;w_\alpha^{-1}) \right\|^q = \int_\Omega \left| h_\alpha\big(x; \, \delta_{\mathbb{M}} u(x)\big) \right|^q w_\alpha^{-1}(x) \, dx
$$

$$
\leq \int_\Omega \left| w_\alpha^{1/q}(x) \Big[g_\alpha(x) + c_\alpha \sum_{\beta \in \mathbb{M}} |D^\beta u(x)|^{P-1} \, w_\beta^{1/q}(x) \Big] \right|^q w_\alpha^{-1}(x) \, dx
$$

$$
\leq (m+1)^{q-1} \left(\left\| g_\alpha; \, L^q(\Omega) \right\|^q + c_\alpha^q \sum_{\beta \in \mathbb{M}} \left\| D^\beta u(x); \, L^P(\Omega;w_\beta) \right\|^P \right)
$$

$$
\leq c_2 \left(1 + \left\| u; \, W^{k,P}(\Omega;S) \right\|^P \right)
$$

where c_2 is a fixed constant depending on the c_α's and on the L^q-norm of the functions g_α.

(ii) According to formula (49.13), to find a weak solution $\hat{u}$ of the Dirichlet problem means to find a function $u \in W_0^{k,P}(\Omega;S)$ such that $a(u + u_0, v) = \langle F,v \rangle$, i.e., $\langle \tilde{T}(u + u_0), v \rangle = \langle F,v \rangle$ for every $v \in W_0^{k,P}(\Omega;S)$. If we denote

$$
(50.12) \qquad Tu = \tilde{T}(u + u_0) \, ,
$$

then obviously T is an operator from the space $X = W_0^{k,P}(\Omega;S)$ into its dual space X^*. Consequently, our boundary value problem reduces to the problem of finding a function $u \in X$ such that $\langle Tu,v \rangle = \langle F,v \rangle$ for every $v \in X$, i.e., to the equation

$$
(50.13) \qquad Tu = F \quad \text{on} \quad X
$$

with a given $F \in X^*$.

If we verify that the operator T is bounded, demicontinuous, monotone and coercive, then the existence of a solution $u \in X$ of equation (50.13) — and consequently, also the existence of a weak solution of the Dirichlet problem — will be guaranteed by the Browder Theorem 50.1. Therefore, let us verify the assumptions of Theorem 50.1.

(ii) The *boundedness* of the operator $\tilde{T}$ follows immediately from formula (50.11).

(iii) The *demicontinuity* of the operator T is a consequence of the continuity of the Nemyckij operator $H_\alpha(u)$ from point (i) of this proof. Indeed, if $u_n \to u$ in X, then $\langle Tu_n - Tu, v \rangle \to 0$ for every $v \in X$ since by the Hölder inequality we have

$$\left| \langle Tu_n - Tu, v \rangle \right|$$

$$= \left| \sum_{\alpha \in \mathbb{K}} \int_\Omega \left[h_\alpha(x; \delta_{\mathbb{K}} u_n + \delta_{\mathbb{K}} u_0) - h_\alpha(x; \delta_{\mathbb{K}} u + \delta_{\mathbb{K}} u_0) \right] D^\alpha v \, w_\alpha^{1/p - 1/q} \, dx \right|$$

$$\leq \sum_{\alpha \in \mathbb{K}} \left\| h_\alpha(\cdot; \delta_{\mathbb{K}} u_n + \delta_{\mathbb{K}} u_0) - h_\alpha(\cdot; \delta_{\mathbb{K}} u + \delta_{\mathbb{K}} u_0); L^q(\Omega; w_\alpha^{-1}) \right\| \cdot$$

$$\cdot \left\| D^\alpha v; L^p(\Omega; w_\alpha) \right\|$$

and the first norms in the last sum tend to zero for $n \to \infty$.

(iv) The *monotonicity* of T follows from condition (50.5), where we take $\xi_\alpha = D^\alpha u(x) + D^\alpha u_0(x)$, $\eta_\alpha = D^\alpha v(x) + D^\alpha u_0(x)$ and then integrate the resulting inequality over Ω :

$$(50.14) \quad \begin{aligned} &\langle Tu - Tv, \, u - v \rangle \\ &= \sum_{\alpha \in \mathbb{K}} \int_\Omega \left[a_\alpha(x; \delta_{\mathbb{K}} u + \delta_{\mathbb{K}} u_0) - a_\alpha(x; \delta_{\mathbb{K}} v + \delta_{\mathbb{K}} u_0) \right] (D^\alpha u - D^\alpha v) \, dx \geq 0. \end{aligned}$$

(v) The *coercivity* of T follows from condition (50.6): If we take there $\xi = \delta_{\mathbb{K}} u(x) + \delta_{\mathbb{K}} u_0(x)$ and then integrate the resulting inequality over Ω , we obtain

$$\langle Tu, \, u + u_0 \rangle$$

$$= \sum_{\alpha \in \mathbb{K}} \int_\Omega a_\alpha(x; \delta_{\mathbb{K}} u + \delta_{\mathbb{K}} u_0)(D^\alpha u + D^\alpha u_0) \, dx$$

$$(50.15) \quad \geq c_1 \sum_{\alpha \in \mathbb{K}} \int_\Omega |D^\alpha u(x) + D^\alpha u_0(x)|^p \, w_\alpha(x) \, dx = c_1 \| u + u_0; \, W^{k,p}(\Omega; S) \|^p$$

$$\geq c_1 \left| \, \| u; \, W^{k,p}(\Omega; S) \| - \| u_0; \, W^{k,p}(\Omega; S) \| \, \right|^p$$

$$\geq c_1 \| u; \, W^{k,p}(\Omega; S) \|^p \left| 1 - \| u_0; \, W^{k,p}(\Omega; S) \| / \| u; \, W^{k,p}(\Omega; S) \| \right|^p .$$

Further, from (50.11) we have

$$\left| \langle Tu, u_0 \rangle \right| \leq \| Tu; \, [W^{k,p}(\Omega; S)]^* \| \, \| u_0; \, W^{k,p}(\Omega; S) \|$$

$$\leq c_3 \left(1 + \| u + u_0; \, W^{k,p}(\Omega; S) \|^p \right)^{1/q} \| u_0; \, W^{k,p}(\Omega; S) \|$$

$$(50.16) \quad \leq c_4 \left(1 + \| u; \, W^{k,p}(\Omega; S) \|^{p-1} + \| u_0; \, W^{k,p}(\Omega; S) \|^{p-1} \right) \| u_0; \, W^{k,p}(\Omega; S) \|$$

$$= c_5 + c_6 \| u; \, W^{k,p}(\Omega; S) \|^{p-1}$$

(note that u_0 is given and $q = p/(p-1)$). Formulas (50.15) and (50.16) yield

$$\langle Tu, u \rangle = \langle Tu, \, u + u_0 \rangle - \langle Tu, u_0 \rangle \geq c_1 \| u; \, W^{k,p}(\Omega; S) \|^p \left| 1 - \frac{\| u_0; W^{k,p}(\Omega; S) \|}{\| u; \, W^{k,p}(\Omega; S) \|} \right|^p -$$

$$- \left(c_5 + c_6 \| u; \ W^{k,p}(\Omega;S) \|^{p-1} \right)$$

$$= \| u; \ W^{k,p}(\Omega;S) \|^p \left[c_1 \left| 1 - \frac{\| u_0; W^{k,p}(\Omega;S) \|}{\| u; \ W^{k,p}(\Omega;S) \|} \right|^p - \frac{c_5}{\| u; W^{k,p}(\Omega;S) \|^p} - \frac{c_6}{\| u; W^{k,p}(\Omega;S) \|} \right]$$

and consequently,

$$\frac{<Tu,u>}{\| u; W^{k,p}(\Omega;S) \|} \to \infty \quad \text{for} \quad \| u; \ W^{k,p}(\Omega;S) \| \to \infty \ .$$

So, the *existence* of at least one weak solution of the Dirichlet problem is proved. The *uniqueness* follows by contradiction if we assume that the inequality in (50.5) is strict : Analogously as in (50.14), for t w o solutions u , u^* we obtain the inequality $<Tu - Tu^*, \ u - u^*> \ > \ 0$ while $Tu = Tu^* = F$.

<u>50.6. REMARKS.</u> There are many possibilities how to generalize the foregoing results by weakening the assumptions on the operator $\mathscr{N}$ and using stronger tools. Let us mention some of these possibilities.

(i) The Browder Theorem 50.1 is one of the simpler tools of the theory of· monotone operators. If we use some deeper results (the so-called Leray-Lions Theorem, the concept of pseudomonotonicity etc. - see, e.g., J. L. LIONS [1] or S. FUČÍK, A. KUFNER [1]), we can generalize conditions (49.9), (50.5) and (50.6) in a direction usual if classical Sobolev space are considered (the so-called *monotonicity of the main part* of the operator $\mathscr{N}$ together with some *compact* imbeddings for the weighted spaces involved etc.).

(ii) Conditions (49.9), (50.5) and (50.6) are of a certain "algebraic" form and can be sometimes replaced by conditions involving directly the (weighted) spaces and, consequently, having a certain less restrictive "integral" form - analogously as in the linear case, where we used the ellipticity condition in an "algebraic" form - see (46.1) - as well as in an "integral" one - see (45.14).

(iii) In this section, we have made no assumptions about the domain Ω and about the weights w_α [except the requirement $w_\alpha \in W(\Omega)$]. On the other hand, we can weaken our assumptions provided we have more information about the weighted spaces considered [i.e., about the weight functions w_α : estimates of the type of imbedding theorems as (45.11) and (45.15) in the linear case, etc.]. Let us mention two of such generalizations :

(iii-1) If there is a subset $\mathbb{K}_1 \subset \mathbb{K}$ such that $\| u; \ W^{k,p}(\Omega;S) \|$
$\leq c_0 \left(\sum_{\alpha \in \mathbb{K}_1} \| D^\alpha u; \ L^p(\Omega;w_\alpha) \|^p \right)^{1/p}$ for every $u \in W_0^{k,p}(\Omega;S)$, then we can modify the coercivity condition (50.6) summing only over $\mathbb{K}_1$ (instead of $\mathbb{K}$) on the right hand side of (50.6).

(iii-2) If the set Ω has a *finite measure*, then (50.6) can be replaced by

$$\sum_{\alpha \in \text{Ж}} a_\alpha(x;\xi)\xi_\alpha \geq c_1 \sum_{\alpha \in \text{Ж}_1} |\xi_\alpha|^P w_\alpha(x) - c_2$$

with $c_1 > 0$, $c_2 \geq 0$.

(iv) Obviously, other boundary value problems can be handled similarly as the Dirichlet problem.

Section 18 . *E l l i p t i c   b o u n d a r y   v a l u e   p r o b l e m s*

In this Section, we want to show shortly how the approach from [I] and rom Chapter IV can be extended to the nonlinear case.

§ 51 .   F o r m u l a t i o n   a n d   s o m e   e x i s t e n c e
        r e s u l t s

51.1. THE "CLASSICAL" APPROACH. Let us again consider the nonlinear differential operator $\mathcal{N}$ from (49.1), i.e.

(51.1) $$(\mathcal{N}u)(x) = \sum_{|\alpha|\leq k} (-1)^{|\alpha|} D^\alpha a_\alpha\big(x;\, \delta_k u(x)\big)$$

with

(51.2) $$\delta_k u(x) = \{D^\beta u(x);\ |\beta| \leq k\} .$$

Further, let us consider the c l a s s i c a l   (= nonweighted) Sobolev space $W^{k,P}(\Omega)$ and its subspace $W_0^{k,P}(\Omega)$.

A function $\hat{u} \in W^{k,P}(\Omega)$ is called a *weak solution of the Dirichlet prob-*
lem for the operator $\mathcal{N}$ - with a given "right hand side" $F \in \big[W_0^{k,P}(\Omega)\big]^*$
and with given "boundary data" $u_0 \in W^{k,P}(\Omega)$ - if

(51.3) $$\hat{u} - u_0 \in W_0^{k,P}(\Omega)$$

and

51.4) $$a(\hat{u},v) = <F,v> \quad \text{for every} \quad v \in W_0^{k,P}(\Omega) ,$$

where

(51.5) $$a(u,v) = \sum_{|\alpha|\leq k} \int_\Omega a_\alpha\big(x;\, \delta_k u(x)\big) D^\alpha v(x)\, dx .$$

This formulation is, in fact, the formulation of Subsection 49.5, where we choose for Ж the set $\{\alpha;\ |\alpha| \leq k\}$ and for the weight functions the "tri-
vial weights" $w_\alpha(x) \equiv 1$, $|\alpha| \leq k$. According to the results of Section 17,

the existence of a weak solution is guaranteed if the coefficients $a_\alpha(x;\xi)$ satisfy

(a)   the Carathéodory condition on $\Omega \times \mathbf{R}^\kappa$ [see Subsection 49.4 (i)] and further,

(b)   the growth condition, monotonicity condition and coercivity condition, respectively :

$$(51.6) \qquad |a_\alpha(x;\xi)| \le g_\alpha(x) + c_\alpha \sum_{|\alpha| \le k} |\xi_\beta|^{p-1} \; , \quad |\alpha| \le k$$

with $g_\alpha \in L^q(\Omega)$, $q = p/(p-1)$, and $c_\alpha \ge 0$;

$$(51.7) \qquad \sum_{|\alpha| \le k} \big(a_\alpha(x;\xi) - a_\alpha(x;\eta) \big)(\xi_\alpha - \eta_\alpha) \ge 0 \; ;$$

$$(51.8) \qquad \sum_{|\alpha| \le k} a_\alpha(x;\xi)\xi_\alpha \ge c_0 \sum_{|\alpha| \le k} |\xi_\alpha|^p$$

with $c_\alpha \ge 0$. Here $x \in \Omega$ and $\xi, \eta \in \mathbf{R}^\kappa$.

The classical approach consists in the transformation of the boundary value to an *operator equation*

$$(51.9) \qquad Tu = F$$

with $T : V \to V^*$ [$V = W_0^{k,p}(\Omega)$ for the case of the Dirichlet problem], and in the solution of this operator equation with help of the theory of monotone operators [for the construction of the operator T see Subsection 50.5 (ii)].

<u>51.2.   THE "WEIGHTED" APPROACH.</u>   Our aim is to extend the method described in Subsection 51.1 to the case of a w e i g h t e d Sobolev space $W^{k,p}(\Omega;S)$. We ask whether it is possible to obtain assertions about the *existence* of a weak solution of the Dirichlet problem in the space $W^{k,p}(\Omega;S)$ *preserving*, roughly speaking, *the fundamental assertions* (a), (b) from Subsection 51.1, which concern the differential operator $\mathcal{N}$ [i.e., taking into account e l l i p t i c operators in the sense of (51.6) – (51.8)], and changing in an appropriate manner only the assumptions about the data F and u_0 .

This is an extension of the approach described in Chapter IV for the linear differential operator $\mathcal{L}$ to the nonlinear differential operator $\mathcal{N}$. We are interested in the answer to the question *for w h a t weight functions this approach is possible*.

For simplicity, let us consider a s p e c i a l c o l l e c t i o n S: $w_\alpha(x) = w(x)$ for *all* $|\alpha| \le k$ with $w \in W(\Omega)$, i.e., $S = \{w,w,\ldots,w\}$. We denote the corresponding weighted space $W^{k,p}(\Omega;S)$ by

$$(51.10) \qquad W^{k,p}(\Omega;w) \; ,$$

and we are now able to prove the following important assertion about the operator T from (51.9):

(i) *Let the coefficients* $a_\alpha(x;\xi)$ *of the differential operator* $\mathcal{N}$ *from (51.1) satisfy the Carathéodory condition and the growth condition (51.6) with* $g_\alpha \in L^q(\Omega;w)$. *Let* $u_0 \in W^{k,p}(\Omega;w)$. *Then*

$$(51.11) \qquad a(u + u_0, v) = <Tu,v>,$$

where T *is an operator from* $V_1 = W_0^{k,p}(\Omega;w)$ *into* $V_2^* = [W_0^{k,p}(\Omega;\ w^{1-p})]^*$. *The right hand side in (51.11) expresses the duality on the space* V_2 , *and (51.11) takes place for all* $u \in V_1$, $v \in V_2$.

P r o o f : It follows from (51.6) that

$$|a(u,v)| \leq \sum_{|\alpha|\leq 1} \int_\Omega |a_\alpha(x;\ \delta_k u(x))| \ |D^\alpha v(x)| \ dx$$

$$\leq \sum_{|\alpha|\leq 1} \left(\int_\Omega |g_\alpha(x)| \ |D^\alpha v(x)| dx + c_\alpha \sum_{|\beta|\leq k} \int_\Omega |D^\beta u(x)|^{p-1} \ |D^\alpha v(x)| \ dx \right) .$$

Using the Hölder inequality with $p > 1$, $q = \dfrac{p}{p-1}$, we obtain with respect to the fact that $p/q = p - 1$ and $q(p - 1) = p$

$$\int_\Omega |g_\alpha| \ |D^\alpha v| \ dx = \int_\Omega |g_\alpha| w^{1/q} \ |D^\alpha v| \ w^{-1/q} \ dx$$

$$\leq \left(\int_\Omega |g_\alpha|^q \ w \ dx \right)^{1/q} \left(\int_\Omega |D^\alpha v|^p \ w^{-p/q} \ dx \right)^{1/p}$$

$$= \|g_\alpha;\ L^q(\Omega;w)\| \cdot \|D^\alpha v;\ L^p(\Omega;w^{1-p})\| \ ;$$

$$\int_\Omega |D^\beta u|^{p-1} \ |D^\alpha v| \ dx = \int_\Omega |D^\beta u|^{p-1} \ w^{1/q} \ |D^\alpha v| \ w^{-1/q} \ dx$$

$$\leq \left(\int_\Omega |D^\beta u|^{q(p-1)} \ w \ dx \right)^{1/q} \left(\int_\Omega |D^\alpha v|^p \ w^{-p/q} \ dx \right)^{1/p}$$

$$= \|D^\beta u;\ L^p(\Omega;w)\|^{p/q} \|D^\alpha u;\ L^p(\Omega;w^{1-p})\|$$

and consequently,

$$|a(u,v)| \leq \left(c_3 + c_4 \|u;\ W^{k,p}(\Omega;w)\|^{p-1} \right) \|v;\ W^{k,p}(\Omega;w^{1-p})\| \ .$$

This means that for every fixed $u \in W^{k,p}(\Omega;w)$ $a(u,v)$ is a continuous linear functional on $W^{k,p}(\Omega;w^{1-p})$, and assertion (i) follows as in point (i) of Subsection 50.5.

51.3. <u>REMARK.</u> Now, we are able to formulate the Dirichlet problem (in a weak sense) in the weighted space $W^{k,p}(\Omega;w)$. The formulation is left to the reader; we can almost word by word repeat the definition from Subsection 51.1, seeking the weak solution $\hat{u}$ in $W^{k,p}(\Omega;w)$ and replacing the classical Sobolev spaces $W^{k,p}(\Omega)$ and $W_0^{k,p}(\Omega)$ in formulas (51.3) and (51.4) by the weighted spaces $W_0^{k,p}(\Omega;w)$ and $W^{k,p}(\Omega;w^{1-p})$, respectively. Naturally, we have to assume that $u_0 \in W^{k,p}(\Omega;w)$ and $F \in [W_0^{k,p}(\Omega;w^{1-p})]^*$.

Analogously as in Subsection 51.1, the Dirichlet problem can be transformed to the *operator equation*

$$Tu = F$$

but this time for

$$T : V_1 \to V_2^*$$

with *t w o   d i f f e r e n t* Banach spaces

$$V_1 = W_0^{k,p}(\Omega;w) \quad , \quad V_2 = W_0^{k,p}(\Omega;w^{1-p}) \quad .$$

(These two spaces coincide if $w(x) \equiv 1$, i.e., in the nonweighted case.)

The situation is similar to the   l i n e a r   case where we investigated the   b i l i n e a r   form on the product $H_1 \times H_2$ of two different Hilbert spaces. But while in that case we were able to solve our problems with help of Lemma 37.1, which was an extension of the Lax–Milgram Theorem 39.5, here we have at our disposal *no* appropriate extension of the theory of monotone operators to the case of operators T acting from one space V_1 into   a n – o t h e r , V_2^* .

Nevertheless, J. VOLDŘICH succeeded in obtaining some existence results at least for the case of power type weights, i.e., for spaces $W^{k,p}(\Omega;w)$ with

$$w(x) = [\mathrm{dist}(x,M)]^\varepsilon = d_M^\varepsilon(x) \quad , \quad M \subset \partial\Omega \quad ,$$

which we have denoted by

$$W^{k,p}(\Omega;d_M,\varepsilon) \quad .$$

Since $w^{1-p} = d_M^{\varepsilon(1-p)} = d_M^{-\varepsilon(p-1)}$, we have in this case

$$V_2 = W_0^{k,p}\big(\Omega;\ d_M,-\varepsilon(p-1)\big)$$

[and, naturally, $V_1 = W_0^{k,p}(\Omega;d_M,\varepsilon)$].

We cannot describe here J. VOLDŘICH's rather sophisticated method; he was able to find a certain   r e f i n e m e n t   of the *method of pseudomonotone operators,* which turned out to be a suitable tool for deriving an existence theorem in this particular case. In fact, he considered only second order equations ($k = 1$) and the case $M = \partial\Omega$, but his results can be extended to higher order equations and to more general sets M . Let us now formulate his results without proofs; all details can be found in J. VOLDŘICH [1], [3].

<u>51.4.</u> <u>THE DIRICHLET PROBLEM.</u> Let us consider the (formal) Dirichlet problem

$$(51.12) \qquad - \sum_{i=1}^{N} \frac{\partial}{\partial x_i} a_i(x;u,\nabla u) + a_0(x;u,\nabla u) = F \quad \text{in } \Omega \ , \quad u = u_0 \quad \text{on } \partial\Omega$$

where Ω is a bounded domain in $\mathbf{R}^N$ with Lipschitzian boundary $\partial\Omega$. Suppose that $u_0 \in W^{1,P}(\Omega;d_M,\varepsilon)$ and $F \in \left[W_0^{1,P}(\Omega; d_M, -\varepsilon(p-1))\right]^*$ with $p > 1$, $M = \partial\Omega$ and $\varepsilon \in \mathbf{R}$.

A function $\hat{u} \in W^{1,2}(\Omega;d_M,\varepsilon)$ is called a *weak solution of the Dirichlet problem* (51.12) if

$$\hat{u} - u_0 \in W_0^{1,P}(\Omega;d_M,\varepsilon)$$

and

$$\sum_{i=1}^{N} \int_\Omega a_i(x;\hat{u},\nabla\hat{u}) \frac{\partial v}{\partial x_i} \ dx + \int_\Omega a_0(x;\hat{u},\nabla\hat{u}) \ v \ dx = <F,v>$$

for every $v \in W_0^{1,P}(\Omega;d_M,-\varepsilon(p-1))$.

<u>51.5.</u> <u>THEOREM</u> (existence of a weak solution). *Let the functions* $a_i(x;\xi_0,\xi)$ *with* $\xi = (\xi_1,\ldots,\xi_N)$, $i = 0,1,\ldots,N$, *be defined on* $\Omega \times \mathbf{R} \times \mathbf{R}^N$ *and satisfy the Carathéodory condition as well as the following inequalities :*

$$(51.13) \qquad |a_i(x;\xi_0,\xi)| \leq g_i(x) + c_i\left(|\xi_0|^{P-1} + |\xi|^{P-1}\right) \ , \quad i = 0,1,\ldots,N \ ,$$

with $g_i \in L^q(\Omega;d_M,\varepsilon)$, $p > 1$, $q = p/(p-1)$, $c_i > 0$;

$$(51.14) \qquad \sum_{i=1}^{N} \left(a_i(x;\xi_0,\xi) - a_i(x;\xi_0,\xi)\right)(\xi_i - \eta_i) > 0$$

for a.e. $x \in \Omega$ *and all* $\xi_0 \in \mathbf{R}$, $\xi, \eta \in \mathbf{R}^N$, $\xi \neq \eta$;

$$(51.15) \qquad \sum_{i=1}^{N} a_i(x;\xi_0,\xi)\xi_i + a_0(x;\xi_0,\xi)\xi_0 \geq c_0|\xi|^P - \tilde{c}_0|\xi_0|^{P-\delta} - s(x)$$

with $s \in L^1(\Omega;d_M,\varepsilon)$, $c_0, \tilde{c}_0 > 0$, $\delta \in (0,p-1)$.

Then there is an open interval I *containing the origin and such that for* $\varepsilon \in I$ *the Dirichlet problem* (51.12) *has at least one weak solution* $\hat{u} \in W^{1,P}(\Omega;d_M,\varepsilon)$ *provided* $u_0 \in W^{1,P}(\Omega;d_M,\varepsilon)$ *and* $F \in \left[W_0^{1,P}(\Omega;d_M,-\varepsilon(p-1))\right]^*$.

<u>51.6.</u> <u>REMARK.</u> The assumptions (51.13) – (51.15) are a little more general than the assumptions (51.6) – (51.8). Further generalization is possible if we consider the homogeneous Dirichlet problem

$$(51.16) \qquad - \sum_{i=1}^{N} \frac{\partial}{\partial x_i} b_i(x;u,\nabla u) + b_0(x;u,\nabla u) = F \quad \text{in } \Omega \ , \quad u = 0 \quad \text{on } \partial\Omega \ .$$

Obviously, this problem is equivalent to the nonhomogeneous Dirichlet problem (51.12) : it suffices to take $b_i(x;\xi_0,\xi) = a_i\big(x;\ \xi_0 + u_0(x),\ \xi + \nabla u(x_0)\big)$.

In view of the homogeneous boundary condition, we look for a weak solution of (51.16) - i.e., for a function u such that

$$a(u + u_0,\ v) = \langle F,v \rangle \quad \text{for every} \quad v \in W_0^{1,P}\big(\Omega;d_M,-\varepsilon(p-1)\big)$$

- in the space $W_0^{1,P}(\Omega;d_M,\varepsilon)$, and we can derive an existence theorem completely analogous to Theorem 51.5 assuming that, instead of (51.13) - (51.15), the following conditions are satisfied :

(i) There are positive constants c and γ and a positive function $g \in L^q(\Omega;d_M,\varepsilon)$ such that

$$|b_0(x;\xi_0,\xi)| \le g(x)d_M^{-1}(x) + c\big(|\xi_0|^{p-1}\ d_M^{-P}(x) + |\xi|^{p-1}\ d_M^{-1}(x)\big)\ ,$$

$$|b_i(x;\xi_0,\xi)| \le g(x) + c\big(|\xi_0|^{p-1}\ d_M^{-(p-1)+\gamma}(x) + |\xi|^{p-1}\big)$$

for a.e. $x \in \Omega$ and all $(\xi_0,\xi) \in \mathbf{R}^{N+1}$.

(ii) There are positive functions $s \in L^1(\Omega;d_M,\varepsilon)$ and $\tilde{c}(\omega)$ $(\omega > 0)$ and a positive constant c_1 such that

$$\sum_{i=1}^{N} b_i(x;\xi_0,\xi)\xi_i + b_0(x;\xi_0,\xi)\xi_0 \ge c_1|\xi|^P - \omega|\xi_0|^P\ d_M^{-P}(x) - \tilde{c}(\omega)s(x)$$

for a.e. $x \in \Omega$ and all $(\xi_0,\xi) \in R^{N+1}$.

(iii) Inequality (51.14) holds with b_i instead of a_i .

51.7. __REMARKS.__ (i) The result stated in Theorem 51.5 is similar to that of [I] or of Chapter IV (see, e.g., Theorem 37.11) : Again, the existence of a certain interval is asserted such that for $\varepsilon \in I$, the Dirichlet problem is weakly solvable in $W^{1,P}(\Omega;d_M,\varepsilon)$. Moreover, the weaker condition for b_i - Subsection 51.6 (i) - (iii) - reminds as of the weaker conditions (35.5*) (see the footnote on p. 142) thanks to the factors $d_M^{-\lambda}$.

(ii) The just described "weighted" approach contains the classical approach described in Subsection 51.1 as a special case : Indeed, we obtain the classical approach for $\varepsilon = 0$, where $W^{k,P}(\Omega;d_M,0) = W^{k,P}(\Omega)$; the choice $\varepsilon = 0$ is admissible since $0 \in I$.

(iii) For the l i n e a r differential operator we have

$$a_\alpha(x;\xi) = \sum_{|\beta|\le k} a_{\alpha\beta}(x)\xi_\beta\ ;$$

this indicates that for $a_{\alpha\beta} \in L^\infty(\Omega)$ the growth conditions (51.6) are satisfied with $p = 2$. In this case, we have $V_1 = W_0^{k,2}(\Omega;d_M,\varepsilon)$ and $V_2 =$

$= W_0^{k,2}(\Omega; d_M, -\varepsilon)$ for the spaces mentioned in Subsection 51.2 (i). Consequently, the "linear" approach from [I] (and from Chapter IV) is a special case of our "nonlinear" approach.

Theorem 51.5 states only the *existence* of an interval I of admissible values, but for practical purposes it is necessary to know the interval I at least approximately. Let us give an example.

51.8. <u>EXAMPLE.</u> Let us consider the Dirichlet problem

$$- \sum_{i=1}^{N} \frac{\partial}{\partial x_i}\left(|\nabla u|^{p-2} \frac{\partial u}{\partial x_i}\right) + |u|^{p-2} u = F \quad \text{in } \Omega , \quad u = 0 \quad \text{on } \partial\Omega$$

with $p > 1$ (for $1 < p < 2$, we define $|0|^{p-2} \cdot 0 = 0$). This is a problem of the type (51.12), since we have

$$a_i(x; \xi_0, \xi) = |\xi|^{p-2} \xi_i \quad \text{for } i = 1, \ldots, N , \quad a_0(x; \xi_0, \xi) = |\xi_0|^{p-2} \xi_0 .$$

It can be shown that for this operator, the interval of admissible values of ε is given by the formula

$$(51.17) \qquad\qquad I = \left(\frac{-p + 1}{cp - 1} , \frac{p - 1}{cp + 1}\right)$$

where $c = c(\Omega)$, $c = 1$ for convex domains.

51.9. <u>REMARK.</u> If we take $p = 2$ in Example 51.8, we obtain the l i n e a r Dirichlet problem for the operator $\mathcal{L}u = -\Delta u + u$. For a convex domain Ω , formula (51.17) yields $I = (-1, 1/3)$; this estimate can be improved to $I = (-1, 1)$. On the other hand, if we proceed by the methods described in [I] and in Chapter IV and use Lemma 37.1, we obtain the less satisfactory result $I = (-1/3, 1/3)$ (cf., e.g., Example 38.4). This shows that the "nonlinear" approach gives (sometimes) better estimates for I than the "linear" one.

51.10. <u>UNIQUENESS.</u> In Theorem 51.5 we stated only the e x i s t e n c e of a weak solution, although the inequality sign in the monotonicity condition (51.14) was strict. J. VOLDŘICH [1] has shown that the weak solution $u \in W^{1,p}(\Omega; d_M, \varepsilon)$ is uniquely determined provided $\varepsilon \in I \cap (-\infty, 0)$, while in the case of positive ε's , the problem to find reasonable conditions of the uniqueness is still open.

<u>R E F E R E N C E S</u>

[I] KUFNER, A. : *Weighted Sobolev spaces*. BSB B.G. Teubner Verlagsgesell-
     schaft, Leipzig 1980 (first edition), J.Wiley & Sons, Chichester – New
     York – Brisbane – Toronto – Singapore 1985 (second edition). MR <u>84e</u>:46029

AGMON, S.; DOUGLIS, A.; NIRENBERG, L. :
     [1] *Estimates near the boundary for solutions of elliptic partial diffe-*
         rential equations satisfying general boundary conditions. I. Comm.
         Pure Appl. Mat. <u>12</u>(1959), 623–727. MR <u>23</u> # A 2610.

AUBIN, J. P. :
     [1] *Behavior of the error of the approximate solutions of boundary*
         value problems for linear elliptic operators by Galerkin's and
         finite difference methods. Ann. Scuola Norm. Sup. Pisa (3)<u>21</u> (1967),
         599–637. MR <u>38</u> # 1391.

AGRANOVIČ, M. S.; VIŠIK, M. I. :
     [1] *Elliptic problems with a parameter and parabolic problems of general*
         type. Uspehi Mat. Nauk <u>19</u>(1964), no. 3 (117), 53–161 (Russian). MR
         <u>33</u> # 415.

BABUŠKA, I.; MEJZLÍK, L.; VITÁSEK, E. :
     [1] *Effects of artifical cooling of concrete in a dam during its harde-*
         ning. In: VII Congrès des Grands Barrages, Rome 1961. 1–13.

BACHMAN, G.; NARIČI, L. :
     [1] *Functional analysis*. Academic Press, New York–London 1963.

BEREZANSKIĬ, J. M.; ROĬTBERG, J. A. :
     [1] *A theorem on homeomorphismus and Green's function for general el-*
         liptic boundary problems. Ukrain. Mat. Ž. <u>19</u>(1967), no. 5, 3–32
         (Russian). MR <u>36</u> # 1823.

BLUM, H. :
     [1] *Der Einfluss von Eckensingularitäten bei der numerischen Behandlung*
         der biharmonischen Gleichung. Thesis (Inaugural-Dissertation), Bonn
         1981.

     [2] *Zur Gitterverfeinerung für quasilineare Probleme auf Eckengebieten*.
         Z. Angew. Math. Mech. <u>64</u>(1984), no. 5, 264–266.

BLUM, H.; DOBROWOLSKI, M. :
     [1] *A finite element method for elliptic equations in domains with cor-*
         ners. Manuscript, Univ. of Bonn, 1980.

BLUM, H.; RANNACHER, R.:
     [1] *On the boundary value problem of the biharmonic operator on domains*
         with angular corners. Math. Methods Appl. Sci. <u>2</u>(1980), no. 4,
         566–581. MR <u>82a</u>:35022.

CARLEMAN, T. :
     [1] *Über das Neumann-Poincarésche Problem für ein Gebiet mit Ecken*.
         Thesis, Uppsala 1916.

CIARLET, P. G. :
  [1] *The finite element method for elliptic problems.* North-Holland,
      Amsterdam 1978. MR 58 # 25001 (Russian translation: Mir, Moscow
      1980. MR 82c:65068).

CIARLET, P. G.; SCHULTZ, M. H.; VARGA, R. S. :
  [1] *Numerical methods of high-order accuracy for nonlinear boundary
      value problems. V. Monotone operator theory.* Numer. Math. 13(1969),
      51-77. MR 40 # 3730.

CIARLET, P. G.; RAVIART, P. A. :
  [1] *General Lagrange and Hermite interpolations in R^n with applica-
      tions to finite element methods.* Arch. Rational Mech. Anal. 46(1972),
      177-199. MR 49 # 1730.

  [2] *Interpolation theory over curved elements, with applications to
      finite element methods.* Comput. Methods Appl. Mech. Engrg. 1(1972),
      217-249. MR 51 # 11991.

DOBROWOLSKI, M. :
  [1] *Numerical approximation of elliptic interface and corner problems.*
      Thesis (Habilitationsschrift), Bonn 1981.

  [2] *Nichtlineare Eckenprobleme und finite Elemente Methode.* Z. Angew.
      Math. Mech. 64(1984), no. 5, 270-271. MR 86d:73023.

DUNFORD, N.; SCHWARTZ, J. T. :
  [1] *Linear operators.* Part II. Interscience, New York 1963.

EDMUNDS, D. E.; KUFNER, A.; RÁKOSNÍK, J. :
  [1] *Embeddings of Sobolev spaces with weights of power type.* Z. Anal.
      Anwendungen 4(1985), no. 1, 25-34. MR 86m:46032a.

FIX, G. J.; GULATI, S.; WAKOFF, G. I. :
  [1] *On the use of singular functions with finite element approximations.*
      J. Computational Phys. 13(1973), 209-228. MR 50 # 9010.

FUČÍK, S.; KUFNER, A. :
  [1] *Nonlinear differential equations.* Elsevier Scient. Publ. Comp.,
      Amsterdam - Oxford - New York 1980. MR 81e:35001.

GOCHBERG, I. I.; SIGAL, E. I. :
  [1] *An operator generalization of the logarithmic residue theorem and
      Rouché's theorem.* Mat. Sb. (N.S.) 84(126) (1971), 607-629 (Russian).
      MR 47 # 2409.

GRISVARD, P. :
  [1] *E.D.F. Bulletin de la Direction des études et recherches.* Série C,
      No. 1, 1986, 21-59.

  [2] *Behavior of the solutions of an elliptic boundary value problem in
      a polygonal or polyhedral domain.* Symposium on Numerical Solutions
      of Partial Differential Equations III. B. Hubbard Ed. Academic Press
      1975, 207-274.

  [3] *Elliptic problems in nonsmooth domains.* Pitman, Boston - London -
      Melbourne 1985.

HARDY, G. H.; LITTLEWOOD, J. E.; PÓLYA, G. :
  [1] *Inequalities.* University Press, Cambridge 1952. MR 13 # 727.

KOMEČ, A. I. :
  [1] *Elliptic boundary value problems on manifolds with piecewise smooth
      boundary.* Mat. Sb. (N.S.) 92(134) (1973), 89-134 (Russian). MR 49
      # 9779.

KONDRAT'EV, V. A. :
 [1] *Boundary value problems for elliptic equations on domains with co-
     nical or angular points*. Trudy Moskov. Mat. Obshch. 16(1967), 209-
     -292 (Russian).

 [2] *Asymptotic of solution of the Navier-Stokes equation near the a gu-
     lar point of the boundary*. Prikl. Mat. Meh. 31(1967), 119-123
     (Russian). MR 36 # 1816.

 [3] *The smoothness of the solution of the Dirichlet problem for second
     order elliptic equations in a piecewise smooth domain*. Differencial'-
     nye uravnenija 6(1970), 1831-1843 (Russian). MR 43 #·7766.

 [4] *Singularities of the solution of the Dirichlet problem for a second
     order elliptic equation in the neighbourhood of an edge*. Differen-
     cial'nye uravnenija 13(1977), no. 11, 2026-2032, (Russian). MR 58
     # 6668.

KONDRAT'EV, V. A.; OLEINIK, O. A. :
 [1] *Boundary value problems for partial differential equations in non-
     smooth domains*. Uspehi mat. nauk 38(1983), no. 2(230), 3-76 (Rus-
     sian). MR 85j:35002.

KREJN, S. G. :
 [1] *Linear differential equations in Banach space*. Nauka, Moscow 1967
     (Russian). MR 40 # 508 (English translation: American Math. Society,
     Providence, R. I., 1971. MR 49 # 7548).

KREJN, S. G.; TROFIMOV, V. P. :
 [1] *Holomorphic operator-valued functions of several complex variables*.
     Functional Anal. i Priložen. 3(1969), no. 4, 85-86, (Russian). MR
     41 # 7466.

KUFNER, A. :
 [1] *Boundary value problems in weighted spaces*. EQUADIFF 6 (Proceedings
     of a conference), Lecture Notes in Math. no. 1192, 35-48. Springer-
     -Verlag, Berlin-Heidelberg-New York-Tokyo 1986.

KUFNER, A.; JOHN, O.; FUČÍK, S. :
 [1] *Function spaces*. Academia, Prague & Noordhoff International Publis-
     hing, Leyden 1977. MR 58 # 2189.

KUFNER, A.; OPIC, B. :
 [1] *The Dirichlet problem and weighted spaces*. I. Časopis Pěst. Mat.
     108(1983), no. 4, 381-408. MR 85j:35057.

 [2] *Weighted S. L. Sobolev spaces and the N-dimensional Hardy inequality*.
     Imbedding theorems and their applications to problems of mathemati-
     cal physics, 108-117, Trudy Sem. S. L. Soboleva, No. 1, 1983, Akad.
     Nauk SSSR Sibirsk. Otdel., Inst. Math., Novosibirsk 1983 (Russian).
     MR 85j:46053.

 [3] *Some inequalities in weighted Sobolev spaces*. Constructive theory
     of functions '84 (Varna 1984), 644-648, Bulgar. Acad. Sci., Sofia
     1984.

 [4] *How to define reasonably weighted Sobolev spaces*. Comment. Math.
     Univ. Carolin. 25 (1984), no. 3, 537-554. MR 86i:46036.

 [5] *The Dirichlet problem and weighted spaces*. II. Časopis Pěst. Mat.
     111(1986), no. 3, 242-253.

 [6] *Some remarks on the definition of weighted Sobolev spaces*. In: Par-
     tial differential equations (Proceedings of an international con-
     ference), 120-126 (Russian). Nauka, Novosibirsk 1986.

KUFNER, A.; RÁKOSNÍK, J. :
  [1] *Linear elliptic boundary value problems and weighted Sobolev spaces:
      A modified approach.* Math. Slovaca 34(1984), no. 2, 185-197. MR
      86a:35045.

KUFNER, A.; VOLDŘICH, J. :
  [1] *The Neumann problem in weighted Sobolev spaces.* C. R. Math. Rep.
      Acad. Sci. Canada 7(1985) no. 4, 239-243. MR 86k:35038.

LIONS, J.-L. :
  [1] *Quelques méthodes de résolution des problèmes aux limites non li-
      néaires.* Dunod, Paris 1969.

LIONS, J.-L.; MAGENES, E. :
  [1] *Problèmes aux limites non homogènes et applications.* Vol. 1. Dunod,
      Paris 1968. MR 40 # 512.

LOUIS, A. :
  [1] *Fehlerabschätzungen für Lösungen quasi-linearer elliptischer Diffe-
      rentialgleichungen mittels finiter Elemente.* Thesis, Mainz 1976.

MAZ'JA, V. G.; PLAMENEVSKIĬ, B. A. :
  [1] *Elliptic boundary value problems in a domain with a piecewise smooth
      boundary.* Proceedings of the Symposium on Continuum mechanics and
      Related Problems of Analysis (Tbilisi 1971), Vol. 1 (Russian), 171-
      -181, Mecniereba, Tbilisi 1973. MR 54 # 719.

  [2] *The coefficients in the asymptotic expansions of the solutions of
      elliptic boundary value problems in a cone.* Zap. Naučn.
      Sem. Leningrad. Otdel. Mat. Inst. Steklov (LOMI), 52(1975), 110-
      -127, (Russian). MR 53 # 11220.

  [3] *About boundary value problems for elliptic equations of second order
      in domains with edges.* Vestnik Leningrad. Univ., Ser. Mat. 1975, 1,
      102-108 (Russian).

  [4] *The coefficients in the asymptotics of solutions of elliptic boun-
      dary value problems with conical points.* Math. Nachr. 76(1977),
      29-60, (Russian). MR 58 # 29176.

  [5] *Estimates in L_p and in Hölder classes, and the Miranda-Agmon
      maximum principle for the solutions of elliptic boundary value
      problems in domains with singular points on the boundary.* Math.
      Nachr. 81(1978), 25-82, (Russian). MR 58 # 11886.

  [6] *L_p-estimates of solutions of elliptic boundary value problems in
      domains with ribs.* Trudy Moskov. Mat. Obshch. 37(1978), 49-93,
      (Russian). MR 81b:35027.

  [7] *On the maximum principle for the biharmonic equation in a domain
      with conical points.* Izv. Vyssh. Uchebn. Zaved. Mat. 1981, no. 2,
      52-59, (Russian). MR 84b:35037.

MAZ'JA, V. G.; ROSSMANN, J. :
  [1] *Über die Lösbarkeit und die Asymptotik der Lösungen elliptischer
      Randwertaufgaben in Gebieten mit Kanten.* I, II, III. Preprints
      P-Math-07/84, 30/84, 31/84. Akad. der Wissenschaften der DDR, Inst.
      für Math.

  [2] *Über die Asymptotik der Lösungen elliptischer Randwertaufgaben in
      Gebieten mit Kanten.* Math. Nachr. (to appear).

MELZER, H.; RANNACHER, R. :
  [1] *Spannungskonzentrationen in Eckpunkten der Kirchhoffschen Platte.*
      Der Bauingenieur 1980 and Preprint no. 270 SFB 72, Univ. Bonn 1979.

NEČAS, J. :
  [1] *Les méthodes directes en théorie des équations elliptiques.* Acade-
       mia, Prague & Masson et Cie , Paris 1967. MR 37 # 3168.

  [2] *Sur une méthode pour résoudre les équations aux dérivées partielles
       du type elliptique, voisine de la variationnelle.* Ann. Scuola Norm.
       Sup. Pisa 16(1962), 305-326. MR 29 # 357.

NIKISHKIN, V. A. :
  [1] *Singularities of the solution of the Dirichlet problem for a second
       order equation in the neighborhood of an edge.* Vestnik Moskov. Univ.
       Ser. I. Mat. Meh. No. 2 (1979), 51-62 (Russian). MR 80h:35046.

NITSCHE, J. :
  [1] *Ein Kriterium für die Quasi-Optimalität des Ritzschen Verfahrens.*
       Numer. Math. 11(1968), 346-348. MR 38 # 1823.

OGANESJAN, L. A. :
  [1] *Singularities at corners of the solution of the Navier-Stokes equa-
       tion.* Zap. Naučn. Sem. Leningrad. Otdel. Mat. Inst. Steklov. (LOMI)
       27(1972), 131-144 (Russian). MR 47 # 5465.

ODEN, J. T.; CAREY, G. F. :
  [1] *The Texas Finite Element Series. Finite Elements, Mathematical As-
       pects,* Vol. IV. Prentice-Hall Inc., 1983. MR 86m:65001d.

RÁKOSNÍK, J. :
  [1] *On imbeddings of Sobolev spaces with power-type weights.* Theory of
       approximation of functions. Proceedings of the International Confe-
       rence, Kiev, May 31 – June 5, 1983. Nauka, Moscow 1987, 505-507.

REKTORYS, K. :
  [1] *Variational methods in mathematics, science and engineering.* D. Rei-
       del Publ. Comp., Dordrecht-Boston 1977. MR 58 # 7268b.

REMPEL, S.; SCHULZE, B.-W. :
  [1] *Index theory of elliptic boundary value problems.* Akademie-Verlag,
       Berlin 1982.

ROJTBERG, J. A. :
  [1] *The values on the boundary of the domain of generalized solutions
       of elliptic equations.* Mat. Sb. (N.S.), 86(128) (1971), 248-267
       (Russian). MR 47 # 3813.

ROSSMANN, J. :
  [1] *Das Dirichletproblem für stark elliptische Differentialgleichungen,
       bei denen die rechte Seite* f *zum Raum* $W^{-k}(G)$ *gehört, in Gebieten
       mit konischen Ecken.* Rostock. Math. Kolloq, No. 22(1983), 13-41.
       MR 85f:35075.

  [2] *Elliptische Randwertaufgaben mit Kanten.* Thesis, Wilhelm-Pieck-Uni-
       versität, Rostock 1984.

SÄNDIG, A.-M. :
  [1] *Error estimates for finite element solutions of elliptic boundary
       value problems in non smooth domains* (to appear).

  [2] *Klassische und schwache Lösungen des Dirichletproblems für lineare
       elliptische Gleichungen höherer Ordnung in Gebieten mit konischen
       Ecken.* Math. Ann. 264 (1983), no.2, 189-195. MR 84I:35055.

  [3] *Über das Randverhalten schwacher Lösungen von elliptischen Differen-
       tialgleichungen höherer Ordnung in beschränkten Gebieten im* $\mathbb{R}^N$.
       Math. Nachr. 97 (1980), 147-158. MR 82c:35026.

SCHULZE, B.-W. :
   [1] *Symbolstrukturen und Regularität mit Asymptotik für partielle Diffe-
        rentialgleichungen in Gebieten mit Singularitäten.* Mitt. Math. Ges.
        DDR, Heft 2-3 (1986), 67-85.

   [2] *Regularity with continuous and branching asymptotics on manifolds
        with edges.* Preprint SFB 72, Univ. Bonn 1986.

   [3] *The conormal asymptotics on manifolds with conical singularities and
        edges.* Sem. Anal., Univ. Nantes (to appear 1987).

TOLKSDORF, P. :
   [1] *On the Dirichlet problem for quasilinear equations in domains with
        conical boundary points.* Preprint no. 518 SFB 72, Univ. Bonn 1982.

VOLDŘICH, J. :
   [1] *On the Dirichlet boundary value problem for nonlinear elliptic par-
        tial differential equations in Sobolev power weight spaces.* Časopis
        Pěst. Mat. 110 (1985), no. 3, 250-269.

   [2] *A remark on the solvability of the Dirichlet problem in Sobolev
        spaces with power-type weights.* Comment. Math. Univ. Carolin. 26
        (1985), no. 4, 745-748.

   [3] *Application of Sobolev weight spaces to the solution of elliptic
        boundary value problems.* Thesis, Math. Inst. Acad. Sci., Prague
        1986 (Czech).

WLOKA, J. :
   [1] *Partielle Differentialgleichungen.* B. G. Teubner, Stuttgart 1982.

WHITEMAN, J. R. :
   [1] *Finite element methods for singularities in two and three dimen-
        sions.* Preprint BICOM 81/4, Brunel University.

I N D E X